Monographs in Electrical and Electronic Engineering 22

Series editors: P. Hammond and R. L. Grimsdale

Monographs in Electrical and Electronic Engineering

10. *The theory of linear induction machinery* (1980) Michel Poloujadoff
12. *Energy methods in electromagnetism* (1981) P. Hammond
13. *Low-noise electrical motors* (1981) S. J. Yang
15. *Superconducting rotating electrical machines* (1983) J. R. Bumby
16. *Stepping motors and their microprocessor controls* (1984) T. Kenjo
17. *Machinery noise measurement* (1985) S. J. Yang and A. J. Ellison
18. *Permanent-magnet and brushless DC Motors* (1985) T. Kenjo and S. Nagamori
19. *Metal–semiconductor contacts. Second edition* (1988) E. H. Rhoderick and R. H. Williams
20. *Introduction to power electronics* (1988) Eiichi Ohno
21. *Brushless permanent-magnet and reluctance motor drives* (1989) T. J. E. Miller
22. *Vector control of a.c. machines* (1990) P. Vas

Vector Control of AC Machines

Peter Vas

CLARENDON PRESS · OXFORD
1990

Oxford University Press, Walton Street, Oxford OX2 6DP
Oxford New York Toronto
Delhi Bombay Calcutta Madras Karachi
Petaling Jaya Singapore Hong Kong Tokyo
Nairobi Dar es Salaam Cape Town
Melbourne Auckland
and associated companies in
Berlin Ibadan

Oxford is a trade mark of Oxford University Press

Published in the United States
by Oxford University Press, New York

British Library Cataloguing in Publication Data
Vas, Peter
Vector control of AC machines.
1. *Alternating current machines*
I. Title II. Series
621.3133
ISBN 0–19–859370–8

Library of Congress Cataloging in Publication Data
Vas, Peter.
Vector control of AC machines / Peter Vas.
(Monographs in electrical and electronic engineering; 22)
Includes bibliographical references.
1. Electric machinery—Alternating current—Automatic control.
2. Electric drives. I. Title. II. Series.
TK2796.V29 1990 621.31'042—dc20 89–49453
ISBN 0–19–859370–8

Typeset by Cotswold Typesetting Ltd, Gloucester
Printed in Great Britain by
Bookcraft (Bath) Ltd
Midsomer Norton, Avon

Acknowledgements

The author wishes to acknowledge the support of British Technology Group (B.T.G.), UK, as a result of which a report was presented in 1985 to B.T.G. by the author on the theory and applications of various vector control methods. This report, together with the most recent developments in the field of vector controlled drives, has inspired the author to undertake the writing of the present book. Many engineers have indirectly contributed to this book by their excellent published works which are referenced, and in particular I should like to mention the pioneering work of Professor W. Leonhard and his associates. Thanks are also due to a former colleague, Dr J. E. Brown, for his continuous support during the years the author was at the University of Newcastle-upon-Tyne, UK and for the support of other members of the same university, including Professor A. Jeffrey, Professor M. Harris, and Professor L. Martin. Acknowledgements are also due to Ferranti, NEI Electronics, Lucas, and ASEA BROWN BOVERI for their support during various stages of the research work performed by the author, to Professor P. Hammond for his encouragement, and to Professor P. J. Lawrenson for his earlier help.

The author wishes to thank Oxford University Press for their effective cooperation and their great care in preparing this edition. Suggestions for future improvement of the book will be gratefully received by the author.

To my wife Zsuzsanna,
my son Peter
and also to the memory of
my father and identical twin brother

Contents

Glossary of symbols

$\bar{a}$	complex spatial operator ($e^{j2\pi/3}$)	
$\mathbf{C}_2$	commutator transformation matrix	
f	magnetomotive force (m.m.f.)	At
i_a	instantaneous value of the armature current	A
i_D	instantaneous value of the d.c. link current	A
$\bar{i}_{ms}$ ($\bar{i}_{ms}$)	space phasor of the stator magnetizing currents expressed in the magnetizing-flux-oriented reference frame	A
$\bar{i}_{mr}$ ($\bar{i}_{mr}$)	space phasor of the rotor magnetizing currents expressed in the magnetizing-flux-oriented reference frame	A
$\bar{i}_r$ ($\bar{i}_r$)	space phasor of the rotor currents expressed in the rotor reference frame	A
$\bar{i}'_r$ ($\bar{i}'_r$)	space phasor of the rotor currents expressed in the stator reference frame	A
$\bar{i}_{rg}$ ($\bar{i}_{rg}$)	space phasor of the rotor currents expressed in the general reference frame	A
$\bar{i}_{rm}$ ($\bar{i}_{rm}$)	space phasor of the rotor currents expressed in the magnetizing-flux-oriented reference frame	A
$\bar{i}_{r\psi s}$ ($\bar{i}_{r\psi s}$)	space phasor of the rotor currents expressed in the stator-flux-oriented reference frame	A
i_{ra}, i_{rb}, i_{rc}	instantaneous values of the rotor currents in rotor phases ra, rb, rc respectively	A
i_{rd}, i_{rq}	instantaneous values of direct- and quadrature-axis rotor currents expressed in the stator reference frame	A
$i_{r\alpha}, i_{r\beta}$	instantaneous values of direct- and quadrature-axis rotor current components respectively and expressed in the rotor reference frame	A
i_{rx}, i_{ry}	instantaneous values of the direct- and quadrature-axis rotor current components respectively and expressed in the general or special reference frames	A
$\bar{i}_s$ ($\bar{i}_s$)	space phasor of the stator current expressed in the stator reference frame	A
$\bar{i}'_s$ ($\bar{i}'_s$)	space phasor of the stator currents expressed in the rotor reference frame	A
$\bar{i}_{sg}$ ($\bar{i}_{sg}$)	space phasor of the stator currents expressed in the general reference frame	A
$\bar{i}_{sm}$ ($\bar{i}_{sm}$)	space phasor of the stator currents expressed in the magnetizing-flux-oriented reference frame	A

$\bar{i}_{s\psi r}$ ($\bar{i}_{s\psi r}$)	space phasor of the stator currents expressed in the rotor-flux-oriented reference frame	A
$\bar{i}_{s\psi s}$ ($\bar{i}_{s\psi s}$)	space phasor of the stator currents expressed in the stator-flux-oriented reference frame	A
i_{sA}, i_{sB}, i_{sC}	instantaneous values of the stator currents in stator phases sA, sB, sC respectively	A
i_{sD}, i_{sQ}	instantaneous values of direct- and quadrature-axis stator current components respectively and expressed in the stationary reference frame	A
i_{sd}, i_{sq}	instantaneous values of the direct- and quadrature-axis stator current components respectively and expressed in the rotor reference frame	A
i_{sx}, i_{sy}	instantaneous values of the direct- and quadrature-axis stator current components respectively and expressed in the general or special reference frames (flux and torque producing stator current components respectively)	A
i_{s0}	instantaneous value of zero-sequence stator current component	A
J	polar moment of inertia	kgm^2
L	dynamic inductance	H
L_{DQ}	cross-magnetizing coupling inductance	H
L_F	inductance of filter	H
L_m	magnetizing inductance	H
L_s, L_{sl}	self- and leakage inductances of the stator respectively	H
L_r, L_{rl}	self- and leakage inductances of the rotor respectively	H
L'_s, L'_r	stator and rotor transient inductances respectively	H
$p=d/dt$	differential operator	
p_s, p_r	instantaneous powers of the stator and rotor respectively	W
P	number of pole-pairs	
R_F	resistance of filter	Ohm
R_s, R_r	resistances of a stator and rotor phase winding respectively	Ohm
t	time	s
t_e	instantaneous value of the electromagnetic torque	Nm
t_l	load torque	Nm
T	time lag	s
T_s, T_r	stator and rotor time constants respectively	s
T'_s, T'_r	stator and rotor transient time constants respectively	s
u_D	instantaneous output voltage of rectifier	V
u_{dx}, u_{dy}	instantaneous values of the direct- and quadrature-axis decoupling voltage components expressed in the special reference frames	V

$\bar{u}_s$	space phasor of the stator voltages expressed in the stator reference frame	V
$\bar{u}_{sg}$	space phasor of the stator voltages expressed in the general reference frame	V
$\bar{u}_{sm}$	space phasor of the stator voltages expressed in the magnetizing-flux-oriented reference frame	V
$\bar{u}_{s\psi s}$	space phasor of the stator voltages expressed in the stator-flux-oriented reference frame	V
$\bar{u}_{s\psi r}$	space phasor of the stator voltages expressed in the rotor-flux-oriented reference frame	V
u_{sA}, u_{sB}, u_{sC}	instantaneous values of the stator voltages in phases sA, sB, sC respectively	V
u_{sD}, u_{sQ}	instantaneous values of direct- and quadrature-axis stator voltage components respectively and expressed in the stationary reference frame	V
u_{sx}, u_{sy}	instantaneous values of the direct- and quadrature-axis stator voltage components respectively and expressed in the general or special reference frames	V
$\bar{u}_r$	space phasor of the rotor voltages expressed in the rotor reference frame	V
$\bar{u}'_r$	space phasor of the rotor voltages expressed in the stator reference frame	V
$\bar{u}_{rg}$	space phasor of the rotor voltages expressed in the general reference frame	V
$\bar{u}_{r\psi s}$	space phasor of the rotor voltages expressed in the stator-flux-oriented reference frame	V
u_{rd}, u_{rq}	instantaneous values of the direct- and quadrature-axis rotor voltages respectively and expressed in the stator reference frame	V
u_{rx}, u_{ry}	instantaneous values of the direct- and quadrature-axis rotor voltages respectively and expressed in the general or special reference frames	V
$u_{r\alpha}, u_{r\beta}$	instantaneous values of the direct- and quadrature-axis rotor voltages respectively and expressed in the rotor reference frame	V
u_{r0}	instantaneous value of the zero-sequence rotor voltage	V
W_{mech}	instantaneous value of mechanical output energy	J
$z = e^{sT}$	discrete Laplace variable	
δ	load angle	rad
θ_r	rotor angle	rad
μ_m	phase angle of the magnetizing flux linkage space phasor with respect to the direct-axis of the stator reference frame	rad

ρ_r	phase angle of the rotor flux linkage space phasor with respect to the direct-axis of the stator reference frame	rad
ρ_s	phase angle of stator flux linkage space phasor with respect to the direct-axis of the stator reference frame	rad
σ	resultant leakage constant	
σ_r	rotor leakage constant	
ψ_f	excitation flux linkage	Wb
$\bar{\psi}'_m$	space phasor of magnetizing flux linkages expressed in the stator reference frame	Wb
$\bar{\psi}_{m\psi r}$	space phasor of magnetizing flux linkages expressed in the rotor-flux-oriented reference frame	Wb
$\bar{\psi}_{m\psi s}$	space phasor of magnetizing flux linkages expressed in the stator-flux-oriented reference frame	Wb
ψ_{md}, ψ_{mq}	instantaneous values of the direct- and quadrature-axis magnetizing flux linkage components expressed in the rotor reference frame	Wb
$\bar{\psi}_r$	space phasor of rotor flux linkages expressed in the rotor reference frame	Wb
$\bar{\psi}'_r$	space phasor of the rotor flux linkages expressed in the stator reference frame	Wb
$\bar{\psi}_{rg}$	space phasor of the rotor flux linkages expressed in the general reference frame	Wb
$\bar{\psi}_{r\psi r}$	space phasor of the rotor flux linkages expressed in the rotor-flux-oriented reference frame	Wb
ψ_{ra}, ψ_{rb}, ψ_{rc}	instantaneous values of the flux linkages in rotor phases ra, rb, rc respectively	Wb
ψ_{rx}, ψ_{ry}	instantaneous values of the direct- and quadrature-axis rotor flux linkage components respectively and expressed in the general or special reference frames	Wb
$\bar{\psi}_s$	space phasor of stator flux linkages expressed in the stator reference frame	Wb
$\bar{\psi}'_s$	space phasor of the stator flux linkages expressed in the rotor reference frame	Wb
$\bar{\psi}_{sg}$	space phasor of the stator flux linkages expressed in the general reference frame	Wb
$\bar{\psi}_{sm}$	space phasor of the stator flux linkages expressed in the magnetizing-flux-oriented reference frame	Wb
$\bar{\psi}_{s\psi s}$	space phasor of the stator flux linkages expressed in the stator-flux-oriented reference frame	Wb
ψ_{sA}, ψ_{sB}, ψ_{sC}	instantaneous values of stator flux linkages in stator phases sA, sB, sC respectively	Wb
ψ_{sd}, ψ_{sq}	instantaneous values of the direct- and quadrature-axis stator flux linkages expressed in the rotor reference frame	Wb

ψ_{sD}, ψ_{sQ}	instantaneous values of the direct- and quadrature-axis stator flux linkage components expressed in the stator reference frame	Wb
ψ_{sx}, ψ_{sy}	instantaneous values of the direct- and quadrature-axis stator flux linkage components respectively and expressed in the general or special reference frames	Wb
ω_m	angular speed of the magnetizing-flux-oriented reference frame	rad/s
ω_{mr}	angular speed of the rotor-flux-oriented reference frame	rad/s
ω_{ms}	angular speed of the stator-flux-oriented reference frame	rad/s
ω_r	angular rotor speed	rad/s
ω_{sl}	angular slip frequency	rad/s

Subscripts

A, B	upper and lower cages respectively
b	base
g	general reference frame
i	induced
m	magnetizing
n	normalized
r	rotor
ra, rb, rc	rotor phases
ref	reference
s	stator
sA, sB, sC	stator phases
x	direct-axis component in general reference frame or in special reference frames (fixed to the stator flux linkage, rotor flux linkage or magnetizing flux linkage space phasors respectively)
y	quadrature-axis component in general reference frame or in special reference frames (fixed to the stator flux linkage, rotor flux linkage or magnetizing flux linkage space phasors respectively)

Mathematical symbols

×	cross vector product
*	complex conjugate

1 Introduction

In the past d.c. motors were used extensively in areas where variable-speed operation was required, since their flux and torque could be controlled easily by the field and armature current. In particular, the separately excited d.c. motor has been used mainly for applications where there was a requirement of fast response and four-quadrant operation with high performance near zero speed. However, d.c. motors have certain disadvantages, which are due to the existence of the commutator and the brushes. That is, they require periodic maintenance; they cannot be used in explosive or corrosive environments and they have limited commutator capability under high-speed, high-voltage operational conditions. These problems can be overcome by the application of alternating current motors, which can have simple and rugged structure, high maintainability and economy; they are also robust and immune to heavy overloading. Their small dimension compared with d.c. motors allows a.c. motors to be designed with substantially higher output ratings for low weight and low rotating mass.

Variable-speed a.c. drives have been used in the past to perform relatively undemanding roles in applications which preclude the use of d.c. motors, either because of the working environment or commutator limits. Because of the high cost of efficient, fast switching frequency static inverters, the lower cost of a.c. motors has also been a decisive economic factor in multi-motor systems. However, as a result of the progress in the field of power electronics, the continuing trend is towards cheaper and more effective power converters, and single motor a.c. drives compete favourably on a purely economic basis with the d.c. drives.

Among the various a.c. drive systems, those which contain the cage induction motor have a particular cost advantage. The cage motor is simple and rugged and is one of the cheapest machines available at all power ratings. Owing to their excellent control capabilities, variable speed drives incorporating a.c. motors and employing modern static converters and vector control can well compete with high-performance four-quadrant d.c. drives.

Vector control techniques incorporating fast microprocessors have made possible the application of induction-motor and synchronous-motor drives for high-performance applications where traditionally only d.c. drives were applied. In the past such control techniques would have not been possible because of the complex hardware and software required to solve the complex control problem. As for d.c. machines, torque control in a.c. machines is achieved by controlling the motor currents. However, in contrast to a d.c. machine, in an a.c. machine, both the phase angle and the modulus of the current has to be controlled, or in other words, the current vector has to be controlled. This is the reason for the terminology 'vector control'. Furthermore, in d.c. machines, the orientation of the field flux and armature m.m.f. is fixed by the commutator and the brushes, while in a.c. machines the field flux and the spatial angle of the armature m.m.f. require external control. In the absence of this control, the spatial angles between the various fields in a.c. machines vary with the load and yield unwanted oscillating dynamic response. With vector control of a.c. machines, the torque- and flux-producing current components are decoupled and the transient response characteristics are similar to those of a separately excited d.c. machine and the system will adapt to any load disturbances and/or reference value variations as fast as a d.c. machine. It is the primary aim of the present book to give a unified and detailed treatment of the various forms of vector-controlled drives.

It is expected that with the rapid developments in the field of microelectronics, vector control of various types of a.c. machines will become a commonly used technique when, even though high dynamic performance is not required, servo-like high performance plays a secondary role to reliability and energy (efficiency) savings. It is possible to contribute to the energy savings by the application of intelligent control of the flux- and torque-producing components of the stator currents. In the case of d.c. drives, the power circuits are relatively uniform and in most cases contain a line-commutated thyristor converter or a transistorized chopper for low power applications. However, for a.c. drives, there is much greater variety, due to the different types of converters (voltage source, current source, natural commutation, forced commutation, d.c. link, cycloconverter) which can be combined with various types of a.c. machines.

Following the early works of Blaschke and Hasse, and largely due to the pioneering work of Professor Leonhard, vector control of a.c. machines has become a powerful and frequently adopted technique world-wide. In recent years, numerous important contributions have been made in this field by contributors from many countries, including Canada, Germany, Italy, Japan, UK, and USA.

In the present book the very wide field of vector control of smooth-air-gap and salient-pole a.c. machines is discussed. Four types of vector control are considered: rotor-oriented control, rotor-flux-oriented control, stator-flux-oriented control, and magnetizing-flux-oriented control. Great emphasis is laid on presenting a unified and detailed physical and mathematical treatment which relies on space-phasor theory. The book contains a step-by-step, physical and mathematical development of space-phasor theory and its applications to the various forms of vector control of a.c. machines. The theory developed also covers the operation of smooth-air-gap and salient-pole machines under non-linear magnetic conditions. The monitoring of the machine parameters and the rotor speed are also discussed. The book is intended to cover the needs of students, academics and industrial users who require a deep understanding of the various aspects of vector control.

A brief review of the contents of the book will now be given. In Chapter 2 the space-phasor model of a.c. machines is developed for smooth-air-gap machines. For this purpose the space phasors of various quantities (stator m.m.f.s, stator currents, magnetizing current, stator flux linkages, etc.) are obtained by physical considerations. This is followed by a discussion of the mechanism of electromagnetic torque production in d.c. and a.c. machines and an explanation based on space-phasor theory. The three-phase and various quadrature-phase machine models (corresponding to various reference frames) are presented and the relationship of space-phasor theory to two-axis theory is also described. The voltage equations are formulated in a general reference frame. It is shown that when a special reference frame is used, which is fixed to either the magnetizing flux-linkage space phasor, the rotor flux-linkage space phasor, or the stator flux-linkage space phasor, the expression for the electromagnetic torque contains a flux-producing current component and a torque-producing current, which are in space quadrature, and this expression is similar to the expression for the electromagnetic torque of a separately excited d.c. machine. This principle is utilized in four implementations of vector control (rotor-oriented control, magnetizing-flux-oriented control, rotor-flux-oriented control, and stator-flux-oriented control). In the final part of Chapter 2 the mechanism of electromagnetic torque production in salient-pole synchronous machines is discussed and the expression for the electromagnetic torque is given in the reference frame fixed to the magnetizing flux-linkage space phasor. This chapter is referred to frequently in later sections.

In Chapter 3, the vector control of synchronous machines is discussed. In particular the vector control of permanent-magnet machines with surface-mounted magnets or interior permanent magnets is described. Both rotor-oriented control and stator-flux-oriented control are discussed. Finally, vector control of salient-pole synchronous machines with electrically excited rotor and vector control of reluctance machines are analysed.

In Chapter 4, which is the largest section in the book, three types of vector control are discussed in detail for induction machines. These are the rotor-flux-, the stator-flux-, and the magnetizing-flux-oriented control methods. Two main cases are dealt with: an induction machine, (i) supplied by impressed stator voltages or, (ii) supplied by impressed stator currents. In addition, doubly-fed induction machines either with impressed rotor currents or containing a static converter cascade in the rotor are also subjected to analysis. These sections contain the details of the so-called direct and indirect implementations. Implementations with and without speed sensors and rotor parameter estimator are also described. For the derivation of the control schemes which correspond to the last two cases, model reference adaptive control systems are used. Implementations which utilize the actual or the reference values of the flux- and torque-producing stator current components are also presented. The steady-state stability limit is determined in the case of the stator-flux-, rotor-flux- and magnetizing-flux-oriented control methods. The space-phasor theory of induction machines either with deep rotor bars or employing a double cage, is also presented in Chapter 4, where the magnetizing-flux-oriented control of the double-cage induction machine is discussed in detail. Enough information is given on all the systems described to enable concrete realization of specific systems, using digital, analogue, or hybrid techniques. The required decoupled circuits are also described in detail for both an ideal drive and a non-ideal drive, for which the effects of inverter time delay are considered.

In Chapter 5 the effects of the saturation of the main flux paths are discussed for both smooth-air-gap and salient-pole machines, by considering the equations which are required for the implementations of the various forms of vector control (rotor-flux-oriented control, stator-flux-oriented control, magnetizing-flux-oriented control). The effects of cross-magnetization due to saturation (cross-saturation) are also discussed in detail, in both physical and mathematical terms.

In all the vector control schemes described it is necessary to know certain machine parameters. In Chapter 6 various techniques are presented for the identification of the parameters of induction and synchronous machines. These include techniques which utilize monitored terminal voltages and currents, while the machine is at standstill and where the inverter itself, which is a part of the vector control system, is utilized to generate the signals necessary for parameter estimation. A high-resolution rotor speed estimator, employing an optimal rotor speed observer, is also discussed. Finally load rejection tests are described for the determination of various synchronous machine parameters.

The book contains almost 100 figures and 150 references on the above topics. The author has attempted to give a most up-to-date list of references published before July 1989, when the complete manuscript was released for printing.

2 The space-phasor model of a.c. machines

The principle of the various forms of vector control of a.c. machines can be well understood by comparing the production of electromagnetic torque in d.c. and a.c. machines. However, for this purpose first the space phasors of various quantities (m.m.f.s, currents, flux linkages, etc.) will be introduced by utilizing physical and mathematical considerations. With the help of these, it is possible to formulate the space-phasor models of a.c. machines and to give a detailed explanation of the fundamental principles involved in vector-controlled machines.

2.1 THE SPACE-PHASOR MODEL OF SMOOTH-AIR-GAP MACHINES

For the purpose of understanding and designing vector controlled drives, it is necessary to know the dynamic model of the machine subjected to control. The machine models which are necessary to design control loops are very different to those used for designing the machine. Machine designers must have tolerance levels (e.g. for power levels) which are less than one percent, while control designs are only rough approximations, where even a ten percent error can be considered as acceptable. This is because every control scheme must absorb the changes of the plant parameters, which are due to the changes in the temperature, supply, non-linearity etc. and the effects of the load are only approximately considered. However, a model of the electrical machine which is adequate for designing the control system must preferably incorporate all the important dynamic effects occurring during steady-state and transient operation. It should also be valid for any arbitrary time variation of the voltages and currents generated by the converter which supplies the machine.

Such a model, valid for any instantaneous variation of voltage and current and adequately describing the performance of the machine under both steady-state and transient operation, can be obtained by the utilization of space-phasor theory. This is very closely related to the two-axis theory of electrical machines, but the simplicity and compactness of the space-phasor equations and the very clear physical pictures obtained by its application can yield further advantages. For better understanding, the relationship of the space-phasor equations to the two-axis equations will be emphasized throughout the book.

First the space-phasor quantities (voltages, currents, m.m.f.s, flux densities, flux linkages, etc.) will be introduced by using both mathematical and physical considerations. For simplicity, a smooth-air-gap a.c. machine is considered with symmetrical two-pole, three-phase windings. Figure 2.1 shows the cross-section view of the machine under consideration and the effects of slotting have been neglected. It is also assumed that the permeability of the iron parts is infinite and the flux density is radial in the air-gap. The effects of iron losses and end-effects are also neglected. In Fig. 2.1 the stator and rotor windings are shown as single, multiple-turn full pitch coils situated on the two sides of the air-gap; these, however, represent distributed windings, which at every instant produce sinusoidal m.m.f. waves centred on the magnetic axes of the respective phases. The phase windings are displaced by 120 electrical degrees from each other. In Fig. 2.1 θ_r is the rotor angle, the angle between the magnetic axes of stator winding sA and rotor winding ra. In general, the speed of the rotor is $\omega_r = \mathrm{d}\theta_r/\mathrm{d}t$, and its positive direction is also shown in Fig. 2.1.

2.1.1 The space phasor of the stator m.m.f.s and stator currents

If the stator windings are supplied by a system of three-phase currents $i_{sA}(t)$, $i_{sB}(t)$, *and* $i_{sC}(t)$, which can vary arbitrarily in time, but the neutral point is isolated, there can be no zero-sequence stator currents. Then

$$i_{s0}(t) = i_{sA}(t) + i_{sB}(t) + i_{sC}(t) = 0 \qquad (2.1\text{-}1)$$

where $i_{s0}(t)$ is the instantaneous value of the zero-sequence stator currents. If it

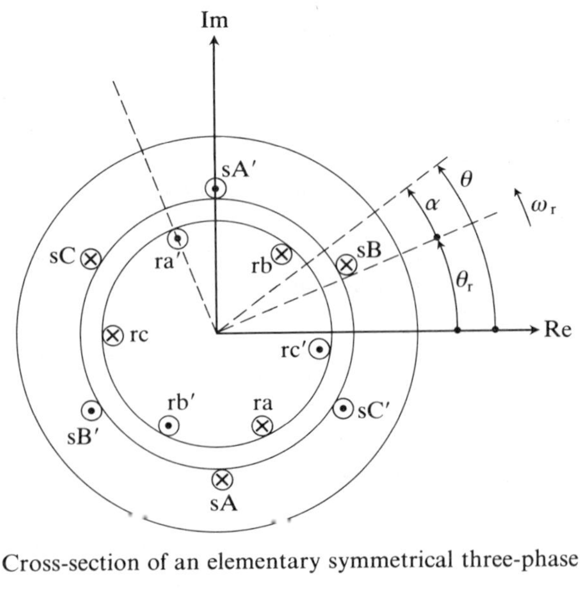

Fig. 2.1. Cross-section of an elementary symmetrical three-phase machine.

is assumed that the stator windings have an equal number of effective turns $N_{se}=N_s k_{ws}$, where N_s and k_{ws} are the number of turns and the winding factor respectively of a stator winding, then the resultant m.m.f. distribution $f_s(\theta, t)$ produced by the stator is as follows, if θ is the angle around the periphery with reference to the axis of stator winding sA, which coincides with the real-axis of the stator denoted by Re in Fig. 2.1,

$$f_s(\theta, t)=N_{se}[i_{sA}(t)\cos\theta+i_{sB}(t)\cos(\theta-2\pi/3) \\ +i_{sC}(t)\cos(\theta-4\pi/3)]. \quad (2.1\text{–}2)$$

By using complex notation, it is possible to put eqn (2.1-2) into the following form

$$f_s(\theta, t)=\tfrac{3}{2}N_{se}\,\mathrm{Re}\{\tfrac{2}{3}[i_{sA}(t)+ai_{sB}(t)+a^2 i_{sC}(t)]e^{-j\theta}\}. \quad (2.1\text{-}3)$$

In eqn (2.1-3), $e^{-j\theta}$ is multiplied by the following quantity:

$$\bar{i}_s(t)=\tfrac{2}{3}[1i_{sA}(t)+ai_{sB}(t)+a^2 i_{sC}(t)]=|\bar{i}_s|e^{j\alpha_s}, \quad (2.1\text{-}4)$$

which is the complex space phasor of the three-phase stator currents in the complex plane in the stationary reference frame fixed to the stator. Furthermore, in eqn (2.1-4) 1, a, and a^2 are spatial operators, $a=e^{j2\pi/3}$ and $a^2=e^{j4\pi/3}$. Although these spatial operators are formally the same as the time operators in the theory of complexors used for the steady-state analysis of sinusoidal voltages and currents, it is very important that the two should not be confused. From this point of view, it would be justifiable to use a different notation for these operators, but historically they have been introduced by using the same notation, which is therefore used throughout the book.

In eqn (2.1-4) $|\bar{i}_s|$ is the modulus of the stator current space phasor and α_s is its phase angle with respect to the real axis of the stationary reference frame fixed to the stator. The real axis of the stator is denoted by sD, corresponding to the terminology: **D**irect-axis of the stator. It should be noted that, since in general all the currents vary in time, both the modulus and the phase angle vary with time. Physically the space phasor of the stator currents determines the instantaneous magnitude and spatial displacement of the peak of the sinusoidal stator m.m.f. distribution produced by the three stator windings. Thus the space phasor of the stator m.m.f.s is defined as follows:

$$\bar{f}_s(t)=N_{se}\bar{i}_s(t)=\bar{f}_{sA}(t)+\bar{f}_{sB}(t)+\bar{f}_{sC}(t), \quad (2.1\text{-}5)$$

where $\bar{f}_{sA}(t)$, $\bar{f}_{sB}(t)$, and $\bar{f}_{sC}(t)$ are the space phasors of the individual phase m.m.f.s.

It follows from eqn (2.1-4) that the three-phase space phasor of the stator currents can be obtained by the addition of the space phasors corresponding to each phase current, $\bar{i}_{sA}=1i_{sA}(t)$; $\bar{i}_{sB}=ai_{sB}(t)$; $\bar{i}_{sC}=a^2 i_{sC}(t)$. In the symmetrical steady state, when the currents are sinusoidal and form a three-phase balanced system,

$$i_{sA}=I_s\cos(\omega_1 t-\phi_s);\quad i_{sB}=I_s\cos(\omega_1 t-\phi_s-2\pi/3);$$
$$i_{sC}=I_s\cos(\omega_1 t-\phi_s-4\pi/3).$$

Thus the space phasor of the stator currents is $\bar{i}_s=I_s e^{j(\omega_1 t-\phi_s)}$ and since I_s = constant, this corresponds to a circle in the complex plane. Thus the locus of the stator current phasor is a circle, which the tip of this space phasor runs around, in the positive direction in space; the angular speed of this space phasor is constant and equal to the synchronous speed. In this case the space phasor of the stator currents is identical to the positive-sequence complexor (complex phasor) of the three-phase currents and this is the reason why in this case the time-vector diagrams are identical to the space-phasor diagrams. However, in other cases, the space phasors and complexors should not be confused. If the currents form an asymmetrical system, in the steady state the locus of the stator-current space phasor will become an ellipse or a straight line. In the transient state the space-phasor locus can have an arbitrary shape.

It should be noted that it is an important consequence of the assumed sinusoidal m.m.f. distribution that the space phasor m.m.f.s of the three phases can be added together. Furthermore, it should also be noted that whilst the m.m.f. waves are measurable, physically existing real quantities, the m.m.f. space phasors are only convenient mathematical abstractions, which of course can be displayed (e.g. in real time, on monitors, oscilloscopes, etc.).

It should be emphasized that it is also possible to introduce the space phasors by utilizing two-axis theory and this was the method originally followed by Park. The space phasor of the stator currents can be defined as a phasor whose real part is equal to the instantaneous value of the direct-axis stator current component, $i_{sD}(t)$, and whose imaginary part is equal to the quadrature-axis stator current component, $i_{sQ}(t)$. Thus, the stator current space vector in the stationary reference frame fixed to the stator can be expressed as

$$\bar{i}_s=i_{sD}(t)+ji_{sQ}(t). \tag{2.1-6}$$

In the technical literature the notation sα and sβ is sometimes used instead of the notation sD, sQ.

In symmetrical three-phase machines, the direct- and quadrature-axis stator currents i_{sD}, i_{sQ} are fictitious, quadrature-phase ('two-phase') current components, which are related to the actual, three-phase stator currents as follows

$$i_{sD}=c[i_{sA}-\tfrac{1}{2}i_{sB}-\tfrac{1}{2}i_{sC}] \tag{2.1-7}$$

and

$$i_{sQ}=c\frac{\sqrt{3}}{2}(i_{sB}-i_{sC}), \tag{2.1-8}$$

where c is a constant. For the so-called classical, non-power invariant form of the phase transformation from three-phase to quadrature-phase components, $c=\frac{2}{3}$. However, for the power-invariant form $c=\sqrt{2/3}$. It follows from the definition of the space phasor of the stator currents (eqn (2.1-4) that, if $c=\frac{2}{3}$, its real-axis component yields

$$\mathrm{Re}(\bar{i}_s)=\mathrm{Re}[\tfrac{2}{3}(i_{sA}+ai_{sB}+a^2i_{sC})]=\tfrac{2}{3}(i_{sA}-\tfrac{1}{2}i_{sB}-\tfrac{1}{2}i_{sC})=i_{sD}, \qquad (2.1\text{-}9)$$

which agrees with eqn (2.1-7) if the non-power invariant form of the phase transformation is used, and its imaginary-axis component yields

$$\mathrm{Im}(\bar{i}_s)=\mathrm{Im}[\tfrac{2}{3}(i_{sA}+ai_{sB}+a^2i_{sC})]=(i_{sB}-i_{sC})/\sqrt{3}=i_{sQ}, \qquad (2.1\text{-}10)$$

which is in agreement with eqn (2.1-8). In a quadrature-phase machine, i_{sD} and i_{sQ} are actual, not transformed currents, which flow in the two stator windings sD and sQ which are in space quadrature.

It should be emphasized that the space phasor does not contain the zero-sequence component and thus if there is a zero-sequence component, an additional definition must be utilized. In general, similarly to eqn (2.1-1), the instantaneous zero-sequence current component of the stator currents is defined as

$$i_{s0}=c_1[i_{sA}(t)+i_{sB}(t)+i_{sC}(t)], \qquad (2.1\text{-}11)$$

where c_1 is a constant; $c_1=\frac{1}{3}$ for the non-power-invariant form, and $c_1=1/\sqrt{3}$ for the power-invariant form.

If the non-power-invariant form of the transformations is used, it is a useful consequence that if there are no zero-sequence components, the projections of a space-phasor quantity on the corresponding phase axes directly yield the instantaneous values of the phase variables of the same quantity. This is shown in Fig. 2.2 for the case of the space phasor of the stator currents.

Mathematically this means that, by utilizing $i_{s0}=0$, and by using the non-power invariant forms of the zero-sequence current component and the space phasor of the stator currents, the following equations are obtained

$$\mathrm{Re}(\bar{i}_s)=\mathrm{Re}[\tfrac{2}{3}(i_{sA}+ai_{sB}+a^2i_{sC})]=\tfrac{2}{3}[i_{sA}-\tfrac{1}{2}i_{sB}-\tfrac{1}{2}i_{sC}]=i_{sA} \qquad (2.1\text{-}12)$$

$$\mathrm{Re}(a^2\bar{i}_s)=\mathrm{Re}[\tfrac{2}{3}(a^2i_{sA}+i_{sB}+ai_{sC})]=i_{sB} \qquad (2.1\text{-}13)$$

$$\mathrm{Re}(a\bar{i}_s)=\mathrm{Re}[\tfrac{2}{3}(ai_{sA}+a^2i_{sB}+i_{sC})]=i_{sC}. \qquad (2.1\text{-}14)$$

If, however, the zero-sequence current is not zero, the phase-variable stator currents are obtained as

$$i_{sA}=\mathrm{Re}(\bar{i}_s)+i_{s0} \qquad (2.1\text{-}15)$$

$$i_{sB}=\mathrm{Re}(a^2\bar{i}_s)+i_{s0} \qquad (2.1\text{-}16)$$

$$i_{sC}=\mathrm{Re}(a\bar{i}_s)+i_{s0} \qquad (2.1\text{-}17)$$

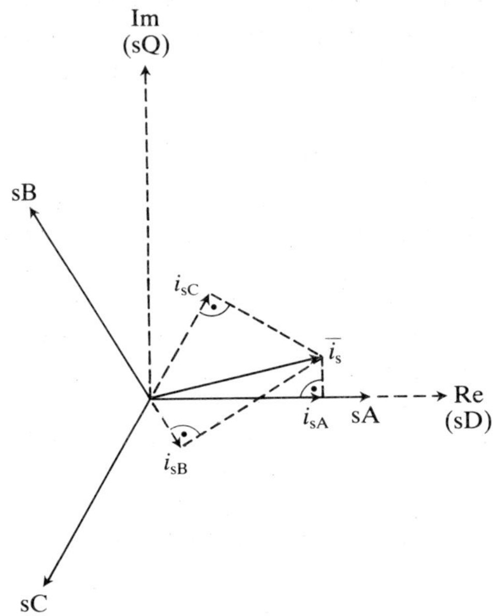

Fig. 2.2. Projections of the stator-current space phasor.

and it follows that the zero-sequence current components are added to the corresponding projections of the space phasor on the corresponding axes.

2.1.2 The space phasor of rotor m.m.f.s and rotor currents

Similar considerations to those for the resultant stator m.m.f. hold for the resultant m.m.f. produced by the symmetrical three-phase sinusoidally distributed rotor windings but the rotor phase ra is displaced from stator phase sA by an angle θ_r and α is the angle around the periphery (also shown in Fig. 2.1) with respect to the axis of rotor winding ra. If it is assumed that the rotor windings have an equal number of effective turns $N_{re}=N_r k_{wr}$, where N_r is the number of turns and k_{wr} the winding factor of a rotor winding, then similarly to eqn (2.1-2), if there is no zero-sequence rotor current, the resultant rotor m.m.f. distribution $f_r(\theta, t)$ produced by the rotor windings carrying currents $i_{ra}(t)$, $i_{rb}(t)$, and $i_{rc}(t)$ is as follows:

$$f_r(\theta, t)=N_{re}[i_{ra}(t)\cos\alpha+i_{rb}(t)\cos(\alpha-2\pi/3)+i_{rc}(t)\cos(\alpha-4\pi/3)]. \qquad (2.1\text{-}18)$$

By introducing complex notation, it is possible to express eqn (2.1-18) as follows:

$$f_r(\theta, t)=\tfrac{3}{2}N_{re}\,\mathrm{Re}\{\tfrac{2}{3}[i_{ra}(t)1+a i_{rb}(t)+a^2 i_{rc}(t)]e^{-j\alpha}\}. \qquad (2.1\text{-}19)$$

In eqn (2.1-19) the complex quantity multiplied by $e^{-j\alpha}$ is the rotor current space phasor $\bar{i}_r$,

$$\bar{i}_r=\tfrac{2}{3}[i_{ra}(t)+ai_{rb}(t)+a^2i_{rc}(t)]=|\bar{i}_r|e^{j\alpha_r}, \qquad (2.1\text{-}20)$$

expressed in the reference frame fixed to the rotor (the real-axis of this reference frame is denoted by rα and its imaginary-axis by rβ, as shown in Fig. 2.1). The speed of this reference frame is $\omega_r=\mathrm{d}\theta_r/\mathrm{d}t$, where θ_r is the rotor angle (also shown in Fig. 2.1). This definition is similar to that of the stator current space phasor expressed in the stationary reference frame (eqn (2.1-4). The rotor current space phasor determines the instantaneous magnitude and angular displacement of the peak of the sinusoidally distributed rotor m.m.f. produced by the sinusoidally distributed rotor windings.

Thus it is possible to generalize, and it follows that the definitions of the space-phasor quantities in their 'own' reference frames ('natural' reference frames) are similar. Furthermore, it should be noted that it would again be possible to introduce the definition of the rotor-current space phasor by the direct application of two-axis theory. Thus if $i_{r\alpha}$ and $i_{r\beta}$ are the instantaneous values of the direct- and quadrature-axis rotor current components of the three-phase rotor currents, which similarly to eqns (2.1-7) and (2.1-8) are related to the instantaneous values of the actual three-phase currents by

$$i_{r\alpha}=c[i_{ra}-\tfrac{1}{2}i_{rb}-\tfrac{1}{2}i_{rc}] \qquad (2.1\text{-}21)$$

$$i_{r\beta}=c\sqrt{3}/2(i_{rb}-i_{rc}), \qquad (2.1\text{-}22)$$

where $c=\tfrac{2}{3}$ for the non-power invariant, classical form of the transformation, then the definition of the space phasor of the rotor currents in the reference frame fixed to the rotor is as follows

$$\bar{i}_r=i_{r\alpha}+ji_{r\beta}. \qquad (2.1\text{-}23)$$

Thus the real part of $\bar{i}_r$ yields the transformed direct-axis rotor current and the imaginary part gives the transformed quadrature-axis rotor current and all the currents can vary arbitrarily in time. For a machine with quadrature-phase rotor winding, $i_{r\alpha}$ and $i_{r\beta}$ are non-transformed, actual rotor currents which flow in the rotor windings rα and rβ respectively.

It follows from eqns (2.1-19) and (2.1-20) that the instantaneous variation of the rotor m.m.f. wave, which rotates at speed ω_r, can be expressed in terms of the rotor current space phasor and the angles θ and θ_r as follows, if it is considered that in accordance with Fig. 2.1, $\alpha=\theta-\theta_r$,

$$f_r(\theta,\theta_r,t)=\tfrac{3}{2}N_{re}\,\mathrm{Re}[\bar{i}_r e^{-j(\theta-\theta_r)}]=\tfrac{3}{2}N_{re}\,\mathrm{Re}(\bar{i}'_r e^{-j\theta}) \qquad (2.1\text{-}24)$$

where

$$\bar{i}'_r=\bar{i}_r e^{j\theta_r}=|\bar{i}_r|e^{j\alpha'_r} \qquad (2.1\text{-}25)$$

is the space phasor of the rotor currents expressed in the stationary reference frame. Substitution of eqn (2.1-20) into eqn (2.1-25) finally gives

$$\bar{i}'_r = |\bar{i}_r| e^{j\alpha'_r} = |\bar{i}_r| e^{j(\alpha_r + \theta_r)}. \quad (2.1\text{-}26)$$

The fact that $\bar{i}'_r$ is the space phasor of the rotor currents expressed in the stationary reference frame can be proved as follows. The relationship between the reference frame fixed to the stator (whose direct and quadrature axes are denoted by sD and sQ respectively) and the reference frame fixed to the rotor, which rotates at angular speed ω_r is shown in Fig. 2.3.

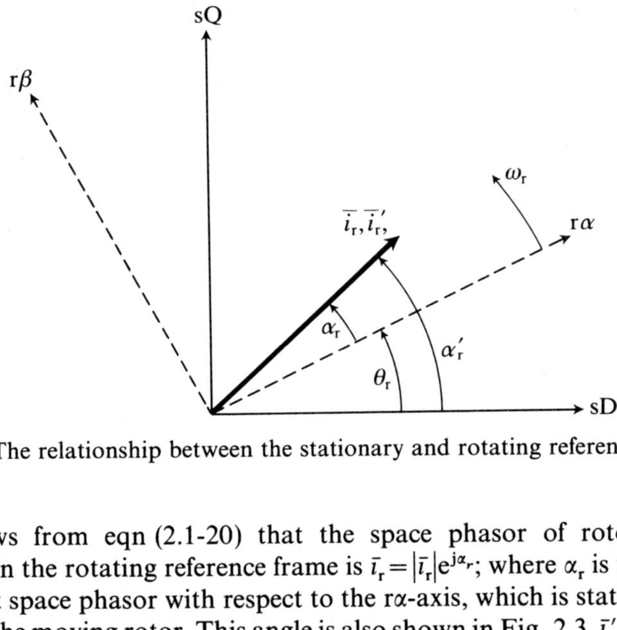

Fig. 2.3. The relationship between the stationary and rotating reference frames.

It follows from eqn (2.1-20) that the space phasor of rotor currents expressed in the rotating reference frame is $\bar{i}_r = |\bar{i}_r| e^{j\alpha_r}$; where α_r is the angle of the current space phasor with respect to the rα-axis, which is stationary with respect to the moving rotor. This angle is also shown in Fig. 2.3. $\bar{i}'_r$ is the space phasor of the rotor currents expressed in the reference frame, which is stationary with respect to the stator, and it follows from Fig. 2.3 that $\bar{i}'_r = |\bar{i}_r| e^{j\alpha'_r}$, where $\alpha'_r = \alpha_r + \theta_r$, which is in full agreement with eqn (2.1-26).

2.1.3 The magnetizing-current space phasor

Due to the combined effects of stator and rotor excitations, the resultant m.m.f. wave is equal to the sum of the stator and rotor m.m.f. distributions,

$$f(\theta, \theta_r, t) = f_s(\theta, t) + f_r(\theta, \theta_r, t) \quad (2.1\text{-}27)$$

and by considering eqns (2.1-3), (2.1-4), and (2.1-24), it follows that

$$f(\theta, \theta_r, t) = \tfrac{3}{2} N_{se} [\mathrm{Re}(\bar{i}_s e^{-j\theta}) + N_{re}/N_{se}\, \mathrm{Re}(\bar{i}'_r e^{-j\theta})]. \quad (2.1\text{-}28)$$

Thus

$$f(\theta, \theta_r, t) = \tfrac{3}{2} N_{se} \operatorname{Re}[(\bar{\imath}_s + N_{re}/N_{se} \bar{\imath}'_r) e^{-j\theta}] \qquad (2.1\text{-}29)$$

is obtained. In eqn (2.1-29), the complex current in the brackets multiplied by $e^{-j\theta}$ is the magnetizing-current space phasor expressed in the stationary reference frame (which is fixed to the stator) and therefore it is defined as

$$\bar{\imath}_m = \bar{\imath}_s + (N_{re}/N_{se}) \bar{\imath}'_r. \qquad (2.1\text{-}30)$$

This is the sum of the stator-current space phasor and the rotor current space phasor expressed in the stationary reference frame fixed to the stator. However, if the rotor current space phasor is referred to the stator, the reference factor is $N_{se}/N_{re} = (N_s k_{ws})/(N_r k_{wr})$. When $\bar{\imath}'_r$ is divided by this factor, then $(N_{re}\bar{\imath}_r)/N_{se}$ is obtained, which is the rotor current space phasor referred to the stator and expressed in the stationary reference frame. By a special selection of the turns ratio, e.g. if the reference factor is equal to the ratio of the magnetizing inductance and the rotor self inductance, eqn (2.1-30) yields the so-called rotor magnetizing-current space phasor expressed in the stationary reference frame, $\bar{\imath}_{mr}$ (for more detailed discussion see Section 4.1.1). The rotor magnetizing-current space phasor is extensively utilized when the so-called rotor-flux-oriented control of a.c. machines is employed (e.g. see Sections 2.1.8 and 4.1).

2.1.4 Flux-linkage space phasors

In this section the space phasors of the stator and rotor flux linkages will be obtained in various reference frames.

The stator flux-linkage space phasor in the stationary reference frame

Similarly to the definitions of the stator current and rotor current space phasors, it is possible to define the space phasor of the stator flux linkages $\bar{\psi}_s$ in terms of the instantaneous values of the flux linkages of the three stator windings. Thus in the stationary reference frame fixed to the stator, the total flux-linkage space phasor can be expressed as follows

$$\bar{\psi}_s = \tfrac{2}{3}(\psi_{sA} + a\psi_{sB} + a^2\psi_{sC}), \qquad (2.1\text{-}31)$$

where the instantaneous values of the phase-variable flux linkage components are

$$\psi_{sA} = \bar{L}_s i_{sA} + \bar{M}_s i_{sB} + \bar{M}_s i_{sC} + \bar{M}_{sr} \cos\theta_r i_{ra} + \bar{M}_{sr} \cos(\theta_r + 2\pi/3) i_{rb} + \bar{M}_{sr} \cos(\theta_r + 4\pi/3) i_{rc} \qquad (2.1\text{-}32)$$

$$\psi_{sB} = \bar{L}_s i_{sB} + \bar{M}_s i_{sA} + \bar{M}_s i_{sC} + \bar{M}_{sr} \cos(\theta_r + 4\pi/3) i_{ra} + \bar{M}_{sr} \cos\theta_r i_{rb} + \bar{M}_{sr} \cos(\theta_r + 2\pi/3) i_{rc} \qquad (2.1\text{-}33)$$

$$\psi_{sC}=\bar{L}_s i_{sC}+\bar{M}_s i_{sB}+\bar{M}_s i_{sA}+\bar{M}_{sr}\cos(\theta_r+2\pi/3)i_{ra}$$
$$+\bar{M}_{sr}\cos(\theta_r+4\pi/3)i_{rb}+\bar{M}_{sr}\cos\theta_r i_{rc}. \tag{2.1-34}$$

In these equations $\bar{L}_s$ is the self inductance of a stator phase winding, $\bar{M}_s$ is the mutual inductance between the stator windings, and $\bar{M}_{sr}$ is the maximal value of the stator–rotor mutual inductance. It can be seen that the phase-variable flux linkages contain six flux linkage terms, a self-flux linkage component produced by the stator currents in the stator winding under consideration, two mutual stator flux linkage components due to the other two stator currents and three stator–rotor mutual flux linkage components, which are due to the three rotor currents. Substitution of eqns (2.1-32), (2.1-33), and (2.1-34) into eqn (2.1-31) yields the following space-phasor equation for the stator flux linkages if eqns (2.1-4), (2.1-20), and (2.1-25) are also considered:

$$\bar{\psi}_s=L_s\bar{i}_s+L_m\bar{i}'_r=L_s\bar{i}_s+L_m\bar{i}_r e^{j\theta_r}, \tag{2.1-35}$$

where $L_s=\bar{L}_s-\bar{M}_s$ is the total three-phase stator inductance and L_m is the so-called three-phase magnetizing inductance, $L_m=(\frac{3}{2})\bar{M}_{sr}$.

The stator flux-linkage space phasor describes the modulus and phase angle of the peak of the sinusoidal stator flux distribution in the air-gap. In eqn (2.1-35) there are two flux-linkage space phasor components. The first component, which is equal to $L_s\bar{i}_s$, is the self flux-linkage space phasor of the stator phases, which is caused by the stator currents. The second component, $L_m\bar{i}'_r$, is a mutual flux-linkage space phasor, which is due to the rotor currents and is expressed in the stationary reference frame. It is important to note that eqn (2.1-35) is general, and holds even under non-linear magnetic conditions. Thus it is also valid when the leakage or main flux paths are saturated. In this case L_s and L_m are not constant, but also depend on the currents of the machine. This will be discussed in detail in Chapter 5.

It is possible to give the definition of the stator flux linkages in terms of the direct- and quadrature-axis flux-linkage components ψ_{sD} and ψ_{sQ}:

$$\bar{\psi}_s=\psi_{sD}+j\psi_{sQ}, \tag{2.1-36}$$

where it follows by considering eqn (2.1-35) that the direct-axis stator flux-linkage component is defined as

$$\psi_{sD}=L_s i_{sD}+L_m i_{rd} \tag{2.1-37}$$

and the quadrature-axis stator flux-linkage component is given by

$$\psi_{sQ}=L_s i_{sQ}+L_m i_{rq}. \tag{2.1-38}$$

The relationship between the instantaneous values of the direct- and quadrature-axis flux-linkage components ψ_{sD} and ψ_{sQ} and the instantaneous values of the three-phase stator flux-linkage components ψ_{sA}, ψ_{sB}, and ψ_{sC} is similar to the relationship between the two-axis currents and the three-phase currents described by eqns (2.1-9) and (2.1-10).

In the equations above, i_{sD}, i_{sQ} and i_{rd}, i_{rq} are the instantaneous values of the direct- and quadrature-axis stator and rotor currents respectively, and it is important to note that all the four currents are defined in the stationary reference frame fixed to the stator and they can arbitrarily vary in time. The rotor currents i_{rd} and i_{rq} are related to the rotor currents $i_{r\alpha}$ and $i_{r\beta}$ by eqn (2.1-25), and the latter current components, defined by eqns (2.1-21) and (2.1-22), are the two-axis components of the rotor currents in the reference frame fixed to the rotor. Thus it follows from eqn (2.1-25) that

$$\bar{i}'_r = i_{rd} + ji_{rq} = \bar{i}_r e^{j\theta_r} \tag{2.1-39}$$

and this yields the following transformational relationship between the d, q and α, β components of the rotor currents, if for convenience the matrix form is used, as is usually the case in the generalized theory of electrical machines,

$$\begin{bmatrix} i_{rd} \\ i_{rq} \end{bmatrix} = \begin{bmatrix} \cos\theta_r & -\sin\theta_r \\ \sin\theta_r & \cos\theta_r \end{bmatrix} \begin{bmatrix} i_{r\alpha} \\ i_{r\beta} \end{bmatrix} = \mathbf{C}_2^{-1} \begin{bmatrix} i_{r\alpha} \\ i_{r\beta} \end{bmatrix} \tag{2.1-40}$$

In eqn (2.1-40) the inverse of the co-called commutator transformation matrix ($\mathbf{C}_2$) appears. It is also an advantage of the application of the space phasors, that in contrast to the matrix forms used in generalized machine theory, where the various machine models corresponding to different reference frames are obtained by the application of matrix transformations, complex transformations—e.g. $e^{j\theta_r}$ in eqn (2.1-39)—are used, which result in more compact, more easily manipulable equations. Furthermore, by expressing the space phasor of a given quantity in various reference frames, all the matrix transformations of generalized machine theory can be obtained.

The rotor flux-linkage space phasor in the rotating reference frame fixed to the rotor

The space phasor of the rotor flux linkages expressed in its own (natural) reference frame, i.e. in the reference frame fixed to the rotor, and rotating at speed ω_r, is defined as follows

$$\bar{\psi}_r = \tfrac{2}{3}[\psi_{ra}(t) + a\psi_{rb}(t) + a^2\psi_{rc}(t)], \tag{2.1-41}$$

where $\psi_{ra}(t)$, $\psi_{rb}(t)$, and $\psi_{rc}(t)$ are the instantaneous values of the rotor flux linkages in the rotor phases ra, rb, and rc respectively. In terms of the instantaneous values of the stator and rotor currents they can be expressed as

$$\begin{aligned}\psi_{ra} &= \bar{L}_r i_{ra} + \bar{M}_r i_{rb} + \bar{M}_r i_{rc} + \bar{M}_{sr}\cos\theta_r i_{sA} \\ &\quad + \bar{M}_{sr}\cos(\theta_r + 4\pi/3) i_{sB} + \bar{M}_{sr}\cos(\theta_r + 2\pi/3) i_{sC}\end{aligned} \tag{2.1-42}$$

$$\begin{aligned}\psi_{rb} &= \bar{L}_r i_{rb} + \bar{M}_r i_{ra} + \bar{M}_r i_{rc} + \bar{M}_{sr}\cos(\theta_r + 2\pi/3) i_{sA} \\ &\quad + \bar{M}_{sr}\cos\theta_r i_{sB} + \bar{M}_{sr}\cos(\theta_r + 4\pi/3) i_{sC}\end{aligned} \tag{2.1-43}$$

$$\psi_{rc}=\bar{L}_r i_{rc}+\bar{M}_r i_{ra}+\bar{M}_r i_{rb}+\bar{M}_{sr}\cos(\theta_r+4\pi/3)i_{sA}$$
$$+\bar{M}_{sr}\cos(\theta_r+2\pi/3)i_{sB}+\bar{M}_{sr}\cos\theta_r i_{sC}, \qquad (2.1\text{-}44)$$

where $\bar{L}_r$ is the self-inductance of a rotor winding, $\bar{M}_r$ is the mutual inductance between two rotor phases. It can be seen that all three rotor flux-linkage components $\psi_{ra}(t)$, $\psi_{rb}(t)$, and $\psi_{rc}(t)$, contain three flux-linkage components produced by the rotor currents and three mutual flux-linkage components produced by the stator currents. A considerable simplification is achieved if eqns (2.1-42), (2.1-43), and (2.1-44) are substituted into eqn (2.1-41) and thus the space phasor of rotor flux linkages in the rotor reference frame is obtained as

$$\bar{\psi}_r=L_r\bar{i}_r+L_m\bar{i}'_s, \qquad (2.1\text{-}45)$$

where $L_r=\bar{L}_r-\bar{M}_r$ is the total three-phase rotor inductance and $\bar{i}'_s$ is the space phasor of the stator currents expressed in the reference frame fixed to the rotor. Equation (2.1-45) contains two terms: (i) the space phasor $L_r\bar{i}_r$ is the rotor self-flux linkage space phasor expressed in the rotor reference frame and is solely due to the rotor currents and; (ii) the space phasor $L_m\bar{i}'_s$ is a mutual flux-linkage space phasor, produced by the stator currents and expressed in the same reference frame.

Instead of defining the rotor flux-linkage space phasor in terms of the rotor flux-linkage components corresponding to the three phases, it is possible to define it in terms of its two-axis components $(\psi_{r\alpha}, \psi_{r\beta})$,

$$\bar{\psi}_r=\psi_{r\alpha}+j\psi_{r\beta}, \qquad (2.1\text{-}46)$$

where it follows from eqn (2.1-45) that the direct-axis rotor flux-linkage component can be defined as

$$\psi_{r\alpha}=L_r i_{r\alpha}+L_m i_{sd} \qquad (2.1\text{-}47)$$

and the quadrature-axis rotor flux-linkage component can be given as

$$\psi_{r\beta}=L_r i_{r\beta}+L_m i_{sq}. \qquad (2.1\text{-}48)$$

In eqns (2.1-47) and (2.1-48) $i_{r\alpha}$, $i_{r\beta}$, i_{sd}, and i_{sq} are the direct- and quadrature-axis rotor and stator current components respectively, and all the current components are expressed in the reference frame fixed to the rotor. The relationship of the stator current components i_{sd}, i_{sq} and the stator current components i_{sD}, i_{sQ} will be shown in the next section.

The rotor flux-linkage space phasor in the stationary reference frame

The rotor flux-linkage components in the reference frame fixed to the rotor $(\psi_{r\alpha}, \psi_{r\beta})$ are related to the rotor flux-linkage components expressed in the stationary reference frame (ψ_{rd}, ψ_{rq}) by the same transformation $e^{j\theta_r}$ as given by eqn (2.1-39) for the rotor currents. Thus the following equation holds

$$\bar{\psi}'_r=\psi_{rd}+j\psi_{rq}=\bar{\psi}_r e^{j\theta_r}=(\psi_{r\alpha}+j\psi_{r\beta})e^{j\theta_r} \qquad (2.1\text{-}49)$$

and this can be put into the following matrix form,

$$\begin{bmatrix}\psi_{rd}\\ \psi_{rq}\end{bmatrix}=\begin{bmatrix}\cos\theta_r & -\sin\theta_r\\ \sin\theta_r & \cos\theta_r\end{bmatrix}\begin{bmatrix}\psi_{r\alpha}\\ \psi_{r\beta}\end{bmatrix}. \tag{2.1-50}$$

In eqn (2.1-50) the transformation matrix is the inverse of the commutator transformation matrix given in eqn (2.1-40).

By the substitution of eqn (2.1-45) into eqn (2.1-49) and by considering eqn (2.1-39), the space phasor of the rotor flux linkages in the stationary reference frame can be expressed as

$$\bar{\psi}'_r=(L_r\bar{i}'_r+L_m\bar{i}'_s e^{j\theta_r})=L_r\bar{i}'_r+L_m\bar{i}_s. \tag{2.1-51}$$

This contains two flux-linkage components, a self flux linkage produced by the rotor currents but expressed in the stationary reference frame ($L_r\bar{i}'_r$) and a mutual flux-linkage component produced by the stator currents and also expressed in the stationary reference frame ($L_m\bar{i}_s=L_m\bar{i}'_s e^{j\theta_r}$). Thus the stator currents in the stationary reference frame are related to the stator current components in the rotating reference frame fixed to the rotor by the following complex transformation

$$\bar{i}_s=\bar{i}'_s e^{j\theta_r} \tag{2.1-52}$$

where $\bar{i}_s$ and $\bar{i}'_s$ are expressed in terms of their two-axis components as follows:

$$\bar{i}_s=i_{sD}+ji_{sQ}$$

$$\bar{i}'_s=i_{sd}+ji_{sq}. \tag{2.1-53}$$

It follows from eqn (2.1-52) that the transformed stator current space phasor in the rotating reference frame fixed to the rotor can be obtained from the space phasor of the stator currents, expressed in the stationary reference frame, as

$$\bar{i}'_s=\bar{i}_s e^{-j\theta_r}. \tag{2.1-54}$$

The corresponding two-axis form can be obtained by the substitution of eqn (2.1-53) into eqn (2.1-54),

$$\begin{bmatrix}i_{sd}\\ i_{sq}\end{bmatrix}=\begin{bmatrix}\cos\theta_r & \sin\theta_r\\ -\sin\theta_r & \cos\theta_r\end{bmatrix}\begin{bmatrix}i_{sD}\\ i_{sQ}\end{bmatrix} \tag{2.1-55}$$

and these are the stator currents used in eqns (2.1-47) and (2.1-48). The transformation described by eqn (2.1-54) can also be obtained by considering Fig. 2.4.

It follows from eqn (2.1-4) that the stator current space phasor expressed in the stationary reference frame ($\bar{i}_s$) can be expressed in terms of its modulus ($|\bar{i}_s|$) and its phase angle (α_s), which are also shown in Fig. 2.4, as $\bar{i}_s=|\bar{i}_s|e^{j\alpha_s}$. However, when the space phasor of the stator currents is expressed in the reference frame rotating with the rotor speed ω_r, then it follows from Fig. 2.4 that $\bar{i}'_s=|\bar{i}_s|e^{j\alpha'_s}$, where the angle α'_s is the angle between $\bar{i}'_s$ and the real-axis (sd)

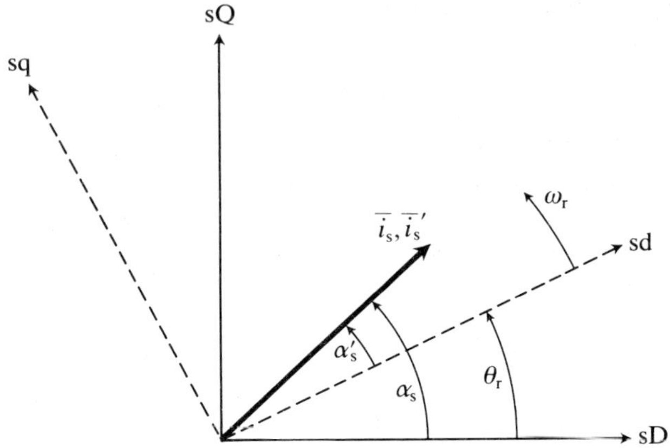

Fig. 2.4. Transformation of the stator-current space phasor.

of the rotating reference frame. However, since it follows from Fig. 2.4 that $\alpha_s' = \alpha_s - \theta_r$, $\bar{i}_s' = |\bar{i}_s| e^{j(\alpha_s - \theta_r)}$ is obtained, which can be expressed as $\bar{i}_s' = \bar{i}_s e^{-j\theta_r}$, and this is in agreement with eqn (2.1-54).

The stator flux-linkage space phasor in the reference frame fixed to the rotor

It should be noted that if, for example, it is necessary to know the space phasor of the stator flux linkages expressed in the rotating reference frame ($\bar{\psi}_s'$) in terms of the stator flux-linkage space phasor expressed in the stationary reference frame ($\bar{\psi}_s$), then similarly to eqn (2.1-54)

$$\bar{\psi}_s' = \bar{\psi}_s e^{-j\theta_r} \tag{2.1-56}$$

holds, or in two-axis form, by considering $\bar{\psi}_s' = \psi_{sd} + j\psi_{sq}$, $\bar{\psi}_s = \psi_{sD} + j\psi_{sQ}$,

$$\begin{bmatrix} \psi_{sd} \\ \psi_{sq} \end{bmatrix} = \begin{bmatrix} \cos\theta_r & \sin\theta_r \\ -\sin\theta_r & \cos\theta_r \end{bmatrix} \begin{bmatrix} \psi_{sD} \\ \psi_{sQ} \end{bmatrix}, \tag{2.1-57}$$

where the transformation matrix is the commutator transformation $\mathbf{C}_2$.

Incidentally, the substitution of eqn (2.1-35) into eqn (2.1-56) yields $\bar{\psi}_s'$ as follows, if eqns (2.1-39) and (2.1-54) are also considered,

$$\bar{\psi}_s' = (L_s\bar{i}_s + L_m\bar{i}_r')e^{-j\theta_r} = L_s\bar{i}_s' + L_m\bar{i}_r, \tag{2.1-58}$$

where $\bar{i}_s'$ and $\bar{i}_r$ are the space phasors of the stator and rotor currents respectively and are expressed in the reference frame fixed to the rotor. By resolving into real and imaginary parts, and using the notation introduced in eqns (2.1-23) and (2.1-53), the direct- and quadrature-axis stator flux linkages in the rotor reference frame are obtained as

$$\psi_{sd} = L_s i_{sd} + L_m i_{r\alpha} \tag{2.1-59}$$

$$\psi_{sq} = L_s i_{sq} + L_m i_{r\beta}, \tag{2.1-60}$$

where i_{sd}, i_{sq}, $i_{r\alpha}$, and $i_{r\beta}$ are the two-axis stator and rotor currents respectively in the reference frame fixed to the rotor.

2.1.5 The space phasors of the stator and rotor voltages

The space phasors of the stator and rotor voltages can be defined similarly to the space-phasor quantities defined in the earlier sections [e.g. eqn (2.1-4)]. Thus the stator voltage space phasor in the stationary reference frame is

$$\bar{u}_s = \tfrac{2}{3}[u_{sA}(t) + a u_{sB}(t) + a^2 u_{sC}(t)] = u_{sD} + j u_{sQ} \qquad (2.1\text{-}61)$$

and the rotor voltage space phasor in the reference frame fixed to the moving rotor is

$$\bar{u}_r = \tfrac{2}{3}[u_{ra}(t) + a u_{rb}(t) + a^2 u_{rc}(t)] = u_{r\alpha} + j u_{r\beta}. \qquad (2.1\text{-}62)$$

In eqns (2.1-61) and (2.1-62) $u_{sA}(t)$, $u_{sB}(t)$, $u_{sC}(t)$ and $u_{ra}(t)$, $u_{rb}(t)$ and $u_{rc}(t)$ are the instantaneous values of the stator and rotor phase voltages respectively and u_{sD}, u_{sQ}, $u_{r\alpha}$ and $u_{r\beta}$ are the corresponding direct- and quadrature-axis components. The relationship between the three-phase and quadrature-phase voltages immediately follows from these equations. For example, for the stator voltages

$$u_{sD} = \mathrm{Re}\{\tfrac{2}{3}[u_{sA}(t) + a u_{sB}(t) + a^2 u_{sC}(t)]\} = \tfrac{2}{3}(u_{sA} - \tfrac{1}{2}u_{sB} - \tfrac{1}{2}u_{sC}) \qquad (2.1\text{-}63)$$

$$u_{sQ} = \mathrm{Im}\{\tfrac{2}{3}\{u_{sA}(t) + a u_{sB}(t) + a^2 u_{sB}(t)]\} = (u_{sB} - u_{sC})/\sqrt{3}. \qquad (2.1\text{-}64)$$

Similarly, the following equations hold for the rotor voltage components

$$u_{r\alpha} = \tfrac{2}{3}(u_{ra} - \tfrac{1}{2}u_{rb} - \tfrac{1}{2}u_{rc}) \qquad (2.1\text{-}65)$$

and

$$u_{r\beta} = (u_{rb} - u_{rc})/\sqrt{3}. \qquad (2.1\text{-}66)$$

It has been emphasized in Section 2.1.1 that the space phasor does not contain the zero-sequence components. Thus if there are zero-sequence voltages, they have to be considered separately in terms of the following stator and rotor zero-sequence components

$$u_{s0} = \tfrac{1}{3}[u_{sA}(t) + u_{sB}(t) + u_{sC}(t)] \qquad (2.1\text{-}67)$$

$$u_{r0} = \tfrac{1}{3}[u_{ra}(t) + u_{rb}(t) + u_{rc}(t)]. \qquad (2.1\text{-}68)$$

It should be noted that if eqns (2.1-63), (2.1-64), and (2.1-67) are put into matrix form,

$$\begin{bmatrix} u_{s0} \\ u_{sD} \\ u_{sQ} \end{bmatrix} = \frac{2}{3} \begin{bmatrix} 1/2 & 1/2 & 1/2 \\ 1 & -1/2 & -1/2 \\ 0 & \sqrt{3}/2 & -\sqrt{3}/2 \end{bmatrix} \begin{bmatrix} u_{sA} \\ u_{sB} \\ u_{sC} \end{bmatrix}, \qquad (2.1\text{-}69)$$

where the transformation matrix is the inverse of the so-called phase transformation matrix ($\mathbf{C}_1$). A similar transformation matrix applies for the rotor voltages. It should also be pointed out that similarly to eqns (2.1-12)–(2.1-14), in the absence of the zero-sequence components, the projections of the space phasor of voltages on the corresponding axes yield the phase voltages, i.e.

$$u_{sA} = \mathrm{Re}(\bar{u}_s) \tag{2.1-70}$$

$$u_{sB} = \mathrm{Re}(a^2\bar{u}_s) \tag{2.1-71}$$

$$u_{sC} = \mathrm{Re}(a\bar{u}_s). \tag{2.1-72}$$

The stator voltage space phasor expressed in the stationary reference frame ($\bar{u}_s$) can be transformed into the stator space phasor expressed in the reference frame fixed to the rotor ($\bar{u}_s'$), similarly to eqn (2.1-54). Thus

$$\bar{u}_s' = \bar{u}_s e^{-j\theta_r} = u_{sd} + ju_{sq}. \tag{2.1-73}$$

If this is resolved into real- and imaginary-axis components, the same transformation matrix ($\mathbf{C}_2$) will appear as in eqn (2.1-57).

The rotor voltage space phasor expressed in the reference frame fixed to the rotor ($\bar{u}_r$) can be expressed in the reference frame fixed to the stator ($\bar{u}_r'$), and in this case the complex transformation to be used is the same as the one used in eqn (2.1-39). Thus

$$\bar{u}_r' = \bar{u}_r e^{j\theta_r} = u_{rd} + ju_{rq}. \tag{2.1-74}$$

When this is put into the two-axis form, the same transformation matrix ($\mathbf{C}_2^{-1}$) will appear as in eqn (2.1-40).

If, for example, the relationship between u_{sA}, u_{sB}, u_{sC} and u_{sd}, u_{sq} is required, it follows from eqns (2.1-73) and (2.1-61) that

$$u_{sd} + ju_{sq} = \tfrac{2}{3}[u_{sA}(t) + au_{sB}(t) + a^2u_{sC}(t)]\,(\cos\theta_r - j\sin\theta_r). \tag{2.1-75}$$

Resolution of eqn (2.1-75) into real and imaginary parts yields

$$u_{sd} = \tfrac{2}{3}[u_{sA}\cos\theta_r + u_{sB}\cos(\theta_r - 2\pi/3) + u_{sC}\cos(\theta_r - 4\pi/3)] \tag{2.1-76}$$

$$u_{sq} = -\tfrac{2}{3}[u_{sA}\sin\theta_r + u_{sB}\sin(\theta_r - 2\pi/3) + u_{sC}\sin(\theta_r - 4\pi/3)], \tag{2.1-77}$$

which define the so-called Park transformation. Incidentally, if the zero-sequence stator voltage component defined by eqn (2.1-67) is combined with eqns (2.1-76) and (2.1-77), the following matrix form is obtained,

$$\begin{bmatrix} u_{s0} \\ u_{sd} \\ u_{sq} \end{bmatrix} = \frac{2}{3}\begin{bmatrix} 1/2 & 1/2 & 1/2 \\ \cos\theta_r & \cos(\theta_r - 2\pi/3) & \cos(\theta_r - 4\pi/3) \\ -\sin\theta_r & -\sin(\theta_r - 2\pi/3) & -\sin(\theta_r - 4\pi/3) \end{bmatrix}\begin{bmatrix} u_{sA} \\ u_{sB} \\ u_{sC} \end{bmatrix}, \tag{2.1-78}$$

2.1.6 The mechanism of electromagnetic torque production

In this section, the mechanism of electromagnetic torque production in d.c. and a.c. machines will be discussed and for simplicity the effects of magnetic non-linearities will be neglected. To enhance the analogy between the mechanism of torque production d.c. and a.c. machines, the space-phasor formulation of the electromagnetic torque will be presented for both types of machines.

Electromagnetic torque production in d.c. machines

Figure 2.5 shows the schematic diagram of a compensated d.c. machine, which for simplicity has a smooth air-gap.

On the stator of the machine there are the field (f) and compensating windings (c) and on the rotor there is the armature winding (a). The current in the field winding i_f produces an excitation flux linkage ψ_f. If current i_a flows in the armature winding, the interaction of the armature current and the excitation flux linkage will result in forces (F) acting on the conductors, as shown in Fig. 2.5. Since the excitation flux linkage is in space quadrature to the armature current, maximal forces are applied to the shaft and therefore the position of the armature winding is optimal for electromagnetic torque production.

It follows from Fig. 2.5 that the armature winding also produces a field, which is superimposed on the field produced by the field winding, but it is in space quadrature with respect to the excitation flux. Thus the resultant field will be displaced from its optimal position. However, this effect can be cancelled by the application of the compensating winding (c), which carries current i_c which is equal to $-i_a$. The currents in the compensating winding and the field winding produce an electromagnetic torque which acts against the armature.

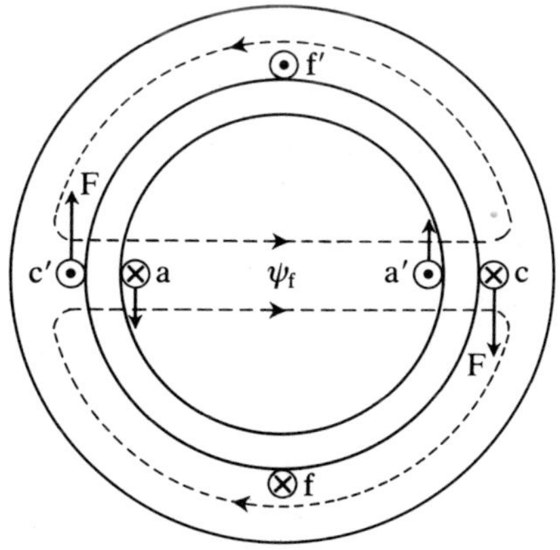

Fig. 2.5. Electromagnetic torque production in a d.c. machine.

Figure 2.6 shows the space phasors of the currents in the field winding, compensating winding and armature winding respectively and also the excitation flux linkage. These are denoted by $\bar{i}_f$, $\bar{i}_c$, $\bar{i}_a$, and $\bar{\psi}_f$ respectively and it should be noted that the stationary reference frame with direct (sD) and quadrature (sQ) axes has been used, so that these space phasors are expressed in the stationary reference frame fixed to the stator of the machine.

It follows that in the d.c. machine described above, the current in the field winding creates the excitation flux linkage, and the interaction of the excitation flux with the currents in the armature winding and compensating winding produce the electromagnetic torque under both steady-state and transient conditions. Under linear magnetic conditions it is possible to express the instantaneous value of the electromagnetic torque (t_e) as a vector (cross) product of the excitation flux linkage and armature current space phasors, and thus the following expression is obtained

$$t_e = c\bar{\psi}_f \times \bar{i}_a. \tag{2.1-79}$$

In eqn (2.1-79) c is a constant and $\times$ denotes the vector product. Since the two space phasors are in space quadrature, eqn (2.1-79) can be put into the following form,

$$t_e = c\psi_f i_a, \tag{2.1-80}$$

where ψ_f and i_a are the moduli of the respective space phasors and are equal to the instantaneous values of the excitation flux linkage and armature current respectively. If the excitation flux is maintained constant, the electromagnetic torque can be controlled by varying the armature current and a change in the

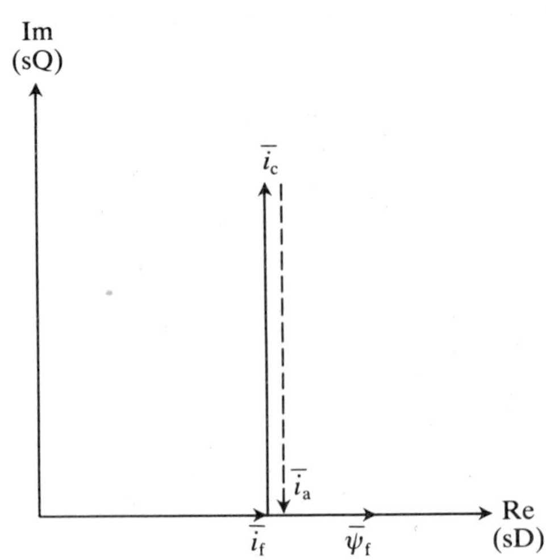

Fig. 2.6. The space phasors in a d.c. machine.

armature current will result in a rapid change in the torque. It is the purpose of vector control of a.c. machines to have a similar technique of rapid torque control, as discussed in the next section.

Electromagnetic torque production in a.c. machines

In this section the analogy between the electromagnetic torque production in d.c. and a.c. machines is described. It will be shown that in general, in symmetrical three-phase or quadrature-phase smooth-air-gap a.c. machines, the developed instantaneous electromagnetic torque can be put into the following vectorial form, which is similar to eqn (2.1-79),

$$t_e = c\bar{\psi}_s \times \bar{\imath}'_r, \qquad (2.1\text{-}81)$$

where under linear magnetic conditions c is a constant and $\bar{\psi}_s$ and $\bar{\imath}'_r$ are the space phasors of the stator flux linkages and rotor currents respectively, expressed in the stationary reference frame. Thus the electromagnetic torque is the cross product of a flux linkage and a current space phasor. In eqn (2.1-81) the stator flux-linkage space phasor and the rotor-current space phasors can be expressed in reference frames other than the stationary reference frame. However, in this case other transformed space phasors will be present in the expression for the electromagnetic torque, e.g. $\bar{\psi}'_s$ and $\bar{\imath}_r$, which correspond to the space phasors of the stator flux linkages and rotor currents expressed in the rotor reference frame.

By utilizing the Euler forms of vector quantities, it is possible to express eqn (2.1-81) as

$$t_e = c|\bar{\psi}_s||\bar{\imath}_r| \sin \gamma, \qquad (2.1\text{-}82)$$

where $|\bar{\psi}_s|$ and $|\bar{\imath}_r|$ are the moduli of the stator flux linkage and rotor current space phasors respectively and γ is the so-called torque angle. It follows that when $\gamma = 90°$, eqn (2.1-82) will take the form of eqn (2.1-80) and maximum torque is obtained. Thus the similarity between the production of the electromagnetic torque in a compensated d.c. machine and in symmetrical, smooth-air-gap a.c. machines has been established. However, it should be noted that whilst in the d.c. machine, where the armature current and main flux distributions are fixed in space—the former is due to the action of the commutator—and where torque control can be established by independently controlling the excitation flux (ψ_f) and armature current (i_a) and where, due to the fixed relationship between the stator field and armature current distribution, maximal torque is produced for given field and armature currents, in an a.c. machine, it is much more difficult to realize this principle, because these quantities are coupled and are stationary with respect to the stator and rotor respectively; they also depend on the modulus, frequency and phase angles of the stator currents. It is a further complication that in machines with a squirrel-cage rotor, it is not possible to monitor the rotor currents by simple means. Of course, there are ways of preparing the rotor cage in advance and using certain transducers which yield signals proportional to the instantaneous values of the rotor currents, but these are only suitable for laboratory work or under very special conditions. The search for simple control schemes, similar to those used for d.c. machines, has led to the development of the co-called vector-controlled schemes, where similarly to d.c. motors, it is possible to obtain two current components, one of which is a flux-producing current component and the other is the torque-producing current component. It will be shown in later sections how the expression for the instantaneous electromagnetic torque can be formulated in terms of these two currents. There will also be a discussion on how they can be independently controlled so as to achieve the required decoupled control.

For completeness, a proof of eqn (2.1-81) will now be presented and this will be discussed in the remainder of the present section. This proof will be based on energy considerations, but for better understanding, in Sections 2.1.7 and 2.1.8 other derivations will also be given, which utilize other concepts. It is possible to obtain an expression of the electromagnetic torque by putting the rate of change of the mechanical output energy ($\mathrm{d}W_{\mathrm{mech}}/\mathrm{d}t$) equal to the mechanical power p_{mech}; the latter is equal to the product of the instantaneous rotor speed and electromagnetic torque,

$$p_{\mathrm{mech}}=\frac{\mathrm{d}W_{\mathrm{mech}}}{\mathrm{d}t}=t_{\mathrm{e}}\omega_r. \tag{2.1-83}$$

For simplicity a two-pole machine is considered, and the expression for the electromagnetic torque is derived from eqn (2.1-83). For this purpose, it is first necessary to formulate an equation for the differential mechanical energy $\mathrm{d}W_{\mathrm{mech}}$.

Since for every machine the principle of conservation of energy must be valid during motion, it follows that the input electrical energy (W_{e}) has to cover the energies related to the stator and rotor losses (W_{loss}), the magnetic energy stored in the field (W_{field}) and the mechanical output energy (W_{mech}), and thus

$$W_{\mathrm{e}}=W_{\mathrm{loss}}+W_{\mathrm{field}}+W_{\mathrm{mech}}. \tag{2.1-84}$$

and the differential mechanical output energy can be obtained as

$$\mathrm{d}W_{\mathrm{mech}}=\mathrm{d}W_{\mathrm{e}}-\mathrm{d}W_{\mathrm{loss}}-\mathrm{d}W_{\mathrm{field}}. \tag{2.1-85}$$

The three different energy components given on the right-hand side of eqn (2.1-85) can be obtained as follows. In general, for a doubly-fed polyphase machine, the differential input electrical energy can be expressed as

$$\mathrm{d}W_{\mathrm{e}}=\tfrac{3}{2}\,\mathrm{Re}(\bar{u}_{\mathrm{s}}\bar{\imath}_{\mathrm{s}}^{*}+\bar{u}_{\mathrm{r}}'\bar{\imath}_{\mathrm{r}}'^{*})\,\mathrm{d}t, \tag{2.1-86}$$

where the stator voltage and current space phasors ($\bar{u}_{\mathrm{s}}$, $\bar{\imath}_{\mathrm{s}}$) are expressed in the stationary reference frame and the rotor voltage and current space phasors ($\bar{u}_{\mathrm{r}}'$, $\bar{\imath}_{\mathrm{r}}'$) are expressed in the reference frame fixed to the stator. Equation

(2.1-86) follows from the physical fact that the total instantaneous power is the sum of the instantaneous power of the stator

$$p_s = \tfrac{3}{2}\,\mathrm{Re}(\bar{u}_s \bar{\imath}_s^*) \tag{2.1-87}$$

and the instantaneous power of the rotor

$$p_r = \tfrac{3}{2}\,\mathrm{Re}(\bar{u}'_r \bar{\imath}'^*_r) = \tfrac{3}{2}\,\mathrm{Re}(\bar{u}_r \bar{\imath}_r^*), \tag{2.1-88}$$

where the asterisk denotes the complex conjugate. Furthermore, the rate of change of the electrical energy must be equal to the total power $p_s + p_r$. Equations (2.1-87) and (2.1-88) can be proved by expressing the stator and rotor space phasor voltages and currents in terms of the instantaneous phase-variable components as defined by eqns (2.1-4), (2.1-20), (2.1-25), (2.1-61), (2.1-62), and (2.1-74) and also by utilizing the assumption that there are no zero-sequence stator and rotor voltages and currents.

The losses are due to heat dissipation across the stator and rotor winding resistances, hysteresis and eddy-current losses within the magnetic material, friction losses between moving parts and either their bearings or the surrounding air, and dielectric losses in the electric fields. However, if only the winding losses are considered, the differential energy related to the stator and rotor losses can be expressed as

$$\mathrm{d}W_{\mathrm{loss}} = \tfrac{3}{2}(R_s|\bar{\imath}_s|^2 + R_r|\bar{\imath}_r|^2)\,\mathrm{d}t, \tag{2.1-89}$$

where R_s and R_r are the stator and rotor resistances respectively and the terms $\tfrac{3}{2}R_s|\bar{\imath}_s|^2$ and $\tfrac{3}{2}R_r|\bar{\imath}_r|^2$ correspond to the ohmic losses across the stator and rotor windings respectively.

The third differential energy component, the differential field energy, can be obtained by considering that the rate of change of the magnetic energy stored in the field must be equal to the input power minus the sum of the stator and rotor losses and the mechanical power. Thus

$$\mathrm{d}W_{\mathrm{field}}/\mathrm{d}t = \tfrac{3}{2}\,\mathrm{Re}(\bar{u}_{si}\bar{\imath}_s^* + \bar{u}'_{ri}\bar{\imath}'^*_r), \tag{2.1-90}$$

where $\bar{u}_{si}$ and $\bar{u}'_{ri}$ are the space phasors of the induced stator and rotor transformer e.m.f.s respectively, both expressed in the stationary reference frame fixed to the stator. The induced stator e.m.f. is due to the rate of change of the stator flux-linkage space phasor

$$\bar{u}_{si} = \frac{\mathrm{d}\bar{\psi}_s}{\mathrm{d}t} \tag{2.1-91}$$

and similarly, the induced rotor e.m.f., which is due to the rate of change of the rotor flux linkages, is defined as

$$\bar{u}'_{ri} = \frac{\mathrm{d}\bar{\psi}'_r}{\mathrm{d}t}. \tag{2.1-92}$$

Substitution of eqns (2.1-91) and (2.1-92) into eqn (2.1-90) yields the following expression for the differential magnetic energy stored in the field,

$$\mathrm{d}W_{\mathrm{field}}=\frac{3}{2}\,\mathrm{Re}\left(\bar{\imath}_{\mathrm{s}}^{*}\,\frac{\mathrm{d}\bar{\psi}_{\mathrm{s}}}{\mathrm{d}t}+\bar{\imath}'^{*}_{\mathrm{r}}\,\frac{\mathrm{d}\bar{\psi}'_{\mathrm{r}}}{\mathrm{d}t}\right)\mathrm{d}t \tag{2.1-93}$$

and it follows that the stored magnetic energy can be expressed as

$$W_{\mathrm{field}}=\tfrac{3}{2}\,\mathrm{Re}(\bar{\imath}_{\mathrm{s}}^{*}\bar{\psi}_{\mathrm{s}}+\bar{\imath}'^{*}_{\mathrm{r}}\bar{\psi}'_{\mathrm{r}})$$

which is the physically expected result. It can be put into the more familiar non-space-phasor form, if the space-phasor quantities are expressed in terms of their phase-variable components.

Thus, by the substitution of eqns (2.1-86), (2.1-89), and (2.1-93) into eqn (2.1-85), the following equation is obtained for the differential mechanical energy

$$\mathrm{d}W_{\mathrm{mech}}=\mathrm{d}W_{\mathrm{mechs}}+\mathrm{d}W_{\mathrm{mechr}} \tag{2.1-94}$$

where $\mathrm{d}W_{\mathrm{mechs}}$ and $\mathrm{d}W_{\mathrm{mechr}}$ are the mechanical energies due to the stator and rotor respectively,

$$\mathrm{d}W_{\mathrm{mechs}}=\tfrac{3}{2}[\mathrm{Re}(\bar{u}_{\mathrm{s}}\bar{\imath}_{\mathrm{s}}^{*})-R_{\mathrm{s}}|\bar{\imath}_{\mathrm{s}}|^{2}-\mathrm{Re}(\bar{\imath}_{\mathrm{s}}^{*}\,\mathrm{d}\bar{\psi}_{\mathrm{s}}/\mathrm{d}t)]\mathrm{d}t \tag{2.1-95}$$

and

$$\mathrm{d}W_{\mathrm{mechr}}=\tfrac{3}{2}[\mathrm{Re}(\bar{u}'_{\mathrm{r}}\bar{\imath}'^{*}_{\mathrm{r}})-R_{\mathrm{r}}|\bar{\imath}'_{\mathrm{r}}|^{2}-\mathrm{Re}(\bar{\imath}'^{*}_{\mathrm{r}}\,\mathrm{d}\bar{\psi}'_{\mathrm{r}}/\mathrm{d}t)]\,\mathrm{d}t. \tag{2.1-96}$$

Since in the stationary reference frame, the stator voltage space phasor $\bar{u}_{\mathrm{s}}$ can only be balanced by the stator ohmic drop ($R_{\mathrm{s}}\bar{\imath}_{\mathrm{s}}$) plus the rate of change of the stator flux linkages ($\mathrm{d}\bar{\psi}_{\mathrm{s}}/\mathrm{d}t$)—see also Section 2.1.7—it follows from eqn (2.1-95) that $\mathrm{d}W_{\mathrm{mechs}}$ must be equal to zero. Furthermore, in the same reference frame, the rotor voltage space phasor $\bar{u}'_{\mathrm{r}}$ must be balanced by the sum of the rotor ohmic voltage drop ($R_{\mathrm{r}}\bar{\imath}'_{\mathrm{r}}$) plus the rate of change of the rotor flux linkages ($\mathrm{d}\bar{\psi}'_{\mathrm{r}}/\mathrm{d}t$) and a rotational voltage $-j\omega_{\mathrm{r}}\bar{\psi}'_{\mathrm{r}}$. (See also Section 2.1.7 where it is shown that in the stationary reference frame, the rotor voltage space-phasor equation must also contain this rotational e.m.f. term, because the rotor is rotating in the direction from stator phase A to stator phase B as shown in Fig. 2.1.) Therefore, by considering eqns (2.1-83), (2.1-94), (2.1-95), and (2.1-96), the mechanical power is obtained as

$$p_{\mathrm{mech}}=\tfrac{3}{2}\,\mathrm{Re}(-j\omega_{\mathrm{r}}\bar{\psi}'_{\mathrm{r}}\bar{\imath}'^{*}_{\mathrm{r}})=\tfrac{3}{2}\omega_{\mathrm{r}}\,\mathrm{Re}(-j\bar{\psi}'_{\mathrm{r}}\bar{\imath}'^{*}_{\mathrm{r}})=-\tfrac{3}{2}\omega_{\mathrm{r}}\bar{\psi}'_{\mathrm{r}}\times\bar{\imath}'_{\mathrm{r}}. \tag{2.1-97}$$

According to eqn (2.1-97), the mechanical power is proportional to the instantaneous rotor speed and to the vectorial product of the rotor flux linkage and current space phasors, and it follows by considering eqn (2.1-83) that the electromagnetic torque can be expressed as

$$t_{\mathrm{e}}=-\tfrac{3}{2}\bar{\psi}'_{\mathrm{r}}\times\bar{\imath}'_{\mathrm{r}}. \tag{2.1-98}$$

For a machine with P pole pairs this has to be multiplied by P (for a two-pole machine $P=1$, for a four-pole machine $P=2$, etc.). It should be noted that in eqn (2.1-98) the rotor flux-linkage and current space phasors $\bar{\psi}'_r$ and $\bar{i}'_r$ are expressed in the stationary reference frame, but since the torque is invariant to the change of the reference frame, the expression $-\frac{3}{2}\bar{\psi}_r \times \bar{i}_r$ is also valid, where $\bar{\psi}_r$ and $\bar{i}_r$ are the space phasors of the rotor flux linkages and currents respectively, but are expressed in the reference frame fixed to the rotor.

It is possible to put eqn (2.1-98) into many other forms and eqn (2.1-81) can be obtained as follows. From eqns (2.1-51) and (2.1-52) the rotor flux-linkage space phasor expressed in the stationary reference frame must contain two flux-linkage components, one of which, $L_r\bar{i}'_r$, is produced by the rotor currents only, where L_r is the self-inductance of a rotor winding, and the other $L_m\bar{i}_s$ is a mutual flux-linkage component produced by the stator currents, where L_m is the magnetizing inductance and $\bar{i}_s$ is the space phasor of the stator currents in the stationary reference frame. Thus eqn (2.1-98) can be put into the following form

$$t_e = -\tfrac{3}{2}(L_r\bar{i}'_r + L_m\bar{i}_s)\times\bar{i}'_r = -\tfrac{3}{2}L_m\bar{i}_s\times\bar{i}'_r, \tag{2.1-99}$$

where the property that a vector product of a vector with itself is zero has been used. Expanding eqn (2.1-99), the following expression is obtained

$$t_e = -\frac{3}{2}L_m\bar{i}_s\times\bar{i}'_r = -\frac{3L_m}{2L_s}(L_s\bar{i}_s + L_m\bar{i}'_r)\times\bar{i}'_r, \tag{2.1-100}$$

since the vector product $\bar{i}'_r\times\bar{i}'_r$ gives zero. In eqn (2.1-100) L_s is the self-inductance of a stator winding and according to eqn (2.1-35), the term $L_s\bar{i}_s + L_m\bar{i}'_r$ is equal to the space phasor of the stator flux linkages expressed in the stationary reference frame ($\bar{\psi}_s$) and contains a self-flux linkage component produced by the stator currents ($L_s\bar{i}_s$) and a mutual flux linkage produced by the rotor currents ($L_m\bar{i}'_r$). Thus the electromagnetic torque produced by the two-pole machine can be expressed as

$$t_e = -\frac{3L_m}{2L_s}\bar{\psi}_s\times\bar{i}'_r. \tag{2.1-101}$$

If the effects of the main flux and leakage flux saturation are neglected, the magnetizing inductance (L_m) and the total stator self-inductance (L_s) are constant and eqn (2.1-101) yields eqn (2.1-81). Further expressions for the electromagnetic torque will be given in Section 2.1.8.

Finally, it should be noted that the general expressions derived for the instantaneous value of the electromagnetic torque are valid for symmetrical three-phase or quadrature-phase machines and no restriction has been made on the time variation of the stator and rotor currents, apart from the fact that there can be no zero-sequence stator or rotor currents. The space phasors used in this book rely on the assumption that the spatial distributions of the flux density and current density are sinusoidal. This is an important assumption frequently adopted in the generalized theory of electrical machines. It would be possible to define harmonic space phasors which correspond to non-sinusoidal flux density and current density distributions, but in this book the effects of space harmonics are incorporated only in the parameters of the machine under consideration and not in extra harmonic equations.

2.1.7 The voltage equations of smooth-air-gap machines in various reference frames

In this section the stator and rotor voltage differential equations, which are valid in both the steady-state and transient operation of smooth-air-gap machines, will be described by using three-phase variables, two-axis variables, and space phasors. The application of space-phasor theory results in a drastic simplification of the voltage equations compared with the phase-variable forms and, while it is possible to give both physical and mathematical analyses of the dynamics concerned in terms of the phase-variable equations, this would lead to unnecessary complications.

For better understanding and to enable a direct comparison to be made between the non-space-phasor and space-phasor forms, the three-phase equations are first briefly introduced and then the two-axis models are presented together with the corresponding matrix forms of the equations. Finally, the space-phasor equations are presented.

The phase-variable voltage equations in the natural reference frame

The phase-variable forms of the three-phase stator and rotor voltage equations are first formulated in their natural reference frames. Thus the stator voltage equations are formulated in the stationary reference frame fixed to the stator and the rotor voltage equations are formulated in the rotating reference frame fixed to the rotor.

The three-phase model: Here the symmetrical three-phase two-pole smooth-air-gap machine with sinusoidally distributed windings, discussed earlier, is considered and the effects of m.m.f. space harmonics are neglected. The schematic of the machine has been shown in Fig. 2.1. It is assumed that the stator and rotor voltages and currents can vary arbitrarily in time. The phase-variable form of the voltage equations is as follows.

In the stationary reference frame the stator voltage equations can be expressed as

$$u_{sA}(t)=R_s i_{sA}(t)+\mathrm{d}\psi_{sA}(t)/\mathrm{d}t \qquad (2.1\text{-}102)$$

$$u_{sB}(t)=R_s i_{sB}(t)+\mathrm{d}\psi_{sB}(t)/\mathrm{d}t \qquad (2.1\text{-}103)$$

$$u_{sC}(t)=R_s i_{sC}(t)+\mathrm{d}\psi_{sC}(t)/\mathrm{d}t, \qquad (2.1\text{-}104)$$

where $u_{sA}(t)$, $u_{sB}(t)$, and $u_{sC}(t)$, $i_{sA}(t)$, $i_{sB}(t)$, and $i_{sC}(t)$ are the instantaneous values of the stator voltages and currents respectively (in the stationary reference frame) and R_s is the resistance of a stator winding. $\psi_{sA}(t)$, $\psi_{sB}(t)$, and $\psi_{sC}(t)$ are the instantaneous values of the stator flux linkages in phases sA, sB, and sC respectively as defined by eqns (2.1-32), (2.1-33), and (2.1-34).

Similar expressions hold for the rotor voltage equations expressed in the reference frame fixed to the rotor,

$$u_{ra}(t) = R_r i_{ra}(t) + \mathrm{d}\psi_{ra}(t)/\mathrm{d}t \tag{2.1-105}$$

$$u_{rb}(t) = R_r i_{rb}(t) + \mathrm{d}\psi_{rb}(t)/\mathrm{d}t \tag{2.1-106}$$

$$u_{rc}(t) = R_r i_{rc}(t) + \mathrm{d}\psi_{rc}(t)/\mathrm{d}t, \tag{2.1-107}$$

where $u_{ra}(t)$, $u_{rb}(t)$, and $u_{rc}(t)$, $i_{ra}(t)$, $i_{rb}(t)$ and $i_{rc}(t)$ are the instantaneous values of the rotor phase voltages and currents respectively, R_r is the resistance of a rotor winding and $\psi_{ra}(t)$, $\psi_{rb}(t)$, and $\psi_{rc}(t)$ are the instantaneous values of the rotor flux linkages in rotor phases ra, rb, and rc respectively, as defined by eqns (2.1-42), (2.1-43), and (2.1-44).

For convenience, the stator and rotor voltage equations [eqns (2.1-102)–(2.1-107), eqns (2.1-32)–(2.1-34)], and eqns (2.1-42)–(2.1-44) of the three-phase machine can be combined into a single matrix equation:

$$\begin{bmatrix} u_{sA} \\ u_{sB} \\ u_{sC} \\ u_{ra} \\ u_{rb} \\ u_{rc} \end{bmatrix} = \begin{bmatrix} R_s + p\bar{L}_s & p\bar{M}_s & p\bar{M}_s & p\bar{M}_{sr}\cos\theta & p\bar{M}_{sr}\cos\theta_1 & p\bar{M}_{sr}\cos\theta_2 \\ p\bar{M}_s & R_s + p\bar{L}_s & p\bar{M}_s & p\bar{M}_{sr}\cos\theta_2 & p\bar{M}_{sr}\cos\theta & p\bar{M}_{sr}\cos\theta_1 \\ p\bar{M}_s & p\bar{M}_s & R_s + p\bar{L}_s & p\bar{M}_{sr}\cos\theta_1 & p\bar{M}_{sr}\cos\theta_2 & p\bar{M}_{sr}\cos\theta \\ p\bar{M}_{sr}\cos\theta & p\bar{M}_{sr}\cos\theta_2 & p\bar{M}_{sr}\cos\theta_1 & R_r + p\bar{L}_r & p\bar{M}_r & p\bar{M}_r \\ p\bar{M}_{sr}\cos\theta_1 & p\bar{M}_{sr}\cos\theta & p\bar{M}_{sr}\cos\theta_2 & p\bar{M}_r & R_r + p\bar{L}_r & p\bar{M}_r \\ p\bar{M}_{sr}\cos\theta_2 & p\bar{M}_{sr}\cos\theta_1 & p\bar{M}_{sr}\cos\theta & p\bar{M}_r & p\bar{M}_r & R_r + p\bar{L}_r \end{bmatrix} \begin{bmatrix} i_{sA} \\ i_{sB} \\ i_{sC} \\ i_{ra} \\ i_{rb} \\ i_{rc} \end{bmatrix}, \tag{2.1-108}$$

where $p = \mathrm{d}/\mathrm{d}t$, which operates on the inductances, since in general they can vary with current. The angles θ, θ_1 and θ_2 are defined as $\theta = \theta_r$, $\theta_1 = \theta_r + 2\pi/3$, and $\theta_2 = \theta_r + 4\pi/3$, where θ_r is the rotor angle shown in Fig. 2.1. The stator self-inductance of one stator phase winding $\bar{L}_s$ can be expressed as the sum of the stator leakage inductance L_{sl} and the stator magnetizing inductance L_{sm}, $\bar{L}_s = L_{sl} + L_{sm}$, and it should be noted that this relationship holds even when the m.m.f. distribution around the periphery is not sinusoidal. For the sinusoidal distribution, the mutual inductance between two stator windings $\bar{M}_s$ can be expressed in terms of the stator magnetizing inductance L_{sm} as

$$\bar{M}_s = L_{sm}\cos(2\pi/3) = -L_{sm}/2.$$

Similarly, it is possible to express the rotor self-inductance of one rotor phase winding $\bar{L}_r$ as the sum of the leakage inductance of a rotor phase winding L_{rl} and the rotor magnetizing inductance L_{rm}, $\bar{L}_r = L_{rl} + L_{rm}$. For a sinusoidal winding distribution the mutual inductance between the rotor windings is $\bar{M}_r = L_{rm}\cos(2\pi/3) = -L_{rm}/2$.

By simple physical considerations of the winding inductances it can be shown that the magnetizing inductance of the stator L_{sm} is related to the maximal value of the mutual inductance between the stator and rotor $\bar{M}_{sr}$ as $L_{sm} = (N_{se}/N_{re})\bar{M}_{sr}$, where N_{se} and N_{re} are the effective number of stator and rotor turns (see Section 2.1.2). Furthermore, the magnetizing inductance of the rotor is related to $\bar{M}_{sr}$ as $L_{rm} = (N_{re}/N_{se})\bar{M}_{sr}$. It also follows that $\bar{M}_{sr}^2 = L_{sm}L_{rm}$.

If the resultant three-phase magnetizing inductance L_m, first introduced in eqn (2.1-35), is utilized, where $L_m = \frac{3}{2}\bar{M}_{sr}$, then it follows from above that the total three-phase stator inductance, L_s [first used in eqn (2.1-35)], takes the form

$$L_s = \bar{L}_s - \bar{M}_s = L_{sl} + L_{sm} + \tfrac{1}{2}L_{sm} = L_{sl} + \tfrac{3}{2}L_{sm}$$

and the total three-phase rotor inductance [first introduced in eqn (2.1-45)] can be expressed as

$$L_r = \bar{L}_r - \bar{M}_r = L_{rl} + L_{rm} + \tfrac{1}{2}L_{rm} = L_{rl} + \tfrac{3}{2}L_{rm}.$$

There are smooth-air-gap machines—e.g. the three-phase squirrel cage induction machine—where instead of coil-wound rotor windings, the rotor currents flow in bars embedded in the rotor slots and connected to end-rings. There are various types of cage rotors, but a so-called single-cage rotor has uniformly distributed bars, it can be considered as a multi-phase winding system, where the number of rotor phases depends on the number of rotor bars and pole-pairs and for a symmetrical rotor construction it is possible to represent this by an equivalent symmetrical three-phase rotor winding and the three-phase model described above can be used for this machine too. If the machine has a so-called double-cage rotor, with an inner and an outer cage, then for modelling purposes, it is still possible to use an equivalent three-phase rotor winding where the rotor parameters (resistance, inductance) contain the effects related to the double cage. However, for certain forms of double cage, or machines with deep bar rotor, it can be advantageous to use a more accurate model. The space-phasor equations for double-cage induction machines or induction machines with deep rotor bars will be discussed in detail in Section 4.3.5.

It is important to note that in eqn (2.1-108) the stator voltages and currents are expressed in the stationary reference frame fixed to the stator, and thus these quantities can be directly monitored on the stator. However, the rotor voltages and currents are expressed in the reference frame fixed to the rotor, so they can be measured on the rotating rotor which of course is almost impossible in the case of a squirrel-cage rotor. Thus, two reference frames have been used for establishing the hybrid model represented by eqns (2.1-108); these are the so-called natural reference frames.

It follows from eqn (2.1-108) that the system of voltage differential equations is non-linear, and contains variable, time-dependent coefficients, since in general the rotor angle is a non-linear function of time and, furthermore, some other parameters can also be variable. For example, under saturated conditions some of the inductance parameters vary with the currents. However, even when all the machine parameters are considered to be constant, the voltage differential equations will contain variable coefficients, since in general the rotor angle varies with time. Furthermore, in eqn (2.1-108), there are 36 terms in the impedance matrix of the machine and the windings are fully coupled.

The complete performance of the machine under transient conditions can be determined by using eqn (2.1-108) together with the equation of motion

$$t_e - t_l = J\frac{d\omega_r}{dt} + D\omega_r, \qquad (2.1\text{-}109)$$

where t_e is the developed electromagnetic torque of the machine, usually given in Newton meters (N m), t_l is the load torque (N m), ω_r is the rotor speed (rad/sec) and is equal to the first time derivative of the electrical rotor angle, $d\theta_r/dt$. The relationship between the electrical and mechanical rotor angles is $\theta_r = P\theta_{rm}$, where θ_{rm} is the mechanical rotor angle, and P is the number of pole pairs. J is the inertia of the rotor, the unit of which is usually given either as kilogram meter^2 (k gm^2) or Joules second^2 (J s^2), or as a quantity called WR^2 expressed in units of pound mass feet^2 (lb mft^2). The quantity $J\omega_r$ is called the inertia constant and is closely related to the kinetic energy of the rotating masses, which can be expressed as $W = \frac{1}{2}J\omega_r^2$. $D\omega_r$ is the damping torque and D is the damping constant which represents dissipation due to windage and friction.

In the voltage equations, significant simplification can be performed if the three-phase variables are replaced by their two-phase equivalents expressed in the same reference frame, since in this case, in the absence of zero-sequence components, there will only be four voltage equations, corresponding to direct- and quadrature-axis stator and rotor voltage equations respectively. Thus there will only be 16 elements in the corresponding new (transformed) impedance matrix. Further simplification can be achieved by using other than the natural reference frames. For example, if instead of expressing the rotor quantities (voltages, currents, flux linkages) in their natural reference frames, they are expressed in the stationary reference frame, the resulting voltage differential equations will contain constant coefficients if the parameters of the machine are considered to be constant. Because of the advantages of these simplifications, they will be briefly discussed below.

The quadrature-phase slip-ring model: To describe the so-called quadrature-phase slip-ring model of smooth-air-gap machines, first the number of phase variables is reduced in the three-phase model, by assuming that there are no zero-sequence voltages and currents on the stator or rotor. Physically this corresponds to using a quadrature-phase machine model instead of the three-phase model. In this so-called quadrature-phase slip-ring model, shown in Fig. 2.7, there is a quadrature-phase stator winding (sD, sQ) and a quadrature-phase rotor winding (rα, rβ); rotor winding rα is displaced from stator winding sD by the angle θ_r.

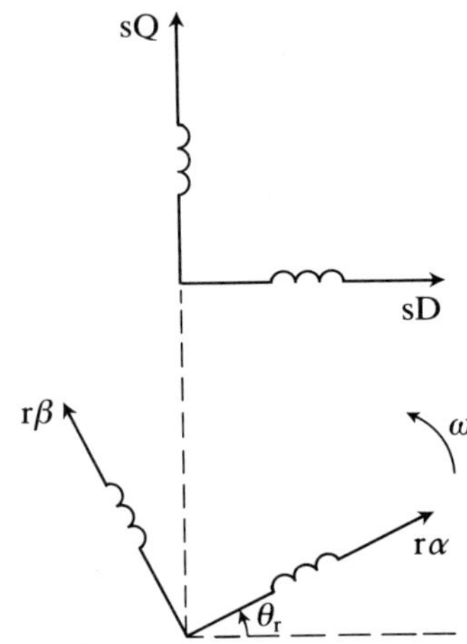

Fig. 2.7. Schematic of the quadrature-phase slip-ring model.

The relationship between the two-axis stator voltage and current components and the corresponding three-phase components can be obtained by considering eqns (2.1-63)–(2.1-66),

$$u_{sD}=\tfrac{2}{3}(u_{sA}-\tfrac{1}{2}u_{sB}-\tfrac{1}{2}u_{sC}) \qquad (2.1\text{-}110)$$

$$u_{sQ}=(u_{sB}-u_{sC})/\sqrt{3} \qquad (2.1\text{-}111)$$

and

$$i_{sD}=\tfrac{2}{3}(i_{sA}-\tfrac{1}{2}i_{sB}-\tfrac{1}{2}i_{sC}) \qquad (2.1\text{-}112)$$

$$i_{sQ}=(i_{sB}-i_{sC})\sqrt{3}. \qquad (2.1\text{-}113)$$

Similar transformations hold for the rotor voltages and currents

$$u_{r\alpha}=\tfrac{2}{3}(u_{ra}-\tfrac{1}{2}u_{rb}-\tfrac{1}{2}u_{rc}) \qquad (2.1\text{-}114)$$

$$u_{r\beta}=(u_{rb}-u_{rc})/\sqrt{3} \qquad (2.1\text{-}115)$$

and

$$i_{r\alpha}=\tfrac{2}{3}(i_{ra}-\tfrac{1}{2}i_{rb}-\tfrac{1}{2}i_{rc}) \tag{2.1-116}$$

$$i_{r\beta}=(i_{rb}-i_{rc})/\sqrt{3}. \tag{2.1-117}$$

Thus with eqns (2.1-110)–(2.1-117), after some algebraic manipulation, eqn (2.1-108) takes the following form:

$$\begin{bmatrix} u_{s\alpha} \\ u_{s\beta} \\ u_{r\alpha} \\ u_{r\beta} \end{bmatrix} = \begin{bmatrix} R_s+pL_s & 0 & pL_m\cos\theta_r & -pL_m\sin\theta_r \\ 0 & R_s+pL_s & pL_m\sin\theta_r & pL_m\cos\theta_r \\ pL_m\cos\theta_r & pL_m\sin\theta_r & R_r+pL_r & 0 \\ -pL_m\sin\theta_r & pL_m\cos\theta_r & 0 & R_r+pL_r \end{bmatrix} \begin{bmatrix} i_{s\alpha} \\ i_{s\beta} \\ i_{r\alpha} \\ i_{r\beta} \end{bmatrix}, \tag{2.1-118}$$

where L_s and L_r are the stator and rotor inductances, introduced in eqns (2.1-35) and (2.1-45) respectively and $L_m=\tfrac{3}{2}\bar{M}_{sr}$ is the magnetizing inductance. In eqn (2.1-118) the stator and rotor variables are expressed in their natural reference frames and, even if the machine parameters are constant, in general the system of voltage differential equations will be time dependent, since the equations contain the rotor angle θ_r which changes with time. If the inductances are constant, the differential operator $p=\mathrm{d}/\mathrm{d}t$ can be moved after the inductance elements.

In the so-called generalized theory of electrical machines, where the various models are obtained by using matrix transformations, the model described by eqn (2.1-118) is usually obtained from the three-phase model described by eqn (2.1-108) by the application of the so-called phase transformation matrix $\mathbf{C}_1$. Thus if $\mathbf{Z}_s$ is the impedance matrix of the three-phase machine described by eqn (2.1-108), the impedance matrix of the quadrature-phase slip ring model described by eqn (2.1-118) can be obtained by performing $\mathbf{Z}_s'=\mathbf{C}_t\mathbf{Z}_s\mathbf{C}$, where $\mathbf{C}=\mathrm{diag}(\mathbf{C}_1;\mathbf{C}_1)$. The inverse of the phase transformation matrix has been defined by eqn (2.1-69). It follows from eqn (2.1-118) that as a consequence of the phase transformation, there are four zero elements in the impedance matrix of the machine and in contrast to the impedance matrix of the three-phase model, which contains 36 non-zero elements, the impedance matrix of the quadrature-phase slip-ring machine contains 12 non-zero elements. However, it is possible to achieve a further reduction in the elements of the impedance matrix, and this is discussed in the following section.

The quadrature-phase commutator model: If the stationary-axis stator quantities of the quadrature-phase slip-ring model are unchanged, but the rotor voltages and currents $u_{r\alpha}$, $i_{r\alpha}$, $u_{r\beta}$, $i_{r\beta}$ are transformed (from the rotor reference frame) into a new reference frame fixed to the stator, the so-called quadrature-phase commutator model with pseudo-stationary rotor windings is obtained. The schematic of this machine is shown in Fig. 2.8; on the stator there are the direct- and quadrature-axis windings denoted by sD, sQ and on the rotor there are the windings denoted by rd and rq respectively.

The following transformations can be obtained from eqns (2.1-40) and (2.1-74) for the rotor voltages and currents

$$u_{r\alpha}=\cos\theta_r u_{rd}+\sin\theta_r u_{rq} \tag{2.1-119}$$

$$u_{r\beta}=-\sin\theta_r u_{rd}+\cos\theta_r u_{rq} \tag{2.1-120}$$

and similarly

$$i_{r\alpha}=\cos\theta_r i_{rd}+\sin\theta_r i_{rq} \tag{2.1-121}$$

$$i_{r\beta}=-\sin\theta_r i_{rd}+\cos\theta_r i_{rq}. \tag{2.1-122}$$

Thus by considering eqns (2.1-118)–(2.1-122), the following transformed set of voltage equations of the commutator model (sometimes also referred to as the commutator primitive model) is obtained

$$\begin{bmatrix} u_{sD} \\ u_{sQ} \\ u_{rd} \\ u_{rq} \end{bmatrix} = \begin{bmatrix} R_s+pL_s & 0 & pL_m & 0 \\ 0 & R_s+pL_s & 0 & pL_m \\ pL_m & \omega_r L_m & R_r+pL_r & \omega_r L_r \\ -\omega_r L_m & pL_m & -\omega_r L_r & R_r+pL_r \end{bmatrix} \begin{bmatrix} i_{sD} \\ i_{sQ} \\ i_{rd} \\ i_{rq} \end{bmatrix}. \tag{2.1-123}$$

In this model the cosine and sine functions of the rotor angle are not present

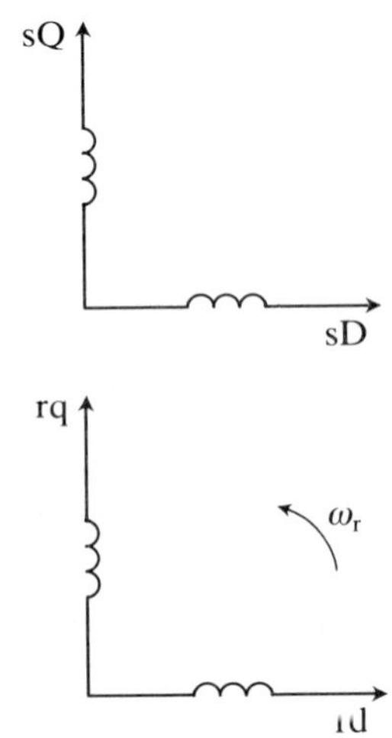

Fig. 2.8. Schematic of the quadrature-phase commutator model.

in the impedance matrix, but the rotor speed ω_r is present in the rotor equations. In the generalized theory of electrical machines, this model is usually obtained from the quadrature-phase slip-ring model by performing $\mathbf{C}_t\mathbf{Z}'_s\mathbf{C}$, where $\mathbf{C}=\text{diag}(\mathbf{I}_2; \mathbf{C}_2)$, $\mathbf{I}_2$ is a 2×2 identity matrix, the matrix $\mathbf{C}_2^{-1}$ has been defined in eqn (2.1-40) and $\mathbf{Z}'_s$ is the impedance matrix of the quadrature-phase slip-ring machine described by eqn (2.1-118). It is also possible to obtain eqn (2.1-123) by physical considerations only, e.g. by considering that in the direct-axis rotor winding, rd, there must be voltages induced by transformer effects, i.e. $p(L_m i_{sD}+L_r i_{rd})$ and rotational voltages due to the rotation of the rotor, i.e. $\omega_r(L_m i_{sQ}+L_r i_{rq})$. Under linear magnetic conditions, the operator p can be moved after the inductances in eqn (2.1-123). Some aspects of the effects of magnetic saturation will be discussed in Chapter 5 and in Section 5.1.1 the quadrature-phase commutator model will be described, where the effects of the saturation of the main flux paths are incorporated in the equations; eqn (2.1-123) is significantly modified.

The space-phasor form of the equations

In the present section the space-phasor forms of the voltage equations of three-phase and quadrature-phase smooth-air-gap machines will be presented. The equations will be expressed both in the stationary reference frames and in a general rotating reference frame, which rotates at a general speed ω_g. The relationship between the space-phasor and matrix forms, given in the previous section, will also be shown.

The space-phasor voltage equations in the stationary reference frame: The stator and rotor voltage differential equations of the smooth-air-gap three-phase machine have been given above in their natural reference frames. By utilizing the definitions of the voltage space phasors [eqns (2.1-61), (2.1-62)], current space phasors [eqns (2.1-4), (2.1-20)] and flux-linkage space phasors [eqns (2.1-31), (2.1-41)], the stator and rotor voltage equations [eqns (2.1-102)–(2.1-104) and eqns (2.1-105)–(2.1-107)] can be put into the following space-phasor form

$$\bar{u}_s = R_s\bar{i}_s + \frac{d\bar{\psi}_s}{dt} \tag{2.1-124}$$

and

$$\bar{u}'_r = R_r\bar{i}'_r + \frac{d\bar{\psi}'_r}{dt} - j\omega_r\bar{\psi}'_r. \tag{2.1-125}$$

In eqns (2.1-124) and (2.1-125) the stator and rotor space phasor flux linkages $\bar{\psi}_s$ and $\bar{\psi}'_r$ are present and these have been defined in terms of the machine inductances and the space phasors of the stator and rotor currents in eqns (2.1-35) and (2.1-51) respectively. ω_r is the instantaneous angular speed of the rotor.

For the sake of a better overview, the definitions of all the space-phasor quantities will be repeated below using the definitions of the space phasors of the three-phase quantities. The space phasors of the stator voltages, currents and flux linkages in the stationary reference frame fixed to the stator are [see eqns (2.1-61), (2.1-4), (2.1-31), and (2.1-35)]:

$$\bar{u}_s = \tfrac{2}{3}[u_{sA}(t) + a u_{sB}(t) + a^2 u_{sC}(t)] = u_{sD} + j u_{sQ} \tag{2.1-126}$$

$$\bar{i}_s = \tfrac{2}{3}[i_{sA}(t) + a i_{sB}(t) + a^2 i_{sC}(t)] = i_{sD} + j i_{sQ} \tag{2.1-127}$$

$$\bar{\psi}_s = \tfrac{2}{3}[\psi_{sA}(t) + a\psi_{sB}(t) + a^2\psi_{sC}(t)] = \psi_{sD} + j\psi_{sQ} = L_s\bar{i}_s + L_m\bar{i}'_r \tag{2.1-128}$$

and similarly the space phasors of the rotor voltages, currents, and flux linkages in the reference frame fixed to the rotor are [see eqns (2.1-62), (2.1-20), (2.1-41), and (2.1-45)]

$$\bar{u}_r = \tfrac{2}{3}[u_{ra}(t) + a u_{rb}(t) + a^2 u_{rc}(t)] = u_{r\alpha} + j u_{r\beta} \tag{2.1-129}$$

$$\bar{i}_r = \tfrac{2}{3}[i_{ra}(t) + a i_{rb}(t) + a^2 i_{rc}(t)] = i_{r\alpha} + j i_{r\beta} \tag{2.1-130}$$

$$\bar{\psi}_r = \tfrac{2}{3}[\psi_{ra}(t) + a\psi_{rb}(t) + a^2\psi_{rc}(t)] = L_r\bar{i}_r + L_m\bar{i}'_s = L_r\bar{i}_r + L_m\bar{i}_s e^{j\theta_r}$$
$$= \psi_{r\alpha} + j\psi_{r\beta}. \tag{2.1-131}$$

The rotor quantities defined above are, in the reference frame fixed to the stator [see eqns (2.1-74), (2.1-39), and (2.1-51)]

$$\bar{u}'_r = \bar{u}_r e^{j\theta_r} = u_{rd} + j u_{rq} \tag{2.1-132}$$

$$\bar{i}'_r = \bar{i}_r e^{j\theta_r} = i_{rd} + j i_{rq} \tag{2.1-133}$$

$$\bar{\psi}'_r = \bar{\psi}_r e^{j\theta_r} = L_r\bar{i}'_r + L_m\bar{i}_s = L_r\bar{i}_r e^{j\theta_r} + L_m\bar{i}_s = \psi_{rd} + j\psi_{rq}. \tag{2.1-134}$$

For completeness, the space phasors of the stator voltages, currents and flux linkages are also given in the rotating reference frame fixed to the rotor [see eqns (2.1-73), (2.1-54), and (2.1-56)] as

$$\bar{u}'_s = \bar{u}_s e^{-j\theta_r} = u_{sd} + j u_{sq} \tag{2.1-135}$$

$$\bar{i}'_s = \bar{i}_s e^{-j\theta_r} = i_{sd} + j i_{sq} \tag{2.1-136}$$

$$\bar{\psi}'_s = \bar{\psi}_s e^{-j\theta_r} = \psi_{sd} + j\psi_{sq}. \tag{2.1-137}$$

The very compact form of eqns (2.1-124) and (2.1-125) makes their application extremely convenient. The first term on the right-hand side of these equations is the space phasor form of the ohmic losses, the second term is a transformer e.m.f., which is the first time derivative of the flux-linkage space phasor of the stator and rotor respectively. Finally, in eqn (2.1-125), the term $-j\omega_r\bar{\psi}'_r$

represents a rotational e.m.f., which is due to the rotation of the rotor and contributes to electromechanical energy conversion. Equations (2.1-124) and (2.1-125) together with the flux linkage equations, eqns (2.1-128), (2.1-131), and (2.1-134), are also valid under saturated conditions.

If the flux-linkage space phasors defined by eqns (2.1-128) and (2.1-134) are substituted into eqns (2.1-124) and (2.1-125), then the space-phasor voltage equations take the following form

$$\bar{u}_s = R_s \bar{i}_s + \mathrm{d}(L_s \bar{i}_s)/\mathrm{d}t + \mathrm{d}(L_m \bar{i}'_r)/\mathrm{d}t \tag{2.1-138}$$

and

$$\bar{u}'_r = R_r \bar{i}'_r + \mathrm{d}(L_r \bar{i}'_r)/\mathrm{d}t + \mathrm{d}(L_m \bar{i}_s)/\mathrm{d}t - j\omega_r(L_r \bar{i}'_r + L_m \bar{i}_s). \tag{2.1-139}$$

These equations can also be put into matrix form:

$$\begin{bmatrix} \bar{u}_s \\ \bar{u}'_r \end{bmatrix} = \begin{bmatrix} R_s & 0 \\ 0 & R_r \end{bmatrix} \begin{bmatrix} \bar{i}_s \\ \bar{i}'_r \end{bmatrix} + \frac{\mathrm{d}}{\mathrm{d}t} \begin{bmatrix} L_s & L_m \\ L_m & L_r \end{bmatrix} \begin{bmatrix} \bar{i}_s \\ \bar{i}'_r \end{bmatrix} - j\omega_r \begin{bmatrix} 0 & 0 \\ L_m & L_r \end{bmatrix} \begin{bmatrix} \bar{i}_s \\ \bar{i}'_r \end{bmatrix}. \tag{2.1-140}$$

The simplicity of these equations should be contrasted with eqn (2.1-108), but while in eqn (2.1-108) the phase-variable voltages and currents appear, in eqn (2.1-140) the space phasors of the voltages and currents are present. If these space phasors are resolved into their real- and imaginary-axis components in accordance with eqns (2.1-126), (2.1-127), (2.1-132), and (2.1-133), then eqn (2.1-140) yields eqn (2.1-123), which corresponds to the quadrature-phase commutator model. Thus it can be seen that if the space-phasor voltage equations are established in the stationary reference frame, they can be used to obtain the equations of the commutator model, without having to perform any matrix transformation. The ability to obtain various models of a machine, in various reference frames, without using matrix transformations, is also an advantage of the application of space-phasors over the application of the conventional generalized matrix theory of electrical machines.

Furthermore, from eqns (2.1-132) and (2.1-133), $\bar{u}_r = \bar{u}'_r e^{-j\theta_r}$ and $\bar{i}_r = i'_r e^{-j\theta_r}$, and the rotor quantities can be expressed in the reference frame fixed to the rotor and eqn (2.1-140) becomes

$$\begin{bmatrix} \bar{u}_s \\ \bar{u}_r \end{bmatrix} = \begin{bmatrix} R_s & 0 \\ 0 & R_r \end{bmatrix} \begin{bmatrix} \bar{i}_s \\ \bar{i}_r \end{bmatrix} + \frac{\mathrm{d}}{\mathrm{d}t} \begin{bmatrix} L_s & L_m e^{j\theta_r} \\ L_m e^{-j\theta_r} & L_r \end{bmatrix} \begin{bmatrix} \bar{i}_s \\ \bar{i}_r \end{bmatrix}. \tag{2.1-141}$$

If all the space-phasor quantities are expressed in terms of their real- and imaginary-axis components, i.e. from eqns (2.1-126), (2.1-127), (2.1-129), and (2.1-130), eqn (2.1-141) yields eqn (2.1-118), which corresponds to the quadrature-phase slip-ring model. Another model can therefore be obtained from the space-phasor equations without utilizing matrix transformations. Of course, the transformation procedure is now 'hidden' in the complex transformations, which contain the terms $e^{j\theta_r}$ and $e^{-j\theta_r}$.

For slip-ring induction machines with short-circuited rotor windings, or induction machines with squirrel-cage rotor, $\bar{u}_r=0$.

The space-phasor voltage equations in the general reference frame: Here the voltage space-phasor equations will be formulated in a general reference frame, which rotates at a general speed ω_g.

It follows from the analysis presented earlier, e.g. from eqn (2.1-136) or Fig. 2.4, that the space phasor of the stator currents in the rotor reference frame is $\bar{i}_s'=\bar{i}_s e^{-j\theta_r}$ where $\bar{i}_s$ is the stator current space phasor in the stator reference frame. Similarly it follows from eqn (2.1-133), that the space phasor of the rotor currents in the stationary reference frame is $\bar{i}_r'=\bar{i}_r e^{j\theta_r}$ where $\bar{i}_r$ is the space phasor of the rotor currents in the rotor reference frame. However, if instead of a reference frame fixed to the rotor, a general reference frame, with direct and quadrature axes x, y rotating at a general instantaneous speed $\omega_g=\mathrm{d}\theta_g/\mathrm{d}t$ is used, as shown in Fig. 2.9, where θ_g is the angle between the direct-axis of the stationary reference frame sD fixed to the stator and the real axis (x) of the general reference frame, then the following equation defines the stator current space phasor in the general reference frame

$$\bar{i}_{sg}=\bar{i}_s e^{-j\theta_g}=i_{sx}+ji_{sy}. \tag{2.1-142}$$

By the substitution of $\theta_g=\theta_r$, eqn (2.1-136) is obtained, which gives the space phasor of the stator currents expressed in the rotor reference frame. Equation (2.1-142) can be proved mathematically by considering that, from Fig. 2.9, in the stator reference frame $\bar{i}_s=|\bar{i}_s|e^{j\alpha_s}$ and in the general reference frame $\bar{i}_{sg}=|\bar{i}_s|e^{j(\alpha_s-\theta_g)}$, thus

$$\bar{i}_{sg}=|\bar{i}_s|e^{j\alpha_s}e^{-j\theta_g}=\bar{i}_s e^{-j\theta_g}.$$

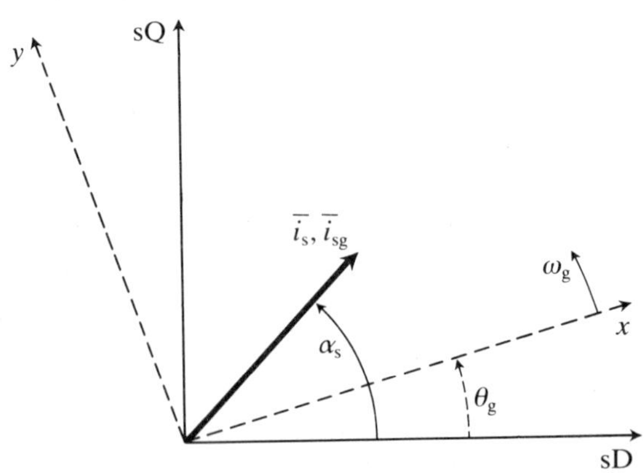

Fig. 2.9. Application of the general reference frame; transformation of the stator quantities.

The stator voltage and flux-linkage space phasors can be similarly obtained in the general reference frame,

$$\bar{u}_{sg}=\bar{u}_s e^{-j\theta_g}=u_{sx}+ju_{sy} \tag{2.1-143}$$

$$\bar{\psi}_{sg}=\bar{\psi}_s e^{-j\theta_g}=\psi_{sx}+j\psi_{sy}, \tag{2.1-144}$$

where $\bar{u}_s$ and $\bar{\psi}_s$ are the space phasors of the stator voltages and stator flux linkages respectively in the stationary reference frame.

Similar considerations hold for the space phasors of the rotor voltages, current and flux linkages. In Fig. 2.10 three reference frames are shown, the reference frame fixed to the rotor, the reference frame fixed to the stator and the general reference frame. The magnetic axes of the rotor winding are rα, rβ and the rotor axis rα is displaced from the direct axis of the stator reference frame by the rotor angle θ_r.

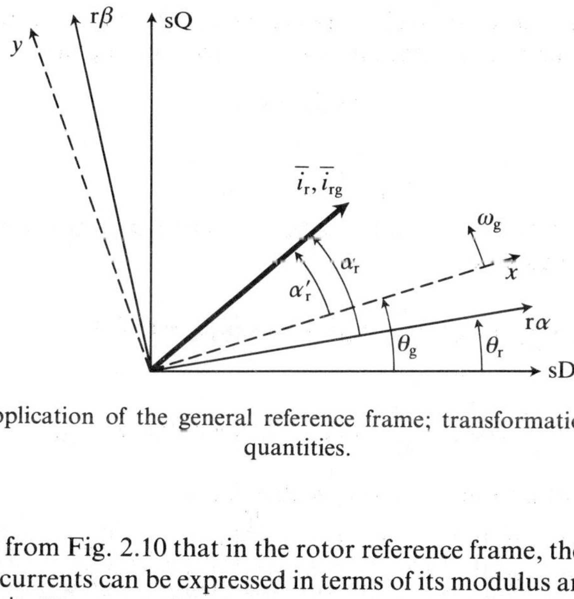

Fig. 2.10. Application of the general reference frame; transformation of the rotor quantities.

It follows from Fig. 2.10 that in the rotor reference frame, the space phasor of the rotor currents can be expressed in terms of its modulus and phase angle α_r as $\bar{i}_r=|\bar{i}_r|e^{j\alpha_r}$. However, since it can be seen that the angle between the real axis (x) of the general reference frame and the real axis of the reference frame rotating with the rotor (rα) is $\theta_g-\theta_r$, in the general reference frame the space phasor of the rotor currents can be expressed as $\bar{i}_{rg}=|\bar{i}_r|e^{j\alpha'_r}$, where $\alpha'_r=\alpha_r-(\theta_g-\theta_r)$. Thus

$$\bar{i}_{rg}=|\bar{i}_r|e^{j\alpha_r}e^{-j(\theta_g-\theta_r)}=\bar{i}_r e^{-j(\theta_g-\theta_r)}=i_{rx}+ji_{ry}. \tag{2.1-145}$$

In the reference frame fixed to the stator, $\theta_g=0$ and $\bar{i}_{rg}=\bar{i}'_r=\bar{i}_r e^{j\theta_r}$ in accordance with eqn (2.1-133). Similarly, the space phasors of the rotor voltages and rotor flux linkages in the general reference frame can be expressed as

$$\bar{u}_{rg}=\bar{u}_r e^{-j(\theta_g-\theta_r)}=u_{rx}+ju_{ry} \tag{2.1-146}$$

and

$$\bar{\psi}_{rg}=\bar{\psi}_r e^{-j(\theta_g-\theta_r)}=\psi_{rx}+j\psi_{ry}. \tag{2.1-147}$$

Substitution of eqns (2.1-142)–(2.1-147) into eqns (2.1-124), (2.1-125), (2.1-128), and (2.1-131) yields the following stator and rotor space-phasor voltage equations in the general reference frame

$$\bar{u}_{sg}=R_s\bar{i}_{sg}+\frac{d\bar{\psi}_{sg}}{dt}+j\omega_g\bar{\psi}_{sg} \tag{2.1-148}$$

$$\bar{u}_{rg}=R_r\bar{i}_{rg}+\frac{d\bar{\psi}_{rg}}{dt}+j(\omega_g-\omega_r)\bar{\psi}_{rg}, \tag{2.1-149}$$

where the stator and rotor flux linkages in the general reference frame can be expressed in terms of the stator and rotor current space phasors as

$$\bar{\psi}_{sg}=L_s\bar{i}_{sg}+L_m\bar{i}_{rg} \tag{2.1-150}$$

and

$$\bar{\psi}_{rg}=L_r\bar{i}_{rg}+L_m\bar{i}_{sg}. \tag{2.1-151}$$

Equations (2.1-148)–(2.1-151) can be combined into the following two complex voltage equations

$$\bar{u}_{sg}=R_s\bar{i}_{sg}+\frac{d}{dt}(L_s\bar{i}_{sg})+\frac{d}{dt}(L_m\bar{i}_{rg})+j\omega_g(L_s\bar{i}_{sg}+L_m\bar{i}_{rg}) \tag{2.1-152}$$

$$\bar{u}_{rg}=R_r\bar{i}_{rg}+\frac{d}{dt}(L_r\bar{i}_{rg})+\frac{d}{dt}(L_m\bar{i}_{sg})+j(\omega_g-\omega_r)(L_r\bar{i}_{rg}+L_m\bar{i}_{sg}), \tag{2.1-153}$$

which can be put into the following matrix form

$$\begin{bmatrix}\bar{u}_{sg}\\ \bar{u}_{rg}\end{bmatrix}=\begin{bmatrix}R_s & 0\\ 0 & R_r\end{bmatrix}\begin{bmatrix}\bar{i}_{sg}\\ \bar{i}_{rg}\end{bmatrix}+\frac{d}{dt}\begin{bmatrix}L_s & L_m\\ L_m & L_r\end{bmatrix}\begin{bmatrix}\bar{i}_{sg}\\ \bar{i}_{rg}\end{bmatrix}+j\omega_g\begin{bmatrix}L_s & L_m\\ L_m & L_r\end{bmatrix}\begin{bmatrix}\bar{i}_{sg}\\ \bar{i}_{rg}\end{bmatrix}$$
$$-j\omega_r\begin{bmatrix}0 & 0\\ L_m & L_r\end{bmatrix}\begin{bmatrix}\bar{i}_{sg}\\ \bar{i}_{rg}\end{bmatrix}. \tag{2.1-154}$$

On the right-hand side of eqn (2.1-153) there are the voltage terms due to the ohmic losses, plus the transformer e.m.f.s, plus the voltages due to the rotation of the general reference frame, plus the true rotational voltages. This last component is only present in the rotor equations and it is this term $-j\omega_r\bar{\psi}_{rg}=-j\omega_r(L_r\bar{i}_{rg}+L_m\bar{i}_{sg})$ which contributes to electromechanical energy conversion. It is also possible to utilize the expression for the rotational voltage to obtain the expressions for the electromagnetic torque in various reference frames, as shown in the next section. In eqns (2.1-152)–(2.1-154), under linear magnetic conditions all the inductance elements can be considered to be constant, and thus the operator d/dt can be moved after the inductances, so it directly operates on the space phasors of the currents.

It is also possible to express all the space-phasor quantities in terms of their real- and imaginary-axis components and thus by utilizing eqns (2.1-142), (2.1-143), (2.1-145), and (2.1-146), eqn (2.1-154) can be put into the following two-axis form,

$$\begin{bmatrix} u_{sx} \\ u_{sy} \\ u_{rx} \\ u_{ry} \end{bmatrix} = \begin{bmatrix} R_s+pL_s & -\omega_g L_s & pL_m & -\omega_g L_m \\ \omega_g L_s & R_s+pL_s & \omega_g L_m & pL_m \\ pL_m & -(\omega_g-\omega_r)L_m & R_r+pL_r & -(\omega_g-\omega_r)L_r \\ (\omega_g-\omega_r)L_m & pL_m & (\omega_g-\omega_r)L_r & R_r+pL_r \end{bmatrix} \begin{bmatrix} i_{sx} \\ i_{sy} \\ i_{rx} \\ i_{ry} \end{bmatrix}, \qquad (2.1\text{-}155)$$

where again $p=\mathrm{d}/\mathrm{d}t$. If $\omega_g=0$, eqn (2.1-155) yields eqn (2.1-123), which corresponds to the quadrature-phase commutator model. If $\omega_g=\omega_1$, where ω_1 is the so-called synchronous speed, and where, say, for an induction machine, $\omega_1-\omega_r=s\omega_1$ is the slip speed and s is the slip, the two-axis equations of the induction machine in the synchronously rotating reference frame are

$$\begin{bmatrix} u_{sx} \\ u_{sy} \\ u_{rx} \\ u_{ry} \end{bmatrix} = \begin{bmatrix} R_s+pL_s & -\omega_1 L_s & pL_m & -\omega_1 L_m \\ \omega_1 L_s & R_s+pL_s & \omega_1 L_m & pL_m \\ pL_m & -s\omega_1 L_m & R_r+pL_r & -s\omega_1 L_r \\ s\omega_1 L_m & pL_m & s\omega_1 L_r & R_r+pL_r \end{bmatrix} \begin{bmatrix} i_{sx} \\ i_{sy} \\ i_{rx} \\ i_{ry} \end{bmatrix}, \qquad (2.1\text{-}156)$$

where u_{sx}, u_{sy} and i_{sx}, i_{sy} are the two-axis components of the stator voltages and currents respectively in the synchronously rotating reference frame. The rotor voltage and current components in the same reference frame are $u_{rx}=u_{ry}=0$ and i_{rx}, i_{ry}. Under linear magnetic conditions, the operator p can be moved after the inductance terms.

The voltage equations are valid under both transient and steady-state conditions. However, under transient conditions, the equation of motion [eqn (2.1-109)],

$$t_e - t_l = J\frac{\mathrm{d}\omega_r}{\mathrm{d}t} + D\omega_r, \qquad (2.1\text{-}157)$$

is also required, where for a machine with P pole-pairs, similarly to eqn (2.1-99), the electromagnetic torque can be expressed as follows, in terms of the stator and rotor current space phasors and their two-axis components formulated in the general reference frame [defined by eqns (2.1-142) and (2.1-145)]:

$$t_e = -\tfrac{3}{2}PL_m \bar{i}_{sg} \times \bar{i}_{rg} = -\tfrac{3}{2}PL_m(i_{sx}i_{ry} - i_{sy}i_{rx}). \qquad (2.1\text{-}158)$$

In eqn (2.1-157) the rotor speed can be expressed as

$$\omega_r = \frac{d\theta_r}{dt}, \qquad (2.1\text{-}159)$$

where θ_r is the rotor angle.

Equation (2.1-158) can also be proved by considering eqns (2.1-98) and (2.1-99). However, in eqn (2.1-98) $\bar{\psi}_r'$ and $\bar{i}_r'$ are the rotor flux linkage and current space phasors formulated in the stationary reference frame fixed to the stator and by considering $\theta_g = 0$, it follows from eqns (2.1-145) and (2.1-147) that they are defined as $\bar{i}_r' = \bar{i}_r e^{j\theta_r}$ and $\bar{\psi}_r' = \bar{\psi}_r e^{j\theta_r}$, where $\bar{i}_r$ and $\bar{\psi}_r$ are space phasors in the rotor reference frame. Similarly, in the general reference frame they are defined as

$$\bar{i}_{rg} = \bar{i}_r e^{-j(\theta_g - \theta_r)} = \bar{i}_r e^{j\theta_r} e^{-j\theta_g} \text{ and } \psi_{rg} = \psi_r e^{-j(\theta_g - \theta_r)} = \psi_r e^{j\theta_r} e^{-j\theta_g}.$$

Thus by also considering eqn (2.1-98), the electromagnetic torque is

$$t_e = -\tfrac{3}{2}\bar{\psi}_r' \times \bar{i}_r' = -\tfrac{3}{2}\bar{\psi}_{rg} e^{j\theta_g} \times \bar{i}_{rg} e^{j\theta_g} = -\tfrac{3}{2}\bar{\psi}_{rg} \times \bar{i}_{rg}, \qquad (2.1\text{-}160)$$

where $\bar{\psi}_{rg}$ has been defined in eqn (2.1-151). Substitution of eqn (2.1-151) into eqn (2.1-160) gives $t_e = -\tfrac{3}{2}L_m \bar{i}_{sg} \times \bar{i}_{rg}$ for the two-pole machine and this agrees with eqn (2.1-158) when $P = 1$.

Another derivation of the electromagnetic torque can be based on the technique used in the conventional generalized matrix theory of electrical machines, where the electromagnetic torque is obtained from the so-called torque matrix,

$$t_e = \tfrac{3}{2} P \mathbf{i}_t^* \mathbf{G} \mathbf{i} \qquad (2.1\text{-}161)$$

where $\mathbf{G}$ is the torque matrix and $\mathbf{i}$ is the column vector of the currents in the general reference frame. From eqn (2.1-156), $\mathbf{i}_t$ is defined as $\mathbf{i}_t = [i_{sx}, i_{sy}\ i_{rx}, i_{ry}]$, where t denotes the transpose, and the torque is defined as

$$\mathbf{G} = \begin{bmatrix} 0 & 0 & 0 & 0 \\ 0 & 0 & 0 & 0 \\ 0 & L_m & 0 & L_r \\ -L_m & 0 & -L_r & 0 \end{bmatrix}, \qquad (2.1\text{-}162)$$

since in eqn (2.1-156) $s\omega_1 = \omega_1 - \omega_r$ and $\mathbf{G}$ contains the terms multiplied by ω_r.

It follows from eqns (2.1-161) and (2.1-162) that

$$t_e = \tfrac{3}{2} P (i_{rx}\psi_{ry} - i_{ry}\psi_{rx}), \qquad (2.1\text{-}163)$$

which can be expressed as a vector product,

$$t_e = -\tfrac{3}{2} P \bar{\psi}_{rg} \times \bar{i}_{rg}$$

which agrees with eqn (2.1-160) if a two-pole machine ($P = 1$) is considered.

It follows that the stator and rotor voltage equations (two complex or four real) expressed in the general reference frame, together with eqns (2.1-157) and (2.1-158) are suitable for the simulation of the transient operation of smooth-air-gap a.c. machines and for designing various forms of control of a.c. drives. With different constraints in the rotor circuit, the general model is applicable to smooth-air-gap synchronous, asynchronous and doubly-fed machines and valid for an arbitrary instantaneous variation of the voltages and currents; the only restriction is that there can be no zero-sequence line-to-neutral voltages and phase currents. Thus the general model is suitable for the analysis of a large number of converter-fed machines, where different voltage or current waveforms are generated by the converter under consideration. If the neutral point of the machine is isolated, but the line-to-neutral voltages are unbalanced and contain a zero-sequence component, the space-phasor equations can still be used, since the voltage asymmetry means only that the neutral of the stator windings has a different potential with respect to the neutral of the supply voltages.

2.1.8 Various expressions for electromagnetic torque

Here different expressions for the electromagnetic torque are obtained, by utilizing various space-phasor quantities (voltages, currents, flux linkages) expressed in different reference frames. It will also be shown that in special reference frames fixed to the rotor flux, the stator flux or the magnetizing flux linkage space phasor, the expression of the electromagnetic torque is similar to the expression for the electromagnetic torque produced by a separately excited d.c. machine. This analogy serves as a basis for various forms of vector control, where the torque control of the a.c. machine is similar to the torque control of the separately excited d.c. machine.

Application of the general reference frame

Further to the expressions for the electromagnetic torque given in Section 2.1.6, it is possible to obtain various other useful expressions of the electromagnetic torque. To give a better physical insight into the processes involved, it is first shown that eqn (2.1-149) can be used directly to obtain a general expression for the electromagnetic torque.

It follows from eqn (2.1-149) that the space-phasor form of the rotational e.m.f. in the rotor windings is $-j\omega_r\bar{\psi}_{rg}=-j\omega_r(\psi_{rx}+j\psi_{ry})$. Thus the real-axis component of this e.m.f. is $u_{rotx}=\omega_r\psi_{ry}$ and the imaginary-axis rotational e.m.f. is $u_{roty}=-\omega_r\psi_{rx}$. It follows that the mechanical power can be expressed as

$$p_{mech}=\tfrac{3}{2}(u_{rotx}i_{rx}+u_{roty}i_{ry})=\tfrac{3}{2}\omega_r(\psi_{ry}i_{rx}-\psi_{rx}i_{ry})=-\tfrac{3}{2}\omega_r\bar{\psi}_{rg}\times\bar{i}_{rg}$$

and when this is divided by the rotor speed ω_r, it yields an expression for the electromagnetic torque which is in agreement with eqn (2.1-160).

It follows from eqn (2.1-160) that the expression for the electromagnetic torque is similar in all reference frames, which is an expected result. For a machine with P pole-pairs, eqn (2.1-160) takes the following form

$$t_e = -\tfrac{3}{2}P\bar{\psi}'_r \times \bar{i}'_r = -\tfrac{3}{2}P\bar{\psi}_{rg} \times \bar{i}_{rg}, \tag{2.1-164}$$

according to which the torque is produced by the interaction of the rotor flux-linkage space phasor and the rotor current space phasor. By a similar argument, it follows that eqn (2.1-99) can be written as

$$t_e = -\tfrac{3}{2}PL_m\bar{i}_s \times \bar{i}'_r = -\tfrac{3}{2}PL_m\bar{i}_{sg} \times \bar{i}_{rg}, \tag{2.1-165}$$

according to which the electromagnetic torque is proportional to the vector product of the stator and rotor current space phasors. It is also possible to obtain eqn (2.1-165) by the substitution of eqn (2.1-151) into eqn (2.1-164). Similarly, eqn (2.1-101) can be written as

$$t_e = -\tfrac{3}{2}P(L_m/L_s)\bar{\psi}_{sg} \times \bar{i}_{rg}, \tag{2.1-166}$$

which could also be obtained from eqns (2.1-150) and (2.1-165).

By the application of the principle of action–reaction, it is possible to write eqn (2.1-164) as

$$t_e = \tfrac{3}{2}P\bar{\psi}_{sg} \times \bar{i}_{sg}. \tag{2.1-167}$$

There are many other ways of giving a rigorous mathematical proof of eqn (2.1-167), e.g. by considering eqn (2.1-165)

$$t_e = \tfrac{3}{2}PL_m\bar{i}_{rg} \times \bar{i}_{sg} \tag{2.1-168}$$

is obtained, which can be expanded as follows, by considering that the vector product $\bar{i}_{sg} \times \bar{i}_{sg} = 0$ and using eqn (2.1-150):

$$t_e = \tfrac{3}{2}P(L_s\bar{i}_{sg} + L_m\bar{i}_{rg}) \times \bar{i}_{sg} = \tfrac{3}{2}P\bar{\psi}_{sg} \times \bar{i}_{sg}, \tag{2.1-169}$$

which agrees with eqn (2.1-167).

It is also possible to split the total stator flux linkages into the sum of a leakage ($\bar{\psi}_{slg}$) and a magnetizing flux linkage ($\bar{\psi}_{mg}$); thus from eqn (2.1-150)

$$\bar{\psi}_{sg} = \bar{\psi}_{slg} + \bar{\psi}_{mg} \tag{2.1-170}$$

where

$$\bar{\psi}_{slg} = L_{sl}\bar{i}_{sg} \tag{2.1-171}$$

is the space phasor of the stator leakage flux linkages in the general reference frame and L_{sl} is the leakage inductance of a stator winding. If the stator and rotor windings have equal numbers of effective turns,

$$\bar{\psi}_{mg} = L_m\bar{i}_{mg}, \tag{2.1-172}$$

which gives the space phasor of the magnetizing flux linkages in the general reference frame, where $\bar{i}_{mg}$ is the magnetizing current space phasor in the general reference frame,

$$\bar{i}_{mg}=\bar{i}_{sg}+\bar{i}_{rg}. \qquad (2.1\text{-}173)$$

Thus substitution of eqns (2.1-170)–(2.1-173) into eqn (2.1-169) yields

$$t_e=\tfrac{3}{2}P(L_{sl}\bar{i}_{sg}+L_m\bar{i}_{mg})\times\bar{i}_{sg}=\tfrac{3}{2}PL_m\bar{i}_{mg}\times\bar{i}_{sg}=\tfrac{3}{2}P\bar{\psi}_{mg}\times\bar{i}_{sg}, \qquad (2.1\text{-}174)$$

according to which the electromagnetic torque can be expressed as the vector product of the magnetizing-current and stator current space phasors. As physically expected, it also follows that the stator leakage fluxes do not contribute to the production of the electromagnetic torque.

By resolving all the space phasors into their real- (x) and imaginary-axis (y) components, it is possible to obtain the electromagnetic torque in terms of the two-axis components of the general reference frame. Thus from eqn (2.1-174), with $\bar{\psi}_{mg}=\psi_{mx}+j\psi_{my}$ and $\bar{i}_{sg}=i_{sx}+ji_{sy}$,

$$t_e=\tfrac{3}{2}P(\psi_{mx}i_{sy}-\psi_{my}i_{sx}). \qquad (2.1\text{-}175)$$

Expansion of eqn (2.1-168) yields the following equation for the electromagnetic torque, if $\bar{i}_{sg}\times\bar{i}_{sg}=0$ and eqn (2.1-151) is utilized:

$$t_e=\frac{3}{2}PL_m\bar{i}_{rg}\times\bar{i}_{sg}=\frac{3}{2}P\frac{L_m}{L_r}\bar{i}_{sg}(L_r\bar{i}_{rg}\times\bar{i}_{sg})$$

$$=\frac{3}{2}P\frac{L_m}{L_r}(L_r\bar{i}_{rg}+L_m\bar{i}_{sg})\times\bar{i}_{sg}=\frac{3}{2}P\frac{L_m}{L_r}\bar{\psi}_{rg}\times\bar{i}_{sg}. \qquad (2.1\text{-}176)$$

By the resolution into direct-axis (x) and quadrature-axis (y) components of the general reference frame and using $\bar{\psi}_{rg}=\psi_{rx}+j\psi_{ry}$, $\bar{i}_{sg}=i_{sx}+ji_{sy}$,

$$t_e=\frac{3}{2}P\frac{L_m}{L_r}(\psi_{rx}i_{sy}-\psi_{ry}i_{sx}), \qquad (2.1\text{-}177)$$

where ψ_{rx} and ψ_{ry} are the two-axis components of the rotor flux-linkage space phasor in the general reference frame, and which from eqn (2.1-151), can be expressed as $\psi_{rx}=L_r i_{rx}+L_m i_{sx}$ and $\psi_{ry}=L_r i_{ry}+L_m i_{sy}$.

The application of special reference frames and the basis for vector control

In Chapter 1 and Section 2.1.6 one aim of the application of vector-controlled a.c. motors is given, which is to be able to implement control schemes which yield high dynamic performance and are similar to those used to control d.c. machines. For this purpose it must first be proved that, under transient conditions, the electromagnetic torque in smooth-air-gap machines is proportional to the product of a flux-producing current component and a torque-producing current component which are in space quadrature. It will be shown in this section that there are three similar ways of achieving this. These are to use reference frames fixed to the stator flux linkage space phasor, to the rotor flux-linkage space phasor, or to the magnetizing flux linkage space phasor (in smooth-air-gap machines the latter is equivalent to fixing the reference frame to the magnetizing current space phasor) to derive an expression for the electromagnetic torque in a form which enables independent control of the torque- and flux-producing current components. The three cases will be discussed in turn.

The application of a special reference frame fixed to the magnetizing flux linkage space phasor: The expression for the electromagnetic torque will be obtained in terms of the modulus of the magnetizing flux-linkage space phasor and the quadrature-axis stator current component, expressed in the reference frame fixed to the magnetizing flux-linkage space phasor.

An induction machine with a symmetrical multiphase, short-circuited rotor winding is assumed, where for simplicity the three-phase stator windings (sA, sB, sC) are replaced by their equivalent quadrature-phase stator windings (sD, sQ). In contrast to Fig. 2.5, which was used to explain the mechanism of torque production in a d.c. machine, with armature winding on the stator, and field and compensating windings on the rotor, in Fig. 2.11 is shown the schematic of an induction machine, with stator windings sD and sQ and a multi-phase short-circuited rotor winding system.

The direct-axis stator current i_{sD} produces the magnetizing flux ψ'_m. If the quadrature-axis current i_{sQ} is suddenly injected into the stator winding sQ—the compensating winding for the d.c. machine—rotor current i'_r will flow in the rotor windings, which at the first instant of time (t_0) is $i'_r = -i_{sQ}$. Thus the space-phasor diagram of the magnetizing flux linkage and stator and rotor currents will be similar to the space-phasor diagram shown in Fig. 2.6 for a d.c. machine. This is shown in Fig. 2.12, where $\bar{i}_s = i_{sD} + ji_{sQ}$ is the space phasor of the stator currents, and $\bar{i}'_r$ is the space phasor of the rotor currents in the stationary reference frame.

From eqn (2.1-173), the sum of the stator and rotor current space phasors

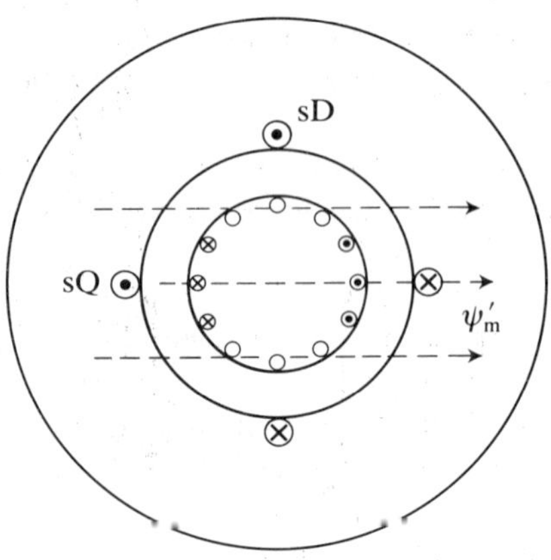

Fig. 2.11. Schematic of the induction machine with quadrature-phase stator windings.

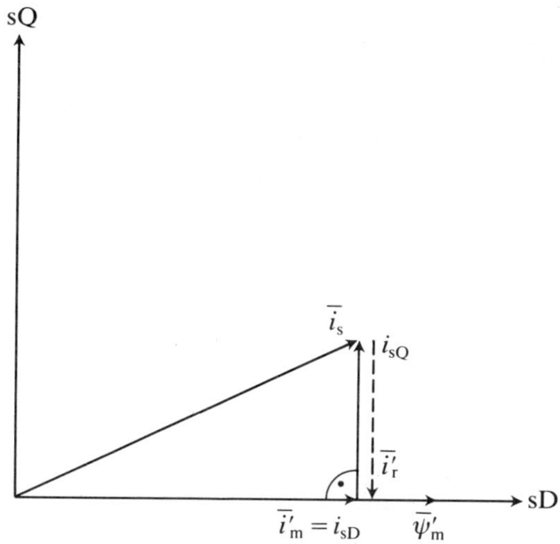

Fig. 2.12. Space phasors of $\bar{i}_s$, $\bar{i}'_r$, and $\bar{\psi}'_m$ at instant t_0.

gives the magnetizing-current phasor if the stator and rotor windings are assumed to have the same number of effective turns. Furthermore, it follows from eqn (2.1-172) that the magnetizing flux linkage space phasor in the stationary reference frame is $\bar{\psi}'_m = L_m \bar{i}'_m$, which is proportional to the magnetizing-current space phasor if the magnetizing inductance L_m is constant, and coaxial with the direct axis of the stationary reference frame. However, at $t > t_0$, the space-phasor diagram must be changed, since the field must change to induce the rotor currents. For simplicity, it is assumed that the rotor is locked in the same position as in Fig. 2.11. The new space-phasor diagram is shown in Fig. 2.13, with the new (changed) magnetizing flux and rotor current space phasors.

It follows from Fig. 2.13 that, in contrast to Fig. 2.12 or Fig. 2.5 (which corresponds to a d.c. machine), the magnetizing flux-linkage space phasor $\bar{\psi}'_m = L_m \bar{i}'_m$ is not coaxial with the direct axis of the stationary reference frame and is not in space quadrature to the quadrature axis of the same reference frame, but it is rotated by angle μ_m, with respect to the direct axis sD. Thus $\bar{\psi}'_m$ is not along the direct axis sD.

If however, there is a fictitious rotation of the stator through angle μ_m in the counter-clockwise direction, the magnetizing flux-linkage space phasor $\bar{\psi}'_m = L_m \bar{i}'_m$ will again be coaxial with the direct-axis stator current and will be in space quadrature to the quadrature-axis stator current, as shown in Fig. 2.14, which is similar to the conditions shown in Fig. 2.5 for the d.c. machine.

In order to keep $\bar{\psi}'_m$ along the direct-axis, if $\bar{\psi}'_m$ is not allowed to rotate, the

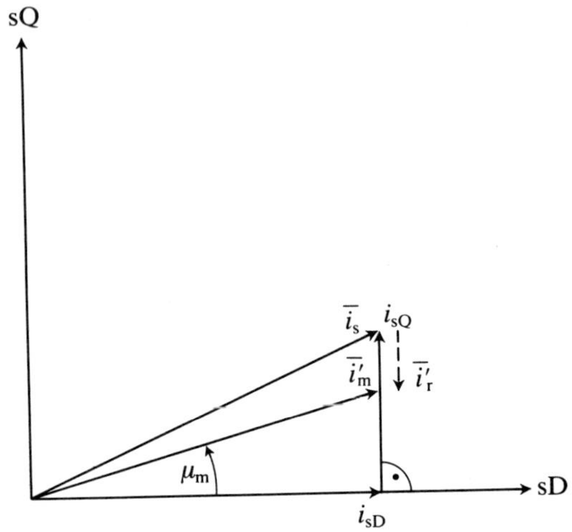

Fig. 2.13. Space phasors of $\bar{i}_s$, $\bar{i}'_r$, and $\bar{\psi}'_m$ at instant $t > t_0$.

stator has to be continuously rotated by the appropriate value of μ_m, until the required orientation is achieved. The conditions are then similar to those discussed for the compensated d.c. machine discussed in Section 2.1.6. In reality, of course, the stator is not rotated and the stator windings remain stationary, and to ensure separation of the stator currents into such two current components, one of which produces the magnetizing flux linkage and the other which produces the electromagnetic torque, a special reference frame (with direct and quadrature axes x, y) has to be used. There are several ways to

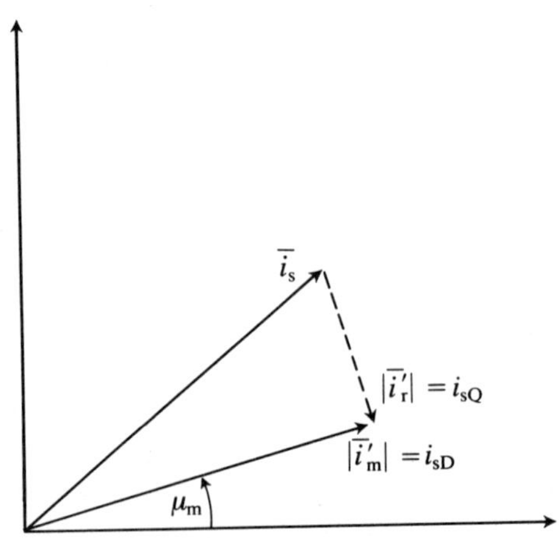

Fig. 2.14. Space phasors of the currents and magnetizing flux linkage when the stator is rotated by the angle μ_m.

choose such a special reference frame. One is to select a reference frame which rotates at the speed of the magnetizing flux-linkage space phasor and has its real-axis coaxial with the magnetizing flux-linkage space phasor, as shown in Fig. 2.15.

It follows from Fig. 2.15 that, if in the stationary reference frame (sD, sQ) the magnetizing flux-linkage space phasor is defined as

$$\bar{\psi}'_{\mathrm{m}} = |\bar{\psi}_{\mathrm{m}}| e^{j\mu_{\mathrm{m}}} = \psi_{\mathrm{mD}} + j\psi_{\mathrm{mQ}} \tag{2.1-178}$$

where $|\bar{\psi}_{\mathrm{m}}|$ and μ_{m} are the modulus and space angle respectively of the magnetizing space phasor and ψ_{mD} and ψ_{mQ} are its two-axis components in the special reference frame, the space phasor of the stator currents in the special reference frame rotating at speed

$$\omega_{\mathrm{m}} = \frac{\mathrm{d}\mu_{\mathrm{m}}}{\mathrm{d}t} \tag{2.1-179}$$

is

$$\bar{i}_{\mathrm{sm}} = \bar{i}_{\mathrm{s}} e^{-j\mu_{\mathrm{m}}} = i_{\mathrm{s}x} + j i_{\mathrm{s}y}, \tag{2.1-180}$$

where $\bar{i}_{\mathrm{s}}$ is the space phasor of the stator currents in the stationary reference frame. Equation (2.1-180) also follows from eqn (2.1-142), by using $\theta_{\mathrm{g}} = \mu_{\mathrm{m}}$ and replacing the subscript g (used to denote the general reference frame) by the subscript m (to denote the reference frame fixed to the magnetizing flux-linkage space phasor). Since the two-axis components of $\bar{i}_{\mathrm{sm}}$ are $i_{\mathrm{s}x}$ and $i_{\mathrm{s}y}$ and the two-axis components of $\bar{i}_{\mathrm{s}}$ are i_{sD} and i_{sQ} (see Fig. 2.15), eqn (2.1-180) can be written as

$$\bar{i}_{\mathrm{sm}} = i_{\mathrm{s}x} + j i_{\mathrm{s}y} = (i_{\mathrm{sD}} + j i_{\mathrm{sQ}}) e^{-j\mu_{\mathrm{m}}} \tag{2.1-181}$$

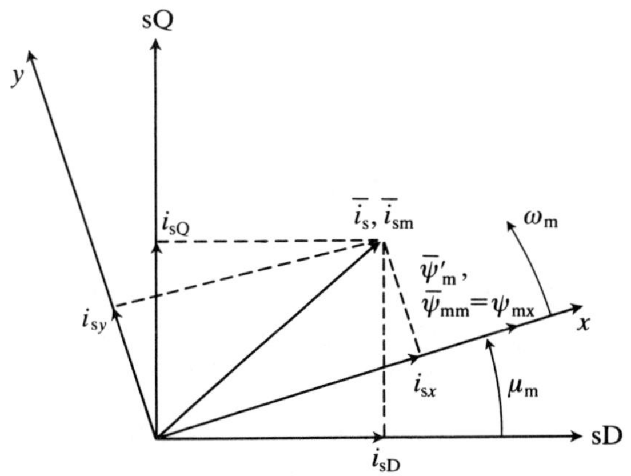

Fig. 2.15. Relationship of stationary (sD, sQ) and special (x, y) reference frames.

which can be put into the matrix form,

$$\begin{bmatrix} i_{sx} \\ i_{sy} \end{bmatrix} = \begin{bmatrix} \cos\mu_m & \sin\mu_m \\ -\sin\mu_m & \cos\mu_m \end{bmatrix} \begin{bmatrix} i_{sD} \\ i_{sQ} \end{bmatrix}. \tag{2.1-182}$$

Equation (2.1-182) shows how the direct- and quadrature-axis stator current components in the special reference frame can be obtained from the direct- and quadrature-axis stator current components of the stationary reference frame if the angle μ_m, which is continuously changing in time, is known. In the transient state all the current components vary arbitrarily in time. In the special reference frame the direct axis is coaxial with the magnetizing flux-linkage space phasor, and thus the quadrature-axis component of the magnetizing flux-linkage space phasor is zero. Therefore it follows from eqn (2.1-174) or eqn (2.1-175) that, in the special magnetizing flux-linkage oriented reference frame, the electromagnetic torque is proportional to the product of the modulus of the magnetizing flux-linkage space phasor $|\bar{\psi}_m|$ and the quadrature-axis component of the stator current space phasor i_{sy},

$$t_e = \tfrac{3}{2}P|\bar{\psi}_m| i_{sy}, \tag{2.1-183}$$

It is possible to expand eqn (2.1-183). From eqns (2.1-172) and (2.1-174), in the special reference frame the magnetizing flux-linkage space phasor $\bar{\psi}_{mm}$, which contains only a direct-axis (x-axis) component ($\bar{\psi}_{mm} = \psi_{mx}$), can be expressed in terms of the magnetizing-current space phasor and the stator and rotor currents as

$$\psi_{mx} = \bar{\psi}_{mm} = L_m \bar{i}_{mm} = L_m(\bar{i}_{sm} + \bar{i}_{rm}) = L_m i_{mx} = |\bar{\psi}_m| = L_m|\bar{i}_{mm}|, \tag{2.1-184}$$

where $\bar{i}_{sm}$ has been defined in eqn (2.1-181) and the space phasor of the rotor currents in the special reference frame ($\bar{i}_{rm}$) is obtained from eqn (2.1-145) with $\theta_g = \mu_m$ as

$$\bar{i}_{rm} = \bar{i}_r e^{-j(\mu_m - \theta_r)} = i_{rx} + j i_{ry}. \tag{2.1-185}$$

Thus by the substitution of eqn (2.1-184) into eqn (2.1-183), the electromagnetic torque can be expressed as

$$t_e = \tfrac{3}{2}P\psi_{mx} i_{sy} = \tfrac{3}{2}PL_m i_{mx} i_{sy}, \tag{2.1-186}$$

where i_{mx} is the main flux-producing current component (direct-axis magnetizing-current component) and i_{sy} is the torque-producing stator current component.

Eqn (2.1-183) forms the basis of the so-called magnetizing flux-oriented control of smooth-air-gap machines, which can be implemented in many different ways, as will be discussed in later sections.

The application of a special reference frame fixed to the rotor flux-linkage space phasor: The expression for the electromagnetic torque will be obtained in terms of the modulus of the rotor flux linkage space phasor and the quadrature-axis stator current component, expressed in a special reference frame fixed to the rotor flux linkage space phasor.

From eqn (2.1-176), or its two-axis form eqn (2.1-177), in the general reference frame, if the effects of saturation are neglected, and thus L_m and L_r are constant, the torque is proportional to the vector product of the rotor flux linkage and the stator current space phasors expressed in the general reference frame. Thus, if such a rotating reference frame is used where the quadrature-axis component of the rotor flux-linkage space phasor is zero ($\psi_{ry}=0$), the electromagnetic torque will be produced by the interaction of the rotor flux linkage in the direct axis (of the special reference frame) and the quadrature-axis component of the stator currents (expressed in the same reference frame). Mathematically this can be described as follows.

According to eqn (2.1-147) or (2.1-134), the rotor flux-linkage space phasor in the stationary ($\omega_g=0$) reference frame can be expressed as

$$\bar{\psi}_r' = \bar{\psi}_r e^{j\theta_r} = \psi_{rd} + j\psi_{rq} = |\bar{\psi}_r| e^{j\rho_r}, \tag{2.1-187}$$

where $|\bar{\psi}_r|$ and ρ_r are the modulus and phase angle of the rotor flux-linkage space phasor in the stationary reference frame. They are shown in Fig. 2.16, which illustrates the relationship between the stator current components in the stationary reference frame and the special reference frame fixed to the rotor flux-linkage space phasor.

The special x, y reference frame shown in Fig. 2.16 rotates at the speed of the rotor flux-linkage space phasor,

$$\omega_{mr} - \frac{d\rho_r}{dt}. \tag{2.1-188}$$

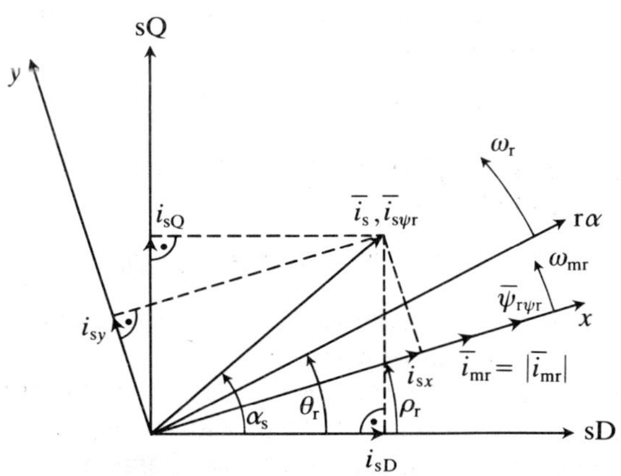

Fig. 2.16. Stator current and rotor flux-linkage space phasors in the stationary reference frame and in the special reference frame fixed to the rotor flux-linkage space phasor.

From eqn (2.1-142) with $\theta_g = \rho_r$, the stator current space phasor in the special reference frame is

$$\bar{i}_{s\psi r} = \bar{i}_s e^{-j\rho_r} = i_{sx} + j i_{sy}, \tag{2.1-189}$$

where $\bar{i}_s$ is the space phasor of the stator currents in the stationary reference frame. It also follows mathematically from eqn (2.1-147) and (2.1-187) that in the special reference frame ($\theta_g = \rho_r$) the rotor flux-linkage space phasor has only a direct-axis component,

$$\bar{\psi}_{r\psi r} = \bar{\psi}_r e^{-j(\rho_r - \theta_r)} = \bar{\psi}_r e^{j\theta_r} e^{-j\rho_r} = \bar{\psi}'_r e^{-j\rho_r} = |\bar{\psi}_r| e^{j\rho_r} e^{-j\rho_r} = |\bar{\psi}_r| = \psi_{rx}, \tag{2.1-190}$$

which is also shown in Fig. 2.16.

Thus by substituting eqns (2.1-189) and (2.1-190) into eqns (2.1-176) or (2.1-177), the electromagnetic torque is given by

$$t_e = \frac{3}{2} P \frac{L_m}{L_r} \psi_{rx} i_{sy}, \tag{2.1-191}$$

where $\psi_{rx} = |\bar{\psi}_r|$ is the modulus of the rotor flux-linkage space phasor and i_{sy} is the quadrature-axis stator current in the reference frame fixed to the rotor flux-linkage space phasor.

The relationship between the stator current components in the stationary reference frame (i_{sD}, i_{sQ}) and the stator current components in the special reference frame (i_{sx}, i_{sy}) can be obtained by considering eqn (2.1-189) as

$$\begin{bmatrix} i_{sx} \\ i_{sy} \end{bmatrix} = \begin{bmatrix} \cos\rho_r & \sin\rho_r \\ -\sin\rho_r & \cos\rho_r \end{bmatrix} \begin{bmatrix} i_{sD} \\ i_{sQ} \end{bmatrix}, \tag{2.1-192}$$

which can also be proved by considering Fig. 2.16.

The rotor flux-linkage space phasor in the general reference frame [defined by eqn (2.1-190)] can also be expressed in terms of the stator and rotor currents by considering eqn (2.1-151)

$$\bar{\psi}_{r\psi r} = \psi_{rx} = |\bar{\psi}_r| = L_r \bar{i}_{r\psi r} + L_m \bar{i}_{s\psi r}, \tag{2.1-193}$$

where the rotor current space phasor in the special reference frame is obtained from eqns (2.1-145) and (2.1-133),

$$\bar{i}_{r\psi r} = i_{rx} + j i_{ry} = \bar{i}_r e^{-j(\rho_r - \theta_r)} = \bar{i}_r e^{j\theta_r} e^{-j\rho_r} = \bar{i}'_r e^{-j\rho_r} = (i_{rd} + j i_{rq}) e^{-j\rho_r}, \tag{2.1-194}$$

where $\bar{i}'_r$ is the rotor current space phasor in the stationary reference frame and $\bar{i}_r$ is the rotor current space phasor in the reference frame fixed to the rotor. It follows from eqn (2.1-194) that in the special reference frame, both of the rotor current components (i_{rx}, i_{ry}) are non-zero. This is of course a physically expected result, since in the special reference frame the quadrature-axis component of the rotor flux linkage component is zero. Thus the quadrature-axis rotor current in the same reference frame can only be zero if there is no mutual rotor flux component due to the quadrature-axis stator current in the special reference frame, i.e. if the quadrature-axis stator current is zero ($i_{sy}=0$), which in general is not the case.

From eqn (2.1-193), the so-called rotor magnetizing current in the special reference frame is defined in terms of the stator and rotor current space phasors given by eqns (2.1-189) and (2.1-194) respectively, as

$$\bar{i}_{mr}=\frac{\bar{\psi}_{r\psi r}}{L_m}=\frac{L_r}{L_m}\bar{i}_{r\psi r}+\bar{i}_{s\psi r}=\bar{i}_{s\psi r}+(1+\sigma_r)\bar{i}_{r\psi r}, \tag{2.1-195}$$

where $\sigma_r=L_{rl}/L_m$ is the rotor leakage factor (L_{rl} and L_r are the rotor leakage inductance and self-inductance respectively). In Fig. 2.16 the rotor magnetizing-current space phasor is also shown. From eqn (2.1-195), the rotor magnetizing current in the special reference frame must also have a component, but only along the real-axis of the special reference frame:

$$\bar{i}_{mr}=i_{mrx}+ji_{mry}=i_{mrx}=|\bar{i}_{mr}|=\frac{|\bar{\psi}_r|}{L_m}=\bar{i}_{s\psi r}+(1+\sigma_r)\bar{i}_{r\psi r}. \tag{2.1-196}$$

Thus, if $\psi_{rx}=|\bar{\psi}_r|$, and $\sigma_r=L_{rl}/L_m$, substitution of eqn (2.1-196) into eqn (2.1-191) yields

$$t_e=\frac{3}{2}P\frac{L_m^2}{L_r}|\bar{i}_{mr}|i_{sy}=\frac{3}{2}P\frac{L_m}{1+\sigma_r}|\bar{i}_{mr}|i_{sy}. \tag{2.1-197}$$

It is a very important feature of eqn (2.1-197) that it shows that the electromagnetic torque can be controlled by independently controlling the flux-producing current component $|\bar{i}_{mr}|$ and the torque producing stator current component i_{sy}. Under linear magnetic conditions L_m, L_r, and the term $\frac{3}{2}PL_m/(1+\sigma_r)$ are constant, and the expression for the torque is similar to that of the separately excited, compensated d.c. machine.

For an induction machine, under linear magnetic conditions, when $|\bar{i}_{mr}|$ is kept constant, $|\bar{i}_{mr}|=i_{sx}=$constant. This will be proved in Section 4.1.1. Thus the rotor flux is solely determined by $|\bar{i}_{mr}|=$constant and the torque is proportional to i_{sy}. That $|\bar{i}_{mr}|=$constant leads to the expression $|\bar{i}_{mr}|=i_{sx}$, can also be proved by considering that the rotor voltage of the x-axis short-circuited rotor winding expressed in the special reference frame must be balanced by a rotor ohmic voltage drop ($R_r i_{rx}$) plus the rate of change of the rotor flux linkage expressed in the special reference frame ($d\psi_{rx}/dt$) and that there is no rotational voltage component due to the quadrature-axis rotor flux linkage (ψ_{ry}), since this component is zero due to the special selection of the reference frame. Under linear conditions $L_m=$constant and, if $|\bar{i}_{mr}|=$constant, it follows that $\psi_{rx}=L_m|\bar{i}_{mr}|=$constant (and thus its rate of change is zero). It follows that $i_{rx}=0$ and thus $|\bar{i}_{mr}|=i_{sx}+(1+\sigma_r)i_{rx}=i_{sx}$.

Torque control schemes of smooth-air-gap a.c. machines based on eqn (2.1-197) have so far found the most widespread applications and, despite the fact that in this case the special reference frame is aligned with the rotor flux-linkage space phasor, and rotates synchronously with the rotor flux-linkage space phasor, and therefore is not fixed to the magnetizing flux-linkage space phasor (as discussed in the previous section), this type of control is often referred to as field-oriented control. However, this terminology can be justified by considering that the rotor flux-linkage space phasor is coaxial with the rotor magnetizing-current space phasor and 'field-orientation' in this sense has the meaning of rotor magnetizing-current orientation. It would be more accurate to use the general terminology of vector control and to distinguish between magnetizing flux (or magnetizing-current)-oriented control (for the control ideas presented in the previous section) and rotor-flux-oriented control or rotor-magnetizing-current-oriented control for the control using the principles developed in the present section. There is also the possibility of stator-flux-oriented control (or stator-magnetizing-current-oriented control), which will be discussed in the next section.

The application of a special reference frame fixed to the stator flux linkage space phasor: It has already been shown [eqn (2.1-174)] that the electromagnetic torque can be expressed in terms of the cross product of the stator flux linkage and current space phasors. From eqn (2.1-167), if a special reference frame attached to the rotating stator flux-linkage space phasor is used, i.e. where the quadrature-axis stator flux-linkage component is zero, in both the steady-state and transient states the electromagnetic torque of a two-pole machine is

$$t_e = \tfrac{3}{2}\psi_{sx} i_{sy}. \tag{2.1-198}$$

ψ_{sx} is the real-axis component of the stator flux-linkage space phasor in the special reference frame rotating at the speed of the stator flux-linkage space phasor (ω_{ms}) and i_{sy} is the imaginary-axis component of the stator current space phasor in the same reference frame. The components of the different space phasors in the special reference frame are related to their corresponding components expressed in the stationary reference frame as follows.

It follows from eqn (2.1-128) that in the stationary reference frame the space phasor of the stator flux linkages can be expressed as

$$\bar{\psi}_s = \psi_{sD} + j\psi_{sQ} = |\bar{\psi}_s| e^{j\rho_s}, \tag{2.1-199}$$

where ψ_{sD} and ψ_{sQ} are the stationary axis components of the stator flux-linkage space phasor, $|\bar{\psi}_s|$ is its modulus, and ρ_s is its phase angle—a spatial angle—with respect to the real axis of the stationary reference frame (sD). The speed of the special reference frame can be expressed as

$$\omega_{ms} = \frac{d\rho_s}{dt}. \tag{2.1-200}$$

Figure 2.17 shows the relationship between the special and the stationary reference frames.

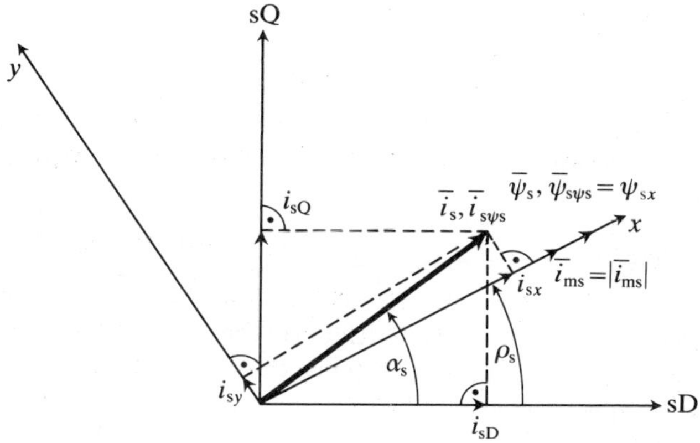

Fig. 2.17. Relationship between the stationary reference frame and the special reference frame fixed to the stator flux linkage space phasor.

From eqns (2.1-142) and (2.1-144), in the special reference frame shown in Fig. 2.17 ($\theta_g = \rho_s$), the space phasors of the stator current and stator flux linkages are

$$\bar{i}_{s\psi s} = \bar{i}_s e^{-j\rho_s} = (i_{sD} + j i_{sQ}) e^{-j\rho_s} = i_{sx} + j i_{sy} \tag{2.1-201}$$

and

$$\bar{\psi}_{s\psi s} = \bar{\psi}_s e^{-j\rho_s} = (\psi_{sD} + j\psi_{sQ}) e^{-j\rho_s} = \psi_{sx} + j\psi_{sx}, \tag{2.1-202}$$

where the two-axis stator current and the stator flux-linkage components in the special reference frame are i_{sx}, i_{sy} and ψ_{sx}, ψ_{sy} respectively and in the stationary reference frame are i_{sD}, i_{sQ} and ψ_{sD}, ψ_{sQ} respectively. However, substitution of eqn (2.1-199) into eqn (2.1-202) yields

$$\bar{\psi}_{s\psi s} = |\bar{\psi}_s| e^{j\rho_s} e^{-j\rho_s} = \psi_{sx} + j0 = \psi_{sx} = |\bar{\psi}_s|, \tag{2.1-203}$$

which shows mathematically that in the special reference frame the space phasor of the stator flux linkages contains only a direct-axis component, which is equal to the modulus of the space phasor (since it has been deliberately aligned with the stator flux-linkage space phasor). For convenience, eqn (2.1-201) can be put into matrix form:

$$\begin{bmatrix} i_{sx} \\ i_{sy} \end{bmatrix} = \begin{bmatrix} \cos\rho_s & \sin\rho_s \\ -\sin\rho_s & \cos\rho_s \end{bmatrix} \begin{bmatrix} i_{sD} \\ i_{sQ} \end{bmatrix}, \tag{2.1-204}$$

which gives the relationship of i_{sD}, i_{sQ} and the transformed currents i_{sx}, i_{sy}.

It is possible to put the equation for the torque in such a form, where the stator flux linkage ψ_{sx} is expressed in terms of the currents. For this purpose the space phasor of the stator flux linkages in the special reference frame [defined by eqn (2.1-201)] is expressed in terms of the stator and rotor currents and from eqn (2.1-150)

$$\bar{\psi}_{s\psi s}=L_s\bar{i}_{s\psi s}+L_m\bar{i}_{r\psi s}, \tag{2.1-205}$$

where $\bar{i}_{s\psi s}$ is defined in eqn (2.1-201) and $\bar{i}_{r\psi s}$ is the space phasor of the rotor currents in the special reference frame. It follows from eqns (2.1-133), (2.1-145), and ($\theta_g=\rho_s$) that

$$\bar{i}_{r\psi s}=i_{rx}+ji_{ry}=\bar{i}_r e^{-j(\rho_s-\theta_r)}=\bar{i}_r e^{j\theta_r}e^{-j\rho_s}=\bar{i}_r' e^{-j\rho_s}=(i_{rd}+ji_{rq})e^{-j\rho_s}, \tag{2.1-206}$$

where $\bar{i}_r$ is the space phasor of the rotor currents in the rotor reference frame and $\bar{i}_r'$ is the space phasor of the rotor currents in the stationary reference frame fixed to the stator. It is now possible to define the so-called stator magnetizing current $\bar{i}_{ms}$ in the special reference frame, which rotates at speed ω_{ms}. From eqns (2.1-205) and (2.1-206)

$$\bar{i}_{ms}=\frac{\bar{\psi}_{s\psi s}}{L_m}=\frac{L_s}{L_m}\bar{i}_{s\psi s}+\bar{i}_{r\psi s}=(1+\sigma_s)\bar{i}_{s\psi s}+\bar{i}_{r\psi s}, \tag{2.1-207}$$

where $\bar{i}_{r\psi s}$ and $\bar{i}_{s\psi s}$ are defined in eqns (2.1-206) and (2.1-201) respectively. Since it follows from the choice of the special reference frame, i.e. that $\bar{\psi}_{s\psi s}$ is aligned with the real axis, and thus has only one component [see eqn (2.1-203)], from eqn (2.1-207), the stator magnetizing-current space phasor will be coaxial with $\bar{\psi}_{s\psi s}$, as shown in Fig. 2.17. From eqn (2.1-207)

$$\bar{i}_{ms}=i_{msx}+ji_{msy}=i_{msx}=|\bar{i}_{ms}|=\frac{\psi_{sx}}{L_m}=\bar{i}_{s\psi s}(1+\sigma_s)+\bar{i}_{r\psi s} \tag{2.1-208}$$

and, if eqn (2.1-208) is substituted into eqn (2.1-198) and in general $P\neq 1$, finally the instantaneous electromagnetic torque can be expressed as

$$t_e=\tfrac{3}{2}PL_m|\bar{i}_{ms}|i_{sy}. \tag{2.1-209}$$

$|\bar{i}_{ms}|$ is the stator flux-producing current component and i_{sy} is the torque-producing stator current component and this equation is again similar to the equation of the electromagnetic torque developed by the separately excited d.c. machine. Thus the torque can be controlled by independently controlling these two currents. It should be noted that in the steady state, these currents are d.c. values.

2.2 ELECTROMAGNETIC TORQUE PRODUCTION AND THE BASIS FOR VECTOR CONTROL IN SALIENT-POLE MACHINES

The mechanism of electromagnetic torque production in salient-pole a.c. machines will be discussed here and it will be assumed that the saliency exists only on one member of the machine, the rotor. For simplicity, a two-pole synchronous machine is assumed. It will be shown that the mechanism of electromagnetic torque production is similar to that shown for smooth-air-gap machines, but owing to the saliency, reluctance torque will also occur.

2.2.1 The electromagnetic torque in the rotor reference frame

A two-pole ($P=1$) salient-pole machine is assumed with symmetrical three-phase, sinusoidally distributed stator windings (sA, sB, sC). Figure 2.18 shows the cross section of the machine; on the rotor there is a field-winding (rF) and there are also quadrature-phase damper windings (rα, rβ), which are short-circuited. θ_r is the rotor angle, the angle between the magnetic axis of stator phase sA (the direct axis of the stationary reference frame sD) and the direct axis of the rotor (rα). It is possible to prove that the equation for the electromagnetic torque in the salient-pole machine can be put into a form similar to eqn (2.1-167):

$$t_e = \tfrac{3}{2}P(\psi_{sx}i_{sy} - \psi_{sy}i_{sx}), \qquad (2.2\text{-}1)$$

where P is the number of pole pairs, ψ_{sx}, ψ_{sy} and i_{sx}, i_{sy} are the direct- (x) and quadrature-axis (y) components of the stator flux linkages and currents respectively in the general reference frame, rotating at the instantaneous speed ω_g. If the reference frame fixed to the rotor ($\omega_g = \omega_r = d\theta_r/dt$) is used

$$t_e = \tfrac{3}{2}P(\psi_{sd}i_{sq} - \psi_{sq}i_{sd}), \qquad (2.2\text{-}2)$$

where ψ_{sd}, ψ_{sq} and i_{sd}, i_{sq} are the direct- and quadrature-axis stator flux linkage and current components expressed in the rotor reference frame.

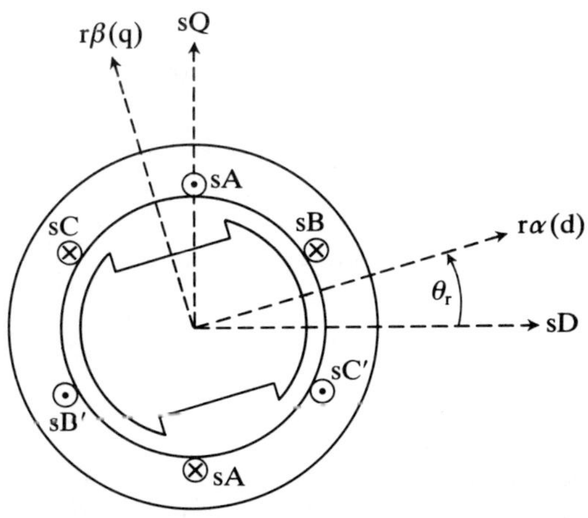

Fig. 2.18. Cross section of the salient-pole machine.

From eqn (2.1-142) the stator currents in the reference frame fixed to the rotor (i_{sd}, i_{sq}) are related to the two-axis currents expressed in the stationary reference frame (i_{sD}, i_{sQ}) as

$$i_{sd}+ji_{sq}=(i_{sD}+ji_{sQ})e^{-j\theta_r}, \tag{2.2-3}$$

The matrix form of which is

$$\begin{bmatrix} i_{sd} \\ i_{sq} \end{bmatrix}=\begin{bmatrix} \cos\theta_r & \sin\theta_r \\ -\sin\theta_r & \cos\theta_r \end{bmatrix}\begin{bmatrix} i_{sD} \\ i_{sQ} \end{bmatrix}. \tag{2.2-4}$$

From physical considerations, the stator flux linkages in the rotor reference frame can be defined in terms of the currents as

$$\psi_{sd}=L_{sd}i_{sd}+L_m i_{rF}+L_{md}i_{r\alpha} \tag{2.2-5}$$

and

$$\psi_{sq}=L_{sq}i_{sq}+L_{mq}i_{r\beta}, \tag{2.2-6}$$

where L_{sd}, L_{sq} are the self-inductances of the stator winding along the d and q axes, L_{md}, L_{mq} are the magnetizing inductances along the same two axes, and owing to saliency $L_{sd}\neq L_{sq}$ and $L_{md}\neq L_{mq}$. i_{rF} and $i_{r\alpha}$, $i_{r\beta}$ are the currents in the field winding and the direct- and quadrature-axis damper windings respectively and are expressed in the reference frame fixed to the rotor.

Substitution of eqns (2.2-5) and (2.2-6) into eqn (2.2-2) yields the electromagnetic torque,

$$t_e=\tfrac{3}{2}P[(L_{sd}-L_{sq})i_{sd}i_{sq}+L_{md}(i_{rF}+i_{r\alpha})i_{sq}-L_{mq}i_{r\beta}i_{sd}], \tag{2.2-7}$$

where $(L_{sd}-L_{sq})i_{sd}i_{sq}$ is the so-called reluctance torque, which is solely due to the saliency and disappears for a smooth-air-gap machine, where $L_{sd}=L_{sq}=L_s$. $L_{md}i_{rF}i_{sq}$ is the field torque, $L_{md}i_{r\alpha}i_{sq}$ is the direct-axis damper-torque component and $L_{mq}i_{r\beta}i_{sd}$ is the quadrature-axis damper-torque component. In the steady-state the damper currents are zero and all the other currents are constant.

It is possible to rearrange the torque equation in terms of the magnetizing flux linkages. Since the total inductances along the direct and quadrature axes can be decomposed into leakage and magnetizing inductances, $L_{sd}=L_{sl}+L_{md}$, $L_{sq}=L_{sl}+L_{mq}$, the flux-linkage components defined by eqns (2.2-5) and (2.2-6) can be written as

$$\psi_{sd}=L_{sl}i_{sd}+\psi_{md} \tag{2.2-8}$$

$$\psi_{sq}=L_{sl}i_{sq}+\psi_{mq}. \tag{2.2-9}$$

The first term on the right-hand side of eqns (2.2-8) and (2.2-9) is a leakage flux component and ψ_{md}, ψ_{mq} are the direct- and quadrature-axis magnetizing flux-linkage components respectively in the rotor reference frame

$$\psi_{md} = L_{md} i_{md} = L_{md}(i_{sd} + i_{rF} + i_{r\alpha}) \tag{2.2-10}$$

$$\psi_{mq} = L_{mq} i_{mq} = L_{mq}(i_{sq} + i_{r\beta}), \tag{2.2-11}$$

where i_{md} and i_{mq} are the magnetizing currents along the two axes. The leakage flux components do not contribute to electromechanical energy conversion, and indeed this can also be seen, if eqns (2.2-8)–(2.2-11) are substituted into eqn (2.2-2) to yield

$$t_e = \tfrac{3}{2} P(\psi_{md} i_{sq} - \psi_{mq} i_{sd}) \tag{2.2-12}$$

or

$$t_e = \tfrac{3}{2} P[(L_{md} - L_{mq}) i_{sd} i_{sq} + L_{md}(i_{rF} + i_{r\alpha}) i_{sq} - L_{mq} i_{r\beta} i_{sd}], \tag{2.2-13}$$

which does not contain the leakage flux linkages. According to eqn (2.2-12) the electromagnetic torque is caused by

(i) the interaction of the direct-axis magnetizing flux-linkage component and the quadrature-axis stator current component and
(ii) by the interaction of the quadrature-axis magnetizing flux-linkage component and the direct-axis stator current component;

the same mechanism exists in smooth-air-gap machines. In eqn (2.2-13) the term $(L_{md} - L_{mq}) i_{sd} i_{sq}$ is reluctance torque and is equal to the reluctance torque component given in eqn (2.2-7). It follows from eqn (2.2-13) that if a smooth air gap is assumed ($L_{md} = L_{mq}$), and if on the rotor only the field winding is excited ($i_{r\alpha} = 0$; $i_{r\beta} = 0$), the electromagnetic torque can be expressed as

$$t_e = \tfrac{3}{2} P L_m i_{rF} i_{sq}. \tag{2.2-14}$$

This expression is similar to the equation for the electromagnetic torque of a synchronous machine with permanent-magnet excitation (see Section 3.1.1).

It can also be seen that in the transient state, if the direct-axis components of the stator currents expressed in the rotor reference frame are absent ($i_{sd} = 0$ or $\gamma = 0$, where γ is the torque angle), there is no reluctance torque and no torque produced by the quadrature-axis damper current. If $i_{sd} = 0$ and $i_{r\alpha} = 0$, the field current (i_{rF}) and the stator current (which lies along the q axis) are in space quadrature, similarly to a separately excited d.c. machine, and 'field-orientation' is achieved, where the stator and field m.m.f.s are in space quadrature. Furthermore, if $i_{sd} = 0$ and the field winding is supplied by a d.c. voltage, then the current in the field winding—expressed in the reference frame fixed to the rotor—must be constant ($i_{rF} = I_F$) and it follows from eqn (2.2-13) that the electromagnetic torque will be

$$t_e = \tfrac{3}{2} P L_{md} I_F i_{sq}. \tag{2.2-15}$$

This expression is again similar to the equation for the torque of a synchronous machine with permanent-magnet excitation (Section 3.1) and it is also similar to the expression for the electromagnetic torque of the compensated, separately excited d.c. machine.

It is an important consequence that, in contrast to an induction machine, where the electromagnetic torque production requires that the stator currents should contain both the excitation (flux-producing) and torque-producing stator current components, in a synchronous machine it is not necessary for the stator currents to contain the excitation component, since excitation is provided by the field currents and torque is produced by the interaction of the field and quadrature-axis (torque-producing) stator current.

If the d axis magnetizing reactance $X_{md} = \omega L_{md}$, in the steady-state $i_{rF} = I_{rF}$ and the e.m.f. $jX_{md}I_{rF}$ is the so-called internal voltage $\bar{E}$, which lies along the q axis, as shown in Fig. 2.19.

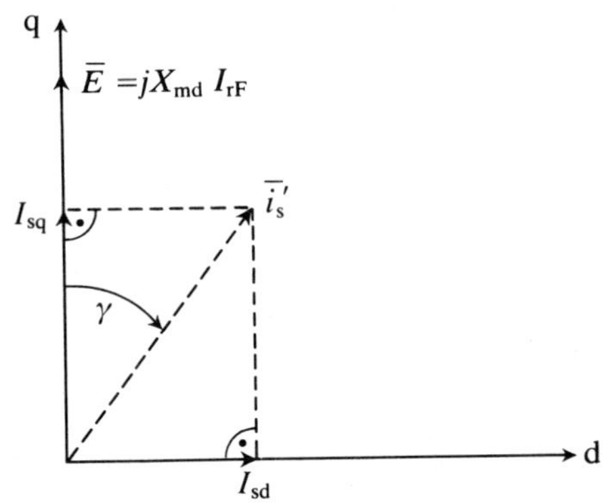

Fig. 2.19. Steady-state vector diagram of the stator currents and internal voltage in the synchronously rotating reference frame (d, q) fixed to the rotor.

In the conventional steady-state theory, the time phase angle between the q axis, where $\bar{E}$ is positioned, and the vector of the steady-state currents $(I_{sd} + jI_{sq})$ expressed in the rotor reference frame, is denoted by γ. In the steady state the torque component $X_{md}I_{rF}I_{sq} = X_{md}I_{rF}|\bar{I}_s| \cos \gamma$ is produced as a result of the interaction of the field current and the quadrature-axis stator current, where $|\bar{I}_s|$ is the modulus of the stator currents. In the transient state, if $i_{r\alpha} = 0$, the torque component $L_{md}i_{rF}i_{sq}$ is present, caused by the interaction of the flux linkage $L_{md}i_{rF}$ and the stator current i_{sq}. If in the transient state the space angle γ is defined as the angle between the stator current space phasor $(i_{sd} + ji_{sq})$ and the quadrature-axis, then the expression $L_{md}i_{rF}i_{sq} = L_{md}i_{rF}|I'_s| \cos \gamma$ is obtained for the torque.

If $\psi_{md} + j\psi_{mq}$ is defined to be the magnetizing flux-linkage space phasor, then as a result of saliency, this vector will not be coaxial with $i_{md} + ji_{mq}$, which could be defined as the magnetizing current space phasor.

2.2.2 The electromagnetic torque in special reference frames

It has been shown in Section 2.1.8 that for smooth-air-gap machines, by suitable orientation of the reference frame, it is possible to obtain expressions for the torque which resemble the expression for the torque of the separately excited, compensated d.c. machine. It is possible to use similar concepts for the salient-pole machine, as will now be briefly discussed.

Application of a special reference frame fixed to the magnetizing flux-linkage space phasor

By considering eqn (2.2-12), where the current and flux components are expressed in the rotor reference frame, it can be shown that in the general reference frame (x, y) rotating at the speed ω_g, for the salient-pole machine discussed in Section 2.2.1, the equation of the torque will be similar:

$$t_e = \tfrac{3}{2}P(\psi_{mx}i_{sy} - \psi_{my}i_{sx}), \tag{2.2-16}$$

where ψ_{mx}, ψ_{my} and i_{sx}, i_{sy} are the direct- and quadrature-axis components of the magnetizing flux linkages and currents respectively in the general reference frame.

If, as in Section 2.1.8, a special reference frame is used, whose real axis is coaxial with the magnetizing flux-linkage space phasor and which rotates with the speed of the magnetizing flux linkage space phasor (ω_m), then the special reference frame, the magnetizing flux-linkage space phasor can be expressed as

$$\bar{\psi}_{mm} = \psi_{mx} + j0 = |\bar{\psi}_m| \tag{2.2-17}$$

since $\psi_{my} = 0$ in the special reference frame. Thus substitution of eqn (2.2-17) into eqn (2.2-16) yields

$$t_e = \tfrac{3}{2}P|\bar{\psi}_m|i_{sy}, \tag{2.2-18}$$

where $|\bar{\psi}_m|$ is the modulus of the magnetizing flux-linkage space phasor and i_{sy} is the quadrature-axis component of the stator current in the special reference frame. If the angle μ_m is the space angle of the magnetizing flux-linkage space phasor with respect to the real axis of the stationary reference frame (sD), then from eqn (2.1-142), i_{sy} can be defined in terms of the two-axis current components (i_{sD}, i_{sQ}) defined in the stationary reference frame as

$$i_{sy} = -\sin\mu_m i_{sD} + \cos\mu_m i_{sQ}, \tag{2.2-19}$$

which also follows from eqn (2.1-182). i_{sy} is in space quadrature to $\bar{\psi}_{mm}$. Equation (2.2-18) serves as the basis of the magnetizing-flux-oriented control of salient-pole synchronous machines.

If the magnetizing-current space phasor in the special reference frame is defined as

$$\bar{i}_{mm} = i_{mx} + ji_{my} \tag{2.2-20}$$

then, in contrast to the smooth-air-gap machine, where the magnetizing flux linkage and magnetizing current space phasors are coaxial, as a result of saliency, $\bar{\psi}_{mm}$ and $\bar{i}_{mm}$ will not be coaxial. Thus whilst for the smooth-air-gap machine the application of a $\bar{\psi}_{mm}$- or $\bar{i}_{mm}$-oriented reference frame has the same effect, this is not the case for the salient-pole machine.

BIBLIOGRAPHY

Blaschke, F. (1972). The principle of field-orientation as applied to the new Transvektor closed-loop control system for rotating-field machines. *Siemens Review* **34**, 217–20.

Jones, C. V. (1967). *The unified theory of electrical machines.* Butterworth, London.

Kovács, K. P. and Rácz, I. (1959). *Transiente Vorgänge in Wechselstrommaschinen*, I–II. Akadémia Kiadó, Budapest.

Leonhard, W. (1985). *Control of electrical drives.* Springer-Verlag, Berlin.

Naunin, D. (1979). The calculation of the dynamic behaviour of electrical machines by space phasors. *Electrical Machines and Electromechanics* **4**, 33–45.

Vas, P. and Brown, J. E. (1985). Real-time monitoring of the electromagnetic torque of multi-phase a.c. machines. *IEEE IAS Meeting, Toronto, 6–11 October*, pp. 732–7.

Vas, P., Brown, J. E., and Shirley, A. (1984). The application of n-phase generalized rotating-field theory to induction machines with arbitrary stator winding connection, *IEEE Transactions on Power Apparatus and Systems* **PAS-103**, 1270–7.

Vas, P., Willems, J. L., and Brown, J. E. (1987). The application of space-phasor theory to the analysis of electrical machines with space harmonics. *Archiv für Elektrotechnik* **69**, 359–63.

3 Vector control of synchronous machines

In this chapter various forms of vector control are described for permanent-magnet synchronous machines (PMSM) with interior or surface magnets and for salient-pole and smooth-air-gap synchronous machines with electrically excited rotor.

3.1 VECTOR CONTROL OF PERMANENT-MAGNET SYNCHRONOUS MACHINES

Synchronous machines with an electrically excited rotor winding have a three-phase stator winding (called the armature) and an electrically excited field winding on the rotor which carries a d.c. current. There can also be damper windings on the rotor. The armature winding is similar to the stator of induction machines. The synchronous machine is a constant-speed machine which always rotates at a synchronous speed which depends on the supply frequency and on the number of poles. The electrically excited field winding can be replaced by permanent magnets. The use of permanent magnets has many advantages including the elimination of brushes, slip-rings and rotor copper losses in the field winding, which leads to higher efficiency. Since the copper and iron losses are then concentrated in the stator, cooling of the machine, through the stator, is more easily achieved. The higher efficiency allows a reduction in the machine frame size. Another advantage of using permanent magnets is, that for the same size machine, the characteristics of the machine can be varied greatly according to the type of magnets chosen and the way they are arranged. Permanent-magnet synchronous machines (PMSM) can be categorized into two main groups:

(i) the brushless d.c. machines fed by current source inverters and having trapezoidal flux distribution and
(ii) synchronous machines having approximately sinusoidal air-gap flux-density distribution, fed by sinusoidal stator currents.

Brushless d.c. machines are not the subject of the present investigation.

Permanent-magnet machines are extensively used in servo drives containing machines with low power for machine-tool (e.g. spindle motors, positioning drives etc.) and robotic applications, but large machines up to 1 MW have also been built; for example, a 1 MW machine has been used in a ship propulsion drive. It is also possible to use such machines in the fields of electricity generation, solar pumping, wind-energy applications etc. In the present book, however, only servo applications are discussed. The design criteria for synchronous servomotors, to be used in machine-tool feed drives, manipulators and industrial robots differ from that of conventional synchronous machines since the following requirements must be met:

- High air-gap flux density;
- High power/weight ratio (greatest possible power/motor mass);
- Large torque/inertia ratio (to enable high acceleration);
- Smooth torque operation (small torque ripples) even at very low speeds (to achieve high positioning accuracy);
- Controlled torque at zero speed;
- High speed operation;
- High torque capability (quick acceleration and deceleration for short time);
- High efficiency and high cos ϕ (low expense for the power supply);
- Compact design.

These requirements can be met well by the permanent-magnet synchronous machine employing vector control. Either the rotor construction can be such that a slim-drum rotor is used with large length/diameter ratio, resulting in a low mechanical time constant (such rotors are sometimes also referred to as hot-dog rotors) or a light-weight, aluminium disc rotor (pancake rotor) is used. Sometimes the disc is manufactured from high-strength plastic material, and the magnets are embedded and encapsulated in this using resin. These machines are also referred to as radial-field machines (they have a long cylindrical rotor) and axial-field machines (with disc rotor). The machines with the disc rotor are usually employed in robotic applications, whilst in servo drives for machine tools the slim-drum rotor is usually applied. The machine with the pancake rotor can yield the same high dynamic performance as the machine with the drum-type rotor, but its power factor, maximum torque and efficiency can be lower. Furthermore, the drum rotor leads to a low-inertia (low mechanical time constant) design, which makes possible a more rapid acceleration. The manufacturing costs of axial-field machines are higher than those for the radial-field machines.

There are various practical implementations of mounting the magnets. In one solution, rare-earth magnets, e.g. samarium–cobalt ($SmCo_5$, Sm_2Co_{17}) or neodymium–iron–boron (NdFeB), which have a high energy product, and thus where the effects of demagnetization are avoided, are mounted on the surface of the solid steel rotor, by using high-strength adhesive material. To obtain high mechanical strength, which is especially important for high-speed operation, either the gaps between the magnets are filled with non-magnetic material and the rotor can be wrapped with high-strength material, e.g. fibre glass, or sometimes screwed pole-shoes are used. At present a disadvantage of using rare-earth magnets is their relatively high price, but with new developments this may change in the future. Furthermore, it is also expected that the characteristics of the NdFeB magnets will improve (at present there can be problems because the field strength decreases with increasing temperature). When the surface-magnet synchronous motor is used for variable frequency operation there are no damper windings on the rotor. In another construction, buried or so-called interior magnets are used.

In the case where the magnet bars are mounted on the rotor surface, the incremental permeability of the magnets relative to the external fields is 1.02–1.2. They have high resistivity, and thus the machine can be considered to have a large effective air gap, which makes the effects of saliency negligible (thus the direct-axis magnetizing inductance is equal to the quadrature-axis magnetizing inductance, $L_{md}=L_{mq}=L_m$). Furthermore, because of the large air-gap, the synchronous inductance ($L_s=L_{sl}+L_m$) is small and therefore the effects of armature reaction are negligible. A further consequence of the large air gap is that the electrical time constant of the stator winding is small. The magnets can have various shapes. Bar-shaped magnets and circumferential segments with angles up to 90° and thickness of a few millimetres are available. Radially magnetized circumferential segments produce a smoother air-gap flux density distribution and fewer torque ripples. These torque ripples can also be reduced by the appropriate design of the stator windings.

When the magnets are buried inside the rotor, a mechanically robust construction is obtained which can be used for high-speed applications since the magnets are physically contained and protected. However, with the interior permanent-magnet construction, the machine cannot be considered to have a uniform air gap. In this case the magnets are mounted inside the steel rotor core and physically there seems to be almost no variation of the surface of the magnet geometry. However, since each magnet is covered by a steel pole piece, this significantly changes the magnetic circuits of the machine, since owing to these iron pole pieces, high-permeance paths are created for the magnetic flux across these poles and also in space quadrature to the magnet flux. Thus there are saliency effects which significantly alter the torque production mechanism of the machine. As described in Section 2.2, in addition to the so-called magnet (or field-alignment) torque, which arises from the interaction of the magnet flux and the quadrature-axis stator current component, there is a reluctance torque component which is due to saliency. It is also a feature of these type of machines that in contrast to conventional, electrically excited salient-pole machines, where the quadrature-axis reactance is smaller than the direct-axis reactance, in synchronous machines with interior permanent magnets on the rotor, the quadrature-axis reactance is larger than the direct-axis reactance. In synchronous machines with buried (interior) permanent magnets, the magnet cost is minimized by the low magnet-weight requirements of the particular design.

One of the requirements to be met for servo applications has been the smooth torque operation. Therefore both the so-called cogging torques (which are due to the slotting) and the pulsating torques, due to space harmonics and time harmonics, must be reduced. The reduction of the cogging torque can be achieved in a number of ways, such as: shaping of the magnets, skewing of the magnets on the rotor with respect to the rotor axis, skewing of the stator, coordinating the design of the number of stator slots, slot opening and magnet dimensions. However, the skewed rotor construction is costly in the case of high-strength ceramic magnets and rare-earth magnets, because it is difficult to manufacture the magnets in complex geometric shapes. Furthermore, the required magnetizing techniques can result in lower flux levels. However, any method used for minimizing the cogging torque must be evaluated on the basis of manufacturing considerations from the point of view of cost, assembly and the equipment required for the process. Under normal circumstances one or two per cent of the rated torque is an acceptable value for the cogging torque in servo applications. However, most of the effects of cogging torque can be eliminated in a converter-fed permanent-magnet synchronous machine drive by a high-performance converter and controller containing an accurate speed or position monitoring device.

It should be noted that whilst in conventional permanent-magnet machines the stator contains teeth, it is possible to construct such permanent-magnet machines where there are no teeth on the stator. In this case the stator windings are assembled outside the machine and are then inserted and fixed in the stator. Such a machine will not cog at low speed and the iron losses are reduced. Furthermore, in this construction more space is available for the stator windings than in the conventional construction. Thus larger size conductor can be used and the current rating of the windings can be higher. Since there is more space peripherally, the radial dimension of the stator windings can be smaller than in the case of a toothed construction. Thus for a toothless machine the rotor diameter can be larger than for a conventional machine of the same frame size. The air gaps of machines with toothless stators are usually several times larger than in the machines which have teeth. This has an adverse effect on the air-gap flux density, which can be partly offset by the application of a rotor with larger diameter and a larger magnet surface.

The reduction of the pulsating torques (ripple torques) is also an important aspect of achieving smooth torque operation. This can be obtained in a number of ways including the minimization of the time harmonic content of the stator currents.

With rare-earth magnets air-gap flux densities up to 1 T (Tesla) can be obtained. However, because of the saturation of the teeth and iron losses, it is not always best to use the strongest magnet material. The teeth can be highly saturated, especially with armature reaction during overload. Despite the large air gap, a field distortion occurs due to saturation, and this can reduce the flux and therefore the induced e.m.f. At high speed, the iron losses are the limiting factor.

3.1.1 Vector control of PMSMs with surface-mounted magnets

It is assumed that the permanent magnets are located on the rotor surface and are of the rare-earth magnet type. The effects of magnetic saturation are neglected and thus for modelling purposes the permanent magnets can be considered as fictitious equivalent constant-current sources (I_{rF}=const.). Thus in the reference frame fixed to the rotor, the rotor current space phasor is

$$\bar{i}_r = cI_{rF} = I_{rf} = \text{constant} \tag{3.1-1}$$

if there are no damper windings on the rotor. The magnets are positioned at angle θ_r relative to the direct axis of the stator, which coincides with the magnetic axis of stator winding sA. In the d, q reference frame fixed to the rotor the flux linkage with the stator windings due to the permanent magnets is

$$\psi_F = L_m \bar{i}_r = L_m I_{rf}. \tag{3.1-2}$$

Since the synchronous reactance is small, ψ_F is almost equal to the modulus of the magnetizing flux-linkage space vector and thus the angle between the magnetizing flux-linkage space vector and ψ_F is small; this angle is the so-called load angle and is denoted by δ. Thus the induced stator magnetizing e.m.f., which is equal to the rate of change of the magnetizing flux linkage, will in the steady state almost be positioned along the quadrature-axis. As a consequence, the terminal voltage will also be almost co-phasal with the quadrature-axis.

It should be noted that the magnet flux linkage ψ_F can be obtained by performing simple tests. If the stator currents are zero and the rotor is rotated at a constant speed (e.g. by a d.c. motor coupled to the shaft), the induced e.m.f.s in the stator windings are proportional to the magnet flux. Thus the magnet flux linkage can be obtained by dividing the terminal voltage by the rotor speed. It follows from eqn (3.1-2) that the equivalent fictitious rotor current can also be determined if the magnetizing inductance and the magnet flux linkage are known. The fictitious rotor current is proportional to the number of pole pairs, the residual flux density of the magnet material (approximately 1.2 T for NdFeB), the radial thickness of the magnets and the sine of half of the electrical angle span of a magnet. Furthermore it is inversely proportional to the recoil permeability of the magnet material (near to unity for NdFeB) and to the equivalent number of sinusoidally distributed stator turns per phase.

The electromagnetic torque of the permanent-magnet synchronous machine with surface mounted magnets and with a symmetrical three-phase stator winding can be obtained physically by considering that the torque is produced by the tendency of the magnets to align with the axis of the stator m.m.f. and now, owing to the lack of saliency, there is no reluctance torque component. Thus the magnet torque has to vary sinusoidally with the angle between the stator m.m.f. axis and the direct axis of the rotor, to which the excitation m.m.f. is frozen. This can also be seen by considering eqns (2.2-14) and (3.1-2) and thus the electromagnetic torque is

$$t_e = \tfrac{3}{2} P L_m I_{rf} i_{sq} = \tfrac{3}{2} P \psi_F i_{sq}, \tag{3.1-3}$$

where i_{sq} is the quadrature-axis component of the stator current space vector

($\bar{i}_s'$) expressed in the reference frame fixed to the rotor, as shown in Fig. 3.1.

It follows from Fig. 3.1, that the space angle of the stator-current space phasor relative to the magnetic axis of stator winding sA is α_s, and since its space angle relative to the direct axis of the reference frame fixed to the rotor is $\alpha_s - \theta_r$, eqn (3.1-3) can be put into the form,

$$t_e = \tfrac{3}{2} P \psi_F |\bar{i}_s| \sin(\alpha_s - \theta_r) \qquad (3.1\text{-}4)$$

where $|\bar{i}_s|$ is the modulus of the stator-current space phasor.

Thus the physically expected result has been obtained and the torque varies with the sine of the angle ($\alpha_s - \theta_r$), the so-called torque angle (β). Since the flux produced by the permanent magnets has been assumed to be constant, the electromagnetic torque can be varied by changing the quadrature-axis stator current expressed in the rotor reference frame (i_{sq}). Thus a constant torque is obtained if i_{sq} = constant. It also follows from eqn (3.1-4) that the maximum torque per stator current is obtained if the torque angle is 90°. Since the electromagnetic torque is a constant times the quadrature-axis stator current component (i_{sq}), it immediately follows that a quick torque response is obtained if the quadrature-axis stator current is changed quickly, e.g. by the application of a current-controlled pulse-width-modulated inverter. It is this feature which makes possible the rotor-oriented control of permanent-magnet synchronous machines.

The control scheme of the rotor-oriented controlled PMSM

The control scheme of the rotor-oriented permanent-magnet synchronous

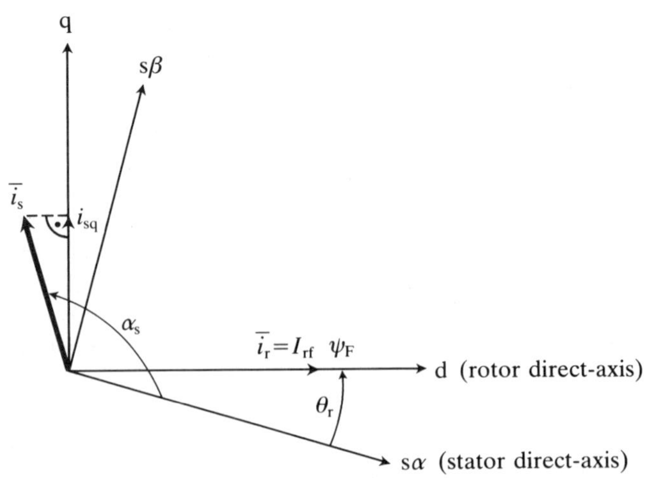

Fig. 3.1. The stator- and rotor-current space phasors and the excitation flux of the PMSM.

machine discussed above is relatively simple. The excitation flux is frozen to the direct axis of the rotor and thus its position can be obtained directly from the rotor shaft by monitoring the rotor angle θ_r or the rotor speed ω_r.

In practice the rotor speed can be monitored by, say, the application of an analogue tachometer and the rotor angle can be monitored by, say, a resolver. Analogue tachometers have about 0.1% accuracy and at low-speed operation of a servo drive, very high-speed resolution is necessary to obtain accurate results; for feed drives and machine tools, for example, a speed resolution of approximately 0.2 min.$^{-1}$ is required. This problem can be solved by using the same digital encoder for sensing the rotor position and rotor speed. It should be noted that at high speeds and high-bandwidth speeds, optical encoders can have limited accuracy and temperature susceptibility, since the monitoring device has to be mounted within the motor enclosure. Resolvers are inherently accurate, but they must be combined with high-resolution digital circuits to achieve and maintain high accuracy over a wide speed range and give no errors due to acceleration. Furthermore, it is also possible to utilize the fact that the rotor speed is equal to the first time derivative of the rotor position, $\omega_r = d\theta_r/dt$. In Section 6.3.1 the estimation of the rotor speed by monitoring the rotor position is described and in Section 6.3.2 a rotor speed estimator is described which employs an optimal rotor-speed observer.

In a squirrel-cage induction motor it may be necessary to use equations, or a corresponding so-called flux model, and certain other monitored quantities (e.g. currents) which yield the position and the modulus of the rotor flux-linkage space phasor, since the rotor flux-linkage space phasor in an induction machine will be displaced from the direct axis of the rotor reference frame by an angle which depends on the value of the slip frequency. However, this will be discussed in detail in Section 4.1.1. It is of course possible to obtain the rotor speed of a synchronous machine by utilizing its voltage equations and the monitored values of the stator voltages and currents. In certain applications, some manufacturers prefer implementations where there are no sensors which involve extra rotating devices for monitoring the rotor speed or rotor angle, but it can be difficult to obtain an accurate representation of the rotor speed or rotor angle by using monitored stator voltages or currents, because this requires very accurate sensing of distorted signals and it may also be necessary to know accurate values of certain machine parameters.

As shown above, in order to produce the largest torque for a given stator current, an optimally efficient operation is achieved by stator current control which ensures that the stator current space phasor contains only a quadrature-axis component (i_{sq}) when expressed in the reference frame fixed to the rotor. This is analogous to the separately excited d.c. machine, where this is achieved by the consecutive switching of the armature coils through the commutator. In the base speed region, for motor operation, the space phasor of the stator currents expressed in the rotor reference frame is therefore $\bar{i}_s' = ji_{sq}$ [see Fig. 3.2(a)], and for braking, it is $\bar{i}_s' = -ji_{sq}$ as shown in Fig. 3.2(b).

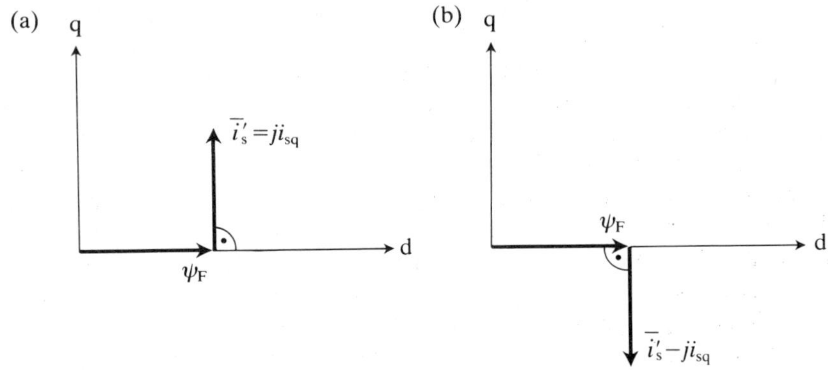

Fig. 3.2. Stator-current space phasors in the case of optimal operation in the base speed region. (a) Motor operation. (b) Braking operation.

This optimal mode of operation is suitable below the base rotor speed, where sufficient voltage is available from the inverter which supplies the stator windings of the machine. However, at higher speeds, above the base speed—constant-horsepower range—the induced e.m.f. increases directly with the rotor speed (the excitation flux is constant due to the permanent magnets), and if a given speed is to be reached, the terminal voltage must also be increased to match the increased stator e.m.f. The increased stator terminal voltage would require an increase in the voltage rating of the inverter used. However, with a given inverter, there is a ceiling voltage which cannot be exceeded. Thus to limit the terminal voltage of the machine to the ceiling voltage of the inverter, field weakening has to be introduced.

In smaller drives, there is no need for field weakening, but for larger drives since, owing to the permanent-magnet construction, it is not possible to achieve direct field weakening, the effect of field weakening can be obtained by controlling the stator currents in such a way that the stator-current space phasor in the rotor reference frame should contain a direct-axis component i_{sd} along the negative direct-axis of the rotor reference frame, in addition to the quadrature-axis stator current component i_{sq}. This is shown in Fig. 3.3, where the same i_{sq} has been maintained as for operation below base speed.

The limiting amplitude of the stator voltages can be obtained by utilizing the steady-state stator equations of the PMSM. It follows from eqn (2.1-141) that in the stationary reference frame, when the effects of magnetic saturation are neglected, the stator voltage equation is

$$\bar{u}_s = R_s\bar{i}_s + L_s\frac{d\bar{i}_s}{dt} + L_m\frac{d}{dt}(\bar{i}_r e^{j\theta_r}). \qquad (3.1\text{-}5)$$

By the substitution of ψ_F defined in eqn (3.1-2), this will take the following form

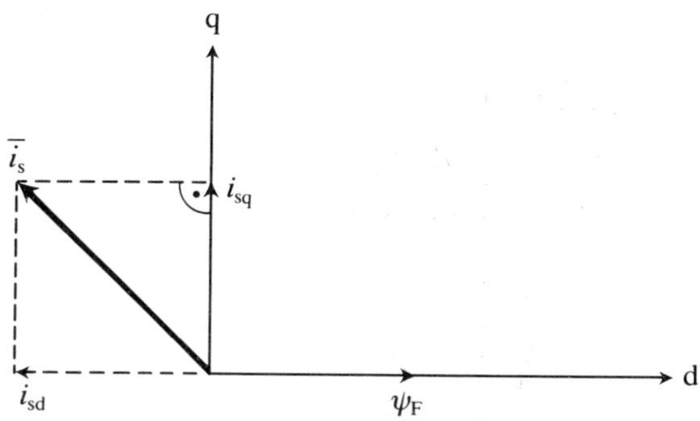

Fig. 3.3. Space phasors of PMSM in the field-weakening range (overspeed range).

$$\bar{u}_s = R_s\bar{i}_s + L_s\frac{d\bar{i}_s}{dt} + \frac{d}{dt}(\psi_F e^{j\theta_r}). \qquad (3.1\text{-}6)$$

By performing the required differentiation, but keeping the excitation flux constant,

$$\bar{u}_s = R_s\bar{i}_s + L_s\frac{d\bar{i}_s}{dt} + j\omega_r\psi_F e^{j\theta_r}. \qquad (3.1\text{-}7)$$

In the steady-state, where $\theta_r = \omega t$, $\omega = \text{const.}$, the space phasor of the stator currents in the stationary reference frame is $\bar{i}_s = c_1\bar{I}_s e^{j\omega t}$, the voltage space phasor in the same reference frame is $\bar{u}_s = c_1\bar{U}_s e^{j\omega t}$, where c_1 is a constant [see the discussion which follows eqn (2.1-5)]. Equation (3.1-7) thus takes the form,

$$\bar{U}_s = R_s\bar{I}_s + j\omega L_s\bar{I}_s + \bar{U}_p = Z_s\bar{I}_s + \bar{U}_p, \qquad (3.1\text{-}8)$$

where

$$\bar{U}_p = \frac{j\omega\psi_F}{c_1} \qquad (3.1\text{-}9)$$

is the phasor of the induced e.m.f.s in the stator winding, $Z_s = R_s + j\omega L_s = R_s + jX_s$ is the impedance of a stator winding and X_s is the synchronous reactance.

By utilizing eqns (3.1-8) and (3.1-9), the two steady-state phasor diagrams shown in Fig. 3.4 can be plotted. The steady-state phasor diagram shown in Fig. 3.4(a) corresponds to the case where there is no direct-axis stator current component ($I_{sd} = 0$), thus $\bar{I}_s = jI_{sq}$ and

$$\bar{U}_s = (R_s + j\omega L_s)jI_{sq} + \bar{U}_p, \qquad (3.1\text{-}10)$$

which holds for operation below base speed (constant-torque region). The

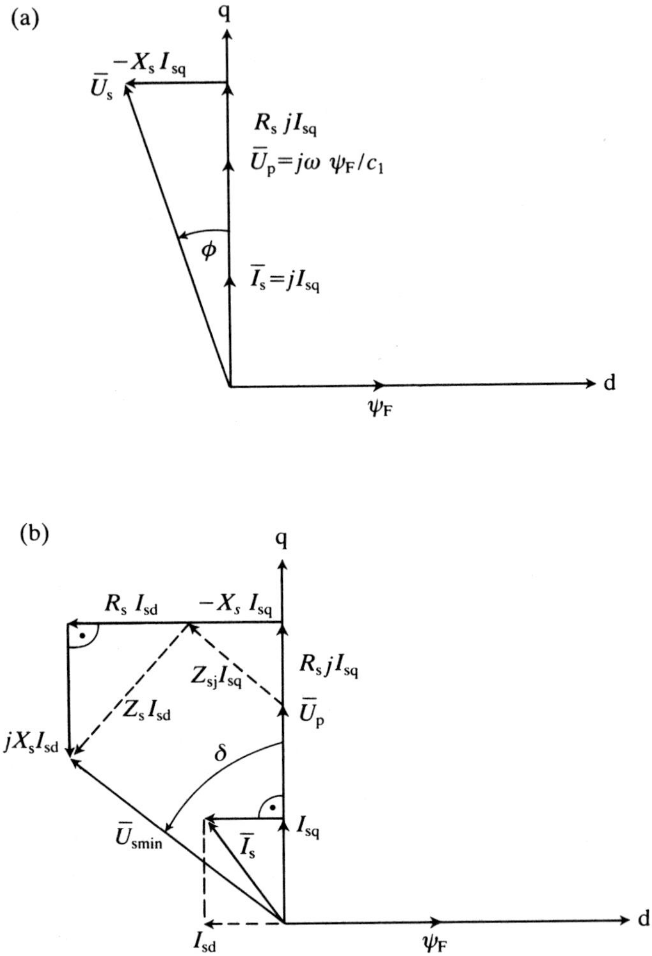

Fig. 3.4. Steady-state phasor diagrams of the PMSM. (a) Operation below base speed. (b) Operation above base speed.

angle between $\bar{U}_p$ and $\bar{U}_s$ is the load angle, and in this case agrees with the displacement angle (ϕ) of the stator current.

In Fig. 3.4(b) the phasors are shown for operation above the base speed (constant-output-power region) and for stator currents where $I_{sd} \neq 0$ ($I_{sd} < 0$). Thus

$$\bar{U}_s = (R_s + j\omega L_s) jI_{sq} + \bar{U}_p + (R_s + j\omega L_s) I_{sd}. \tag{3.1-11}$$

Figure 3.4(b) corresponds to the case where minimum stator voltage ($\bar{U}_{smin}$) is obtained. It follows from Fig. 3.4(b) that the minimum stator voltage is obtained if the stator voltage is adjusted to be in quadrature to the voltage phasor corresponding to the voltage drop along the d-axis, which is equal to $(R_s + j\omega L_s)I_{sd}$.

It is then possible to utilize Fig. 3.4 to obtain the necessary direct-axis stator current component to achieve speeds in the field-weakening range (above base speed). It should be noted that because of the extra current along the d axis, the stator currents are increased, but since the stator currents are limited by the inverter, a corresponding reduction of the maximum quadrature-axis stator current has to be performed, which of course leads to a smaller torque. In other words, when the rotor speed is increased the torque angle is increased for the purposes of field weakening and thus the torque per unit of stator current is reduced.

Owing to the increased stator currents, the copper losses will be increased and the efficiency of the drive will deteriorate. However, field weakening is restricted to short transients at light load. The maximum speed is determined by the current rating of the inverter. It is possible to achieve high-speed operation with light loads and high values of the demagnetizing stator current component. However, in practice, the maximum speed does usually not exceed twice the base speed.

It follows from the equations and vector diagrams presented above that the degree of field weakening with a PMSM depends on the synchronous reactance of the machine, i.e. on the armature reaction of the machine. With the magnets mounted on the surface of the rotor, the air gap can be considered to be relatively large as discussed earlier. So the synchronous reactance is small and therefore the field-weakening range is limited.

In the following discussion it is assumed that the stator windings of the machine are supplied by impressed stator currents obtained from a current-controlled switched transistor inverter with a high switching frequency. This is similar to the application of armature current control in a converter-fed d.c. drive. The main advantage of this type of operation is that because of the impressed stator currents, the effects of stator resistance and leakage inductance on the drive dynamics are eliminated and the motor interactions are simplified. Figure 3.5 shows the schematic diagram of a Pulse-Width-Modulated (PWM) transistorized inverter, using fast current control loops, which supplies the stator windings of the permanent-magnet synchronous machine. Figure 3.5(a) shows a simplified circuit and Fig. 3.5(b) shows a more detailed circuit.

An uncontrolled rectifier produces the constant voltage U across the filter capacitor C_F. The filter establishes a steady d.c. voltage. The constant d.c. voltage source could also be a battery bank. With an appropriately high U fast current control loops can be designed, which will yield machine currents (i_{sA}, i_{sB}, i_{sC}) which coincide with the demanded currents (i_{sAref}, i_{sBref}, i_{sCref}). Thus the stator windings can be considered to be supplied by fictitious current sources. This approximation is valid if the ceiling voltage of the inverter is sufficient.

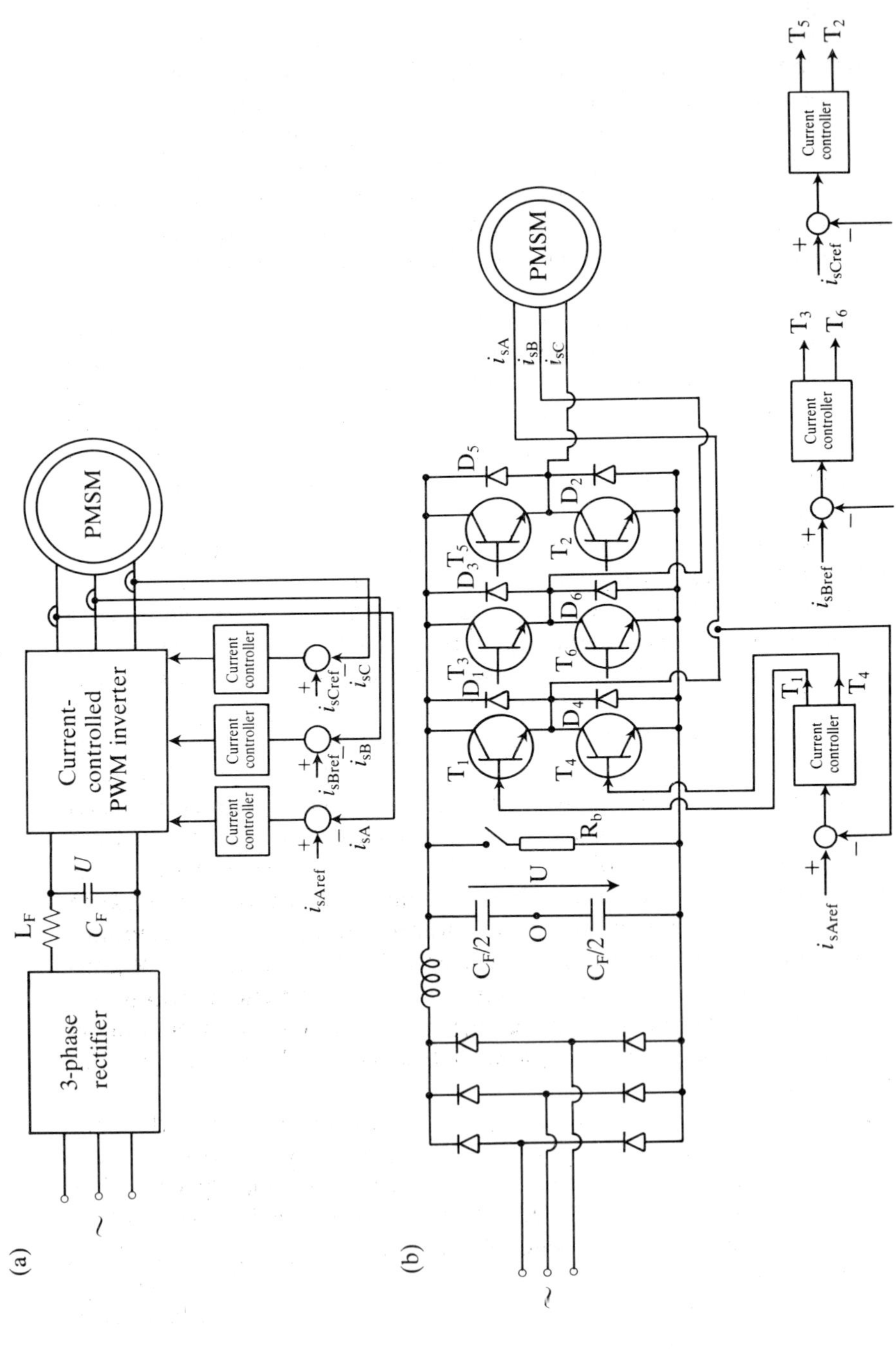

Fig. 3.5. A PWM inverter-fed PMSM. (a) Simplified circuit. (b) More detailed circuit.

It should be noted that inverters used for small drives, under a few kilowatts, are almost always pulse-width-modulated inverters, due to the fact that they avoid the need for a variable d.c. source voltage. Furthermore, they yield superior waveforms and at these power levels it is easy to obtain controlled-turn-off power devices.

The current controllers can be hysteresis or PWM controllers. When a hysteresis current controller is used, upper and lower hysteresis levels are defined relative to the reference value of the current and the inverter is used to ensure that the actual stator current is limited to between the hysteresis bands. By defining narrow hysteresis bands, it is possible to ensure tight control, but this demands a higher switching frequency from the inverter. The actual currents contain harmonics which produce high-frequency torque ripples, but these are filtered out by the inertia of the machine so the speed of the machine can almost be free from the effects of these ripples. The hysteresis current controller reacts almost instantaneously to changes in the reference values of the current, and thus the time lag in the model of this controller is very small. When a PWM current controller is used, the actual current is compared with the reference current, and an error signal is generated. The error signal is compared to a sawtooth wave, and if the resulting error is larger than the sawtooth, the phase voltage is switched positively. However, if it is smaller than the sawtooth wave, it is switched negatively. The advantage of this controller over the hysteresis current controller is that the switching frequency of the inverter is preset and it is easy to determine the time delay between the time the reference changes and when corrective switching action is performed by the current controller.

In Fig. 3.5(b), the current control is implemented by the appropriate firing of the six bipolar transistors $T_1 \ldots T_6$ of a full-bridge inverter which produces adjustable frequency three-phase excitation from the d.c. source. The bipolar transistors can be replaced by any other bipolar or metal-oxide-semiconductor (MOS) power switch devices, e.g. by Metal-Oxide Semiconductor Field-Effect (MOSFET) transistors, which are significantly faster than the bipolar transistors, or by Metal-Oxide-Semiconductor Insulated Gate Transistors (MOSIGT), all of which can be turned both off and on with low-level gating signals (which can be gate voltages or gate currents depending on the specific type of power transistor used). The three-phase full-bridge inverter consists of three half-bridge units (which are phase shifted by 120°). In Fig. 3.5(b) all the transistors are combined with a parallel freewheeling diode to provide circulation paths for the flow of reverse stator currents of the synchronous machine during reactive power flow and regeneration; they also clamp the load voltage to the input level.

For braking, the motor is operated in the regenerating mode and the kinetic energy stored in the system inertia is converted into electrical energy. This can be dissipated across a resistor R_b shown in Fig. 3.5(b). During braking active power flows from the machine to the d.c. link. The braking resistor can be switched on and off within the hysteresis band of the d.c. link voltage to dissipate the braking power.

Implementation of rotor-oriented control of PMSM in polar coordinates: The block diagram of the rotor-oriented controlled permanent-magnet synchronous machine in polar coordinates is shown in Fig. 3.6, where the stator currents (i_{sA}, i_{sB}, i_{sC}) are impressed by fast current loops, e.g. by utilizing a transistorized PWM inverter as discussed above. When the control structure is put into polar form and a microprocessor implementation is used, for algorithmic reasons it can have advantages over an implementation utilizing the Cartesian form.

Since the sum of the stator currents is assumed to be zero, it is sufficient to monitor only two stator currents, $i_{sA}(t)$ and $i_{sB}(t)$. When a microprocessor is used, these currents are converted into their digital forms by using A/D converters. By using rectangular-to-polar (R$\rightarrow$P) conversion, the modulus ($|\bar{i}_s|$) and the phase angle (α_s) of the stator-current space phasor expressed in the stationary reference frame are obtained as shown in Fig. 3.6. It should be noted that the block labelled '3$\rightarrow$2' performs the transformation of the three-phase currents into the quadrature-phase current components, in accordance with eqns (2.1-9) and (2.1-10).

The rotor speed ω_r is measured and the rotor angle θ_r is obtained by integration (denoted by '$1/p$'). It is also possible to monitor the rotor angle directly and to use this to obtain the rotor speed. However, whatever method is used, it is very important to have an extremely accurate knowledge of the rotor position, since rotor-oriented control is based on the accurate knowledge of this quantity. The rotor angle is subtracted from the angle α_s of the stator-current space phasor to obtain $\beta=\alpha_s-\theta_r$, which is the angle of the stator-current space phasor in the rotor reference frame. The sine of this angle is obtained by using a function generator (FG1) and it is then multiplied by the modulus of the stator current space phasor and thus the quadrature-axis stator current component (i_{sq}) is obtained. This is multiplied by the excitation flux (ψ_F) and a signal proportional to the instantaneous value of the electromagnetic torque (t_e) is obtained, in agreement with eqns (3.1-3) and (3.1-4).

The necessary current references to the PWM inverter are generated as follows. A cascaded control structure is shown with superimposed torque, rotor-speed and rotor-position control loops. The difference between the reference value of the rotor angle and the monitored rotor angle serves as an input signal to the rotor position controller, which is a proportional (P) controller. The output of this controller is the rotor speed reference value. The difference between this and the actual monitored rotor speed is the input to the speed controller, which is a proportional-integrating (PI) controller, the output of which is the torque reference signal. The actual value of the electromagnetic torque (t_e) is subtracted from the reference value (t_{eref}) and

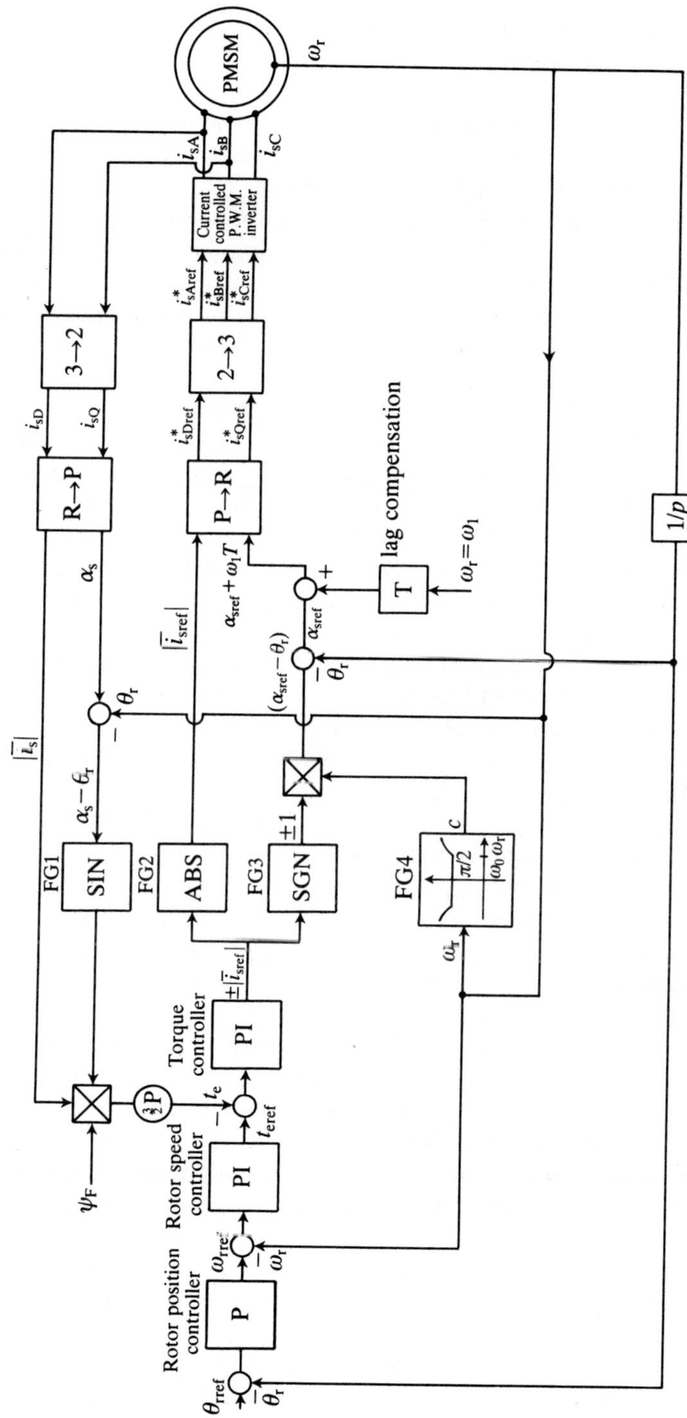

Fig. 3.6. Block diagram of a rotor-oriented controlled PMSM using polar coordinates.

this is the input signal to the torque controller, which is a PI controller. The output of the torque controller is the modulus of the reference value of the stator-current space phasor but multiplied by $+1$ or -1, indicated in Fig. 3.6 by $\pm|\bar{i}_{sref}|$. The reference values of the modulus and phase angle of the stator-current space phasor in the rotor reference frame are $|\bar{i}_{sref}|$ and $(\alpha_{sref}-\theta_r)$ respectively and they are obtained by using three function generators and by utilizing the signal for the rotor speed ω_r and the output signal of the torque controller. The signal $\pm|\bar{i}_{sref}|$ is fed into a function generator (FG2) which takes its absolute value, $|\bar{i}_{sref}|$. The angle $(\alpha_{sref}-\theta_r)$ must be positive or negative, corresponding to a positive or negative output from the torque controller, and therefore a sign function generator (FG3) is used. Thus when $+|\bar{i}_{sref}|$ is the output of the torque controller, the output of FG3 is $+1$ and when $-|\bar{i}_{sref}|$ is the output of the torque controller, the output of FG3 is -1. The factor ± 1 is then multiplied by another factor which is obtained with the fourth function generator FG4. This is equal to $\pi/2$ if the rotor speed is less than the base speed (ω_b), or to a factor greater than $\pi/2$ if the rotor speed is greater than the base speed, in order to obtain a stator-current phasor with a negative i_{sd} component for field-weakening purposes in accordance with the technique described above.

The actual rotor angle (θ_r) is subtracted from the reference value of the angle of the stator-current space phasor in the rotor reference frame, and thus the angle α_{sref} of the stator-current space phasor in the stationary reference frame is obtained. Thus the outputs of the P→R converter are the two-axis stator-current references i^*_{sQref} and i^*_{sDref}, which are transformed into the three-phase stator current references i^*_{sAref}, i^*_{sBref} and i^*_{sCref}. This two-phase to three-phase transformation is indicated by the block labelled '2→3'. The three-phase reference currents are the inputs to the inverter which supplies the permanent-magnet synchronous motor.

The role of the lag compensation showed in Fig. 3.6 will be described in the following section, and it should be noted that i^*_{sQref} and i^*_{sDref} are the reference values of the two-axis stator currents, which are obtained when the time lag of the control loops (T) is not negligible ($T\neq 0$). Similarly, i^*_{sAref}, i^*_{sBref}, and i^*_{sCref} are the reference values of the stator currents when $T\neq 0$.

It follows from the block diagram shown in Fig. 3.6 that there is a substantial amount of signal processing required when the scheme is implemented. With analogue techniques, it would be problematic and uneconomical to realize accurately the required multiplications, to implement the four function generators used to obtain the required modulations by the rotor angle. However, the use of a microprocessor simplifies the required hardware significantly. It is believed that in the future dedicated signal processors will perform the tasks set by the rotor-oriented control scheme described above.

Implementation of rotor-oriented control of PMSM in Cartesian coordin-

ates: The block diagram of the rotor-oriented controlled permanent-magnet synchronous machine using Cartesian coordinates is shown in Fig. 3.7. It is again assumed that the stator currents of the machine are impressed by fast current loops, e.g. by using a current-controlled PWM inverter.

Since there are no zero-sequence stator currents, again only two stator currents are monitored, $i_{sA}(t)$ and $i_{sB}(t)$. These are transformed into direct- and quadrature-axis stator current components $i_{sD}(t)$ and $i_{sQ}(t)$, which are the real- and imaginary-axis components of the stator-current space phasor in the stationary reference frame. The rotor speed is measured and the rotor angle is obtained by integration. However, it is again possible to have another implementation, where the rotor angle is directly monitored. The rotor angle is used to transform the space phasor of the stator currents in the stationary reference frame into the space phasor of the stator currents in the rotor reference frame [$\bar{i}_s' = \bar{i}_s e^{-j\theta_r} = i_{sd} + j i_{sq}$, see eqn (2.1-136)] and thus the current components i_{sd} and i_{sq} are obtained. The excitation flux (ψ_F) is multiplied by the quadrature-axis stator current component (i_{sq}) expressed in the rotor reference frame, and thus the electromagnetic torque (t_e) is obtained in agreement with eqn (3.1-3).

The current references to the inverter are generated as follows, by again using a cascaded control with superimposed torque, speed and rotor control and also using field-weakening.

The difference between the limit stator voltage (U_{smax}) and the peak of the actual stator voltages ($|\bar{U}_s|$) serves as input to the field-weakening controller, which is an integrator and the output of which is the reference value of the stator current in the rotor d-axis (i_{sdref}). The stator current reference in the rotor q-axis (i_{sqref}) is obtained at the output of a PI torque controller, whose input is the difference between the torque reference (t_{eref}) and the actual torque (t_e). The torque reference is obtained at the output of the rotor speed controller, a PI controller, whose output is the difference between the reference rotor speed (ω_{rref}) and the actual rotor speed (ω_r). The reference value of the rotor speed is obtained by the application of a rotor position controller, the input of which is the difference between the rotor angle reference and the actual monitored rotor angle.

The function generator (FG) shown in Fig. 3.7 ensures that if i_{smax} is the limiting value of the stator currents, i_{sq} may have to be reduced and this is achieved by utilizing $(i_{sd}^2 + i_{sq}^2)^{1/2} < i_{smax}$, and thus the function generator performs the operation $(i_{smax}^2 - i_{sd}^2)^{1/2}$.

It has been assumed that the PWM inverter which supplies the machine is current controlled and that the current controllers react instantaneously to the changes in the current commands, i.e. there is no time lag in the current control loops. Thus the transformations cancel and the control is extremely simple and is similar to the control of a converter-fed d.c. machine. However, when this assumption is not valid, that is when the reference value of the stator-current space phasor ($i_{sDref} + j i_{sQref}$) is not equal to the actual stator-current

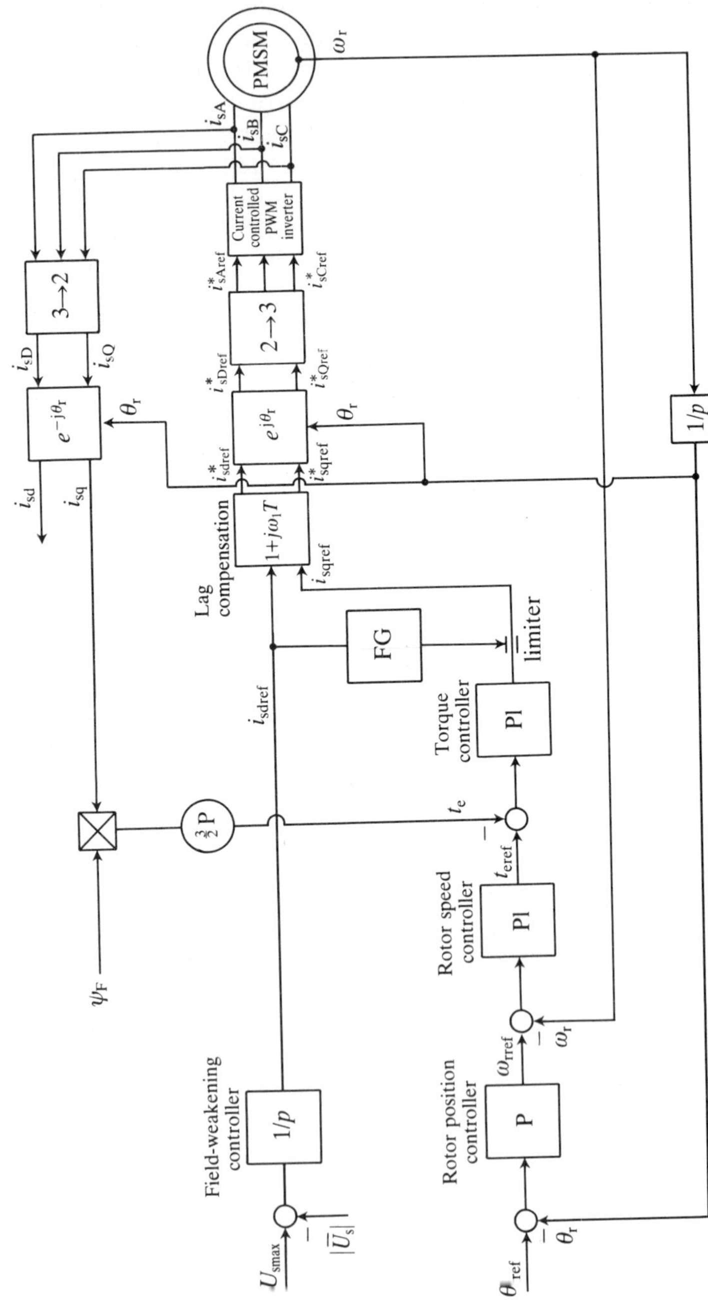

Fig. 3.7. Block diagram of a rotor-oriented controlled PMSM using Cartesian coordinates.

space phasor ($\bar{i}_s = i_{sD} + ji_{sQ}$), owing to the time lag in the current control loops, the following space phasor equation holds

$$\bar{i}_{sref} = T\frac{d\bar{i}_s}{dt} + \bar{i}_s, \tag{3.1-12}$$

where T is the time lag in the control loops. In two-axis form, this is

$$i_{sDref} = T\frac{di_{sD}}{dt} + i_{sD} \tag{3.1-13}$$

$$i_{sQref} = T\frac{di_{sQ}}{dt} + i_{sQ}. \tag{3.1-14}$$

For the rotor-oriented control to work well, it is essential that the required decoupling of the stator current components is achieved, even when this lag is present. A solution to this can be obtained easily if it is assumed that the unwanted coupling caused by the lag is due only to the phase shift of the stator-current space phasor and that its modulus is unchanged. In this case, in Fig. 3.7, in front of the block labelled '$e^{j\theta_r}$', which represents the transformation of the reference values of the two-axis stator current components in the rotor-coordinates into the two-axis stator current components of the stationary reference frame, an extra block labelled '$1+j\omega_1 T$' is added, as shown in Fig. 3.7.

This can also be proved by considering Fig. 3.8. When there is no lag ($T=0$), the relationship between the actual and reference values of the stator-current space phasors in the stationary reference frame is $\bar{i}_s = \bar{i}_{sref}$, and of course a similar relationship must hold in the rotor reference frame $\bar{i}'_s = \bar{i}'_{sref}$, since it follows from eqn (2.1-142) that $\bar{i}'_s = \bar{i}_s e^{-j\theta_r}$, and $\bar{i}'_{sref} = \bar{i}_{sref} e^{-j\theta_r}$.

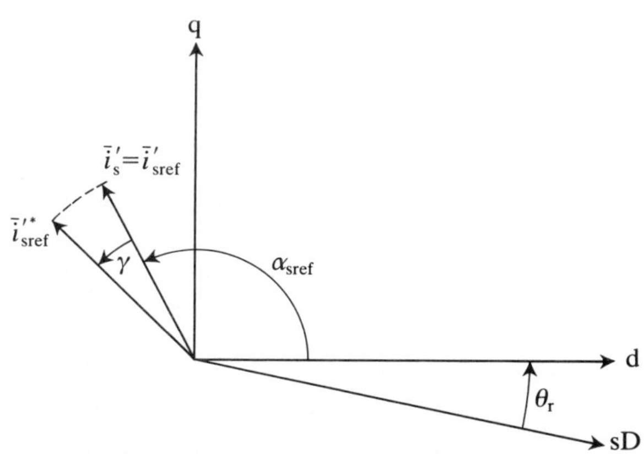

Fig. 3.8. The effects of controller lag.

However, when there is a lag, if a steady-state, where the modulus of the stator-current space phasor is constant ($|\bar{i}'_s|=\text{const.}$), is assumed, the reference stator current phasor in the rotor reference frame ($\bar{i}'^{*}_{\text{sref}}$) will be in a new position, displaced by angle γ from $\bar{i}'_{\text{sref}}$, where $\gamma=\omega_1 T$, but its length is the same as the length of $\bar{i}'_s$. By assuming small values for the lag, and thus of γ,

$$\bar{i}'^{*}_{\text{sref}}=\bar{i}'_{\text{sref}}e^{j\gamma}=\bar{i}'_{\text{sref}}e^{j\omega_1 T}=\bar{i}'_{\text{sref}}(\cos\omega_1 T+j\sin\omega_1 T)\approx\bar{i}'_{\text{sref}}(1+j\omega_1 T). \qquad (3.1\text{-}15)$$

Thus in front of the $e^{j\theta_r}$ block in Fig. 3.7, the block $(1+j\omega_1 T)$ must be inserted, the outputs of which are the two-axis components of $\bar{i}'^{*}_{\text{sref}}(i^{*}_{\text{sdref}}, i^{*}_{\text{sqref}})$ and whose inputs are the two-axis components of $\bar{i}_{\text{sref}}(i_{\text{sdref}}, i_{\text{sqref}})$. Thus, in two-axis form, the equations which define the lag compensation are

$$i^{*}_{\text{sdref}}=i_{\text{sdref}}-\omega_1 T i_{\text{sqref}} \qquad (3.1\text{-}16)$$

$$i^{*}_{\text{sqref}}=i_{\text{sqref}}+\omega_1 T i_{\text{sdref}}. \qquad (3.1\text{-}17)$$

Of course, it is possible to combine the transformation for the lag compensation with the transformation $e^{j\theta_r}$, so that a compound transformation $e^{j\theta_r}e^{j\gamma}=e^{j(\theta_r+\gamma)}$ is present. It can thus be seen that the removal of the unwanted coupling caused by the lag of the converter can be implemented by replacing the angle θ_r by $\theta_r+\gamma=\theta_r+\omega_1 T$ in the $e^{j\theta_r}$ block. This polar form can be utilized directly in Fig. 3.6, if the lag of the inverter is not negligible. In this case, since the angle of $\bar{i}_{\text{sref}}$ relative to the direct axis of the stator (sD) is α_{sref} (as shown in Fig. 3.8), and furthermore, since due to the $e^{j\gamma}$ displacement, the angle of $\bar{i}'^{*}_{\text{sref}}$ relative to the sD axis is $\alpha_{\text{sref}}+\gamma$, in front of the polar-to-rectangular transformation block, $\gamma=\omega_1 T$ has to be added to α_{sref}. Finally, it should be noted, that at higher speeds, the lag of the current control loops can increase, but in this case T can be adapted in the blocks which contain T. In Section 4.1.1 some other aspects of the effects of time delay are discussed.

3.1.2 Vector control of PMSMs with interior magnets

Interior-permanent-magnet synchronous machines (IPMSM) are mechanically robust and thus allow high-speed applications. As discussed in Section 3.1.1, in these machines the effective air gap is small and therefore the effects of armature reaction are significant. This allows the control of the synchronous machine in the constant-torque region as well as in the field-weakening (constant-power) region up to a high speed. Furthermore, in accordance with the discussion presented in Section 3.1.1, in the interior-permanent-magnet machine the quadrature-axis stator inductance (L_{sq}) is smaller than the direct-axis stator inductance (L_{sd}) and this is a distinct difference from the conventional electrically excited salient-pole synchronous machine. In practice there can exist inductance ratios ($L_{\text{sq}}/L_{\text{sd}}$) as high as five. The extra inductance can be useful as it allows the use of lower switching frequencies for the IPM synchronous machine than those used with other inverter-fed machines. The relation $L_{sq} < L_{sd}$ has a direct effect on the electromagnetic torque production and also on the excitation flux requirements of the IPM synchronous machine. Both of these aspects will also be discussed here.

An IPMSM allows economical machine design, because in accordance with eqn (2.2-13), in contrast to the machine with surface-mounted permanent magnets, where the torque is produced by the interaction of the magnet flux and the quadrature-axis stator current, now the electromagnetic torque contains two components. One of these is due to the interaction of the magnet flux and the quadrature-axis stator current and the other one is the so-called reluctance torque component, which is proportional to the difference of the stator inductances in the two axes, $L_{sd} - L_{sq} = L_{md} - L_{mq}$ where L_{md} and L_{mq} are the direct- and quadrature-axis magnetizing inductances respectively. Thus rotor saliency provides a way of reducing the volume of the magnets, which would otherwise be high to obtain the desired rating of the synchronous machine.

The system performance characteristics of the IPM synchronous machine drive can be influenced by the proper adjustment of the rotor parameters of the machine to control the two torque components described above. For example, overexcitation in a permanent-magnet synchronous machine drive endangers the drive electronics when the induced stator voltages (back e.m.f.s) generated by the magnets significantly exceed the source voltage at high speeds. However, the rotor saliency can be utilized to reduce the excitation flux linkage due to the permanent magnets, in order to obtain large speed ranges while reducing the amplitudes of overexcitation.

In the IPM synchronous machine, when the m.m.f. components created by the two-axis stator currents and the rotor magnets are added, because of magnetic saliency, the resultant m.m.f. distribution will be non-sinusoidal, even under linear magnetic conditions. Under linear magnetic conditions, the resultant flux-density distribution is equal to the sum of the component flux-density distributions produced by the permanent magnets (B_m) and the direct- and quadrature-axis components of the stator currents (B_d, B_q). These component distributions, together with the resultant distribution, are shown in Fig. 3.9. The flux-density waveforms shown correspond to operation as a motor and are for rated torque. With a light load or operation as a generator other waveforms apply. In Fig. 3.9, the direct and quadrature axes of the rotor (d, q) are also shown and τ_p is the pole pitch and θ is the angle around the periphery.

Due to the higher permeance along the rotor direct-axis, high flux densities can arise at the edges of the steel pole pieces. Thus, the stator teeth which are opposite the leading edges of these poles tend to saturate as the current excitation level is increased. For given stator currents, saturation of these parts of the stator teeth reduces the magnitude of the fundamental component of the air-gap flux-density distribution around the periphery and also shifts this

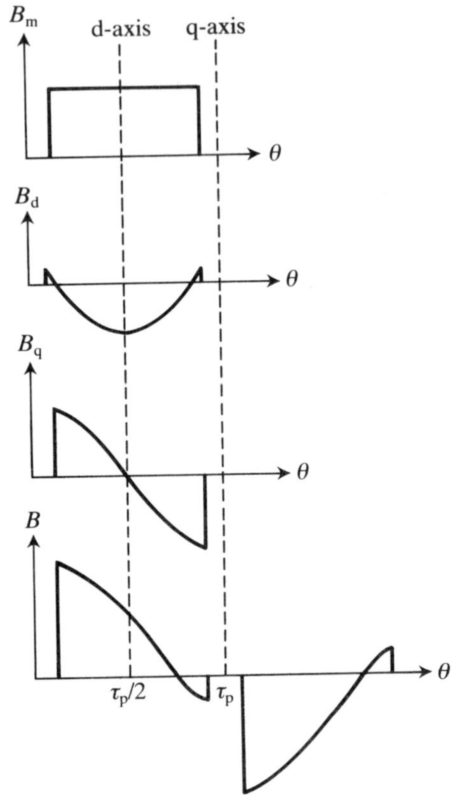

Fig. 3.9. Component and resultant flux density distributions in the IPM synchronous machine.

fundamental component in the direction of the centre of the rotor pole. When viewed from the stator terminals, this reduction of the air-gap flux appears as a reduction in the stator inductances, especially along the quadrature-axis (q). Thus, for a given applied terminal voltage, the maximum torque capability of the machine is greater than the one which could be predicted by the application of a linear model of the machine, which neglects the effects of magnetic saturation.

In the following, the main assumptions (apart from magnetic saliency), are the same as those used for the permanent-magnet synchronous machine with surface-mounted magnets. However, it should be noted that when the effects of magnetic saturation cannot be neglected, there will be a cross-coupling between the direct and quadrature axes not only due to the inherent physical saliency, but also due to the effects of main-flux saturation. This is the phenomenon of cross-saturation. Some aspects of magnetic saturation will be discussed in Chapter 5.

Since the flux-density distribution in the IPM synchronous machine is non-sinusoidal, the most convenient way to produce a smooth, pulsation-free electromagnetic torque is to produce a stator-current m.m.f. distribution which rotates synchronously and which is fixed in space relative to the rotor surface. This requires symmetrical three-phase stator currents, when sinusoidally distributed stator windings are assumed. It should be noted that in contrast to the synchronous motor with surface-mounted magnets, which can be designed for both sinusoidal and square-wave excitation, in the IPM synchronous motor with square-wave excitation, smooth torque operation cannot be obtained since the square waves produce an m.m.f. distribution which discretely shifts along the air-gap periphery only at the instants of switching.

It follows from eqn (2.2-13) that in the absence of damper windings, the electromagnetic torque produced in the IPMSM is

$$t_e = \tfrac{3}{2}P[\psi_F i_{sq} + (L_{sd} - L_{sq})i_{sd}i_{sq}], \tag{3.1-18}$$

where ψ_F is the constant flux linkage produced by the permanent magnets. However, it should be noted that when neodymium–iron–boron magnets are used, the magnet material has approximately 0.1% per °C negative temperature sensitivity, such that with the rise of the rotor temperature, the magnet flux linkage decreases. If the rotor temperature is monitored, it is possible to compensate this flux linkage for the variations of the rotor temperature. There is no simple method to directly monitor the rotor temperature, but, among other methods, it is possible to estimate this parameter by utilizing the monitored stator temperature and the dynamic thermal model of the machine. The control of the instantaneous stator currents provides the possibility of directly controlling the instantaneous electromagnetic torque. In the absence of damper windings, the torque immediately responds to changes in the direct- and quadrature-axis stator currents.

It is possible to put eqn (3.1-18) into the following form,

$$t_e = \tfrac{3}{2}P(\psi_{sd}i_{sq} - \psi_{sq}i_{sd}) = \tfrac{3}{2}P|\bar{\psi}_s|i_{sy}, \tag{3.1-19}$$

where ψ_{sd} and ψ_{sq} are the stator flux linkages in the direct and quadrature axes of the rotor, which in the absence of damper circuits can be expressed in terms of the stator currents and the magnet flux as

$$\psi_{sd} = L_{sd} + \psi_F$$
$$\psi_{sq} = L_{sq}i_{sq}. \tag{3.1-20}$$

In eqn (3.1-19), $|\bar{\psi}_s|$ is the modulus of the stator flux-linkage space phasor,

$$|\bar{\psi}_s| = (\psi_{sd}^2 + \psi_{sq}^2)^{1/2}, \tag{3.1-21}$$

and i_{sd}, i_{sq} are the direct- and quadrature-axis components of the stator currents in the rotor reference frame. i_{sy} is the quadrature-axis component of the stator currents expressed in a special reference frame (x, y) which rotates at the speed of the stator flux-linkage space phasor ($\bar{\psi}_s$). The real axis of this reference frame is aligned with the stator flux-linkage space phasor.

The phasor diagram of the IPMSM in the steady-state is shown in Fig. 3.10, where θ_r is the rotor angle, which is the angle between the direct axis of the stator (sD) and the direct-axis of the rotor (d). Both the magnet flux-linkage (ψ_F) and the stator flux-linkage ($\bar{\psi}_s$) space phasors are shown; δ is the load angle and $\bar{U}_s$ is the phasor of the terminal voltage.

In the special (x, y) reference frame, the two-axis components of the stator-current space phasor are denoted by i_{sx} and i_{sy} respectively. By utilizing Fig. 3.10, it is possible to prove that the electromagnetic torque is proportional to the quadrature-axis stator current component i_{sy}, since it follows from Fig. 3.10 that

$$\sin\delta = \frac{\psi_{sq}}{|\bar{\psi}_s|}$$

$$\cos\delta = \frac{\psi_{sd}}{|\bar{\psi}_s|}. \qquad (3.1\text{-}22)$$

Thus by the substitution of eqn (3.1-22) into eqn (3.1-19), the electromagnetic torque can be expressed as

$$t_e = \tfrac{3}{2}P|\bar{\psi}_s|(i_{sq}\cos\delta - i_{sd}\sin\delta) = \tfrac{3}{2}P|\bar{\psi}_s|i_{sy}. \qquad (3.1\text{-}23)$$

The relationship $i_{sy} = i_{sq}\cos\delta - i_{sd}\sin\delta$ follows from Fig. 3.10, but it also

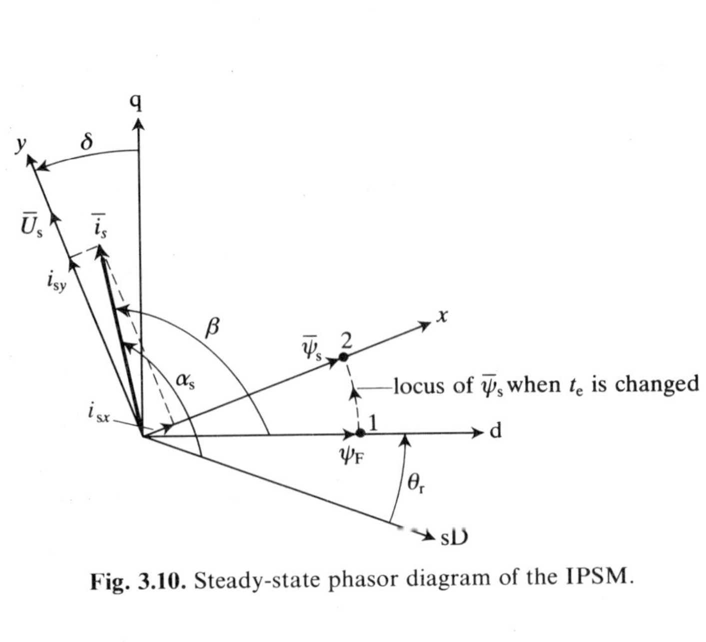

Fig. 3.10. Steady-state phasor diagram of the IPSM.

follows from eqn (2.1.142), since by considering that $\theta_g = \delta + \theta_r$, the transformation

$$i_{sx} + ji_{sy} = \bar{i}_s e^{-j(\theta_r + \delta)} = (i_{sD} + ji_{sQ}) e^{-j(\theta_r + \delta)} = (i_{sd} + ji_{sq}) e^{-j\delta} \qquad (3.1\text{-}24)$$

is obtained. In eqn (3.1-24) the relationship between the stator current phasor in the synchronously rotating reference frame $(i_{sd} + ji_{sq})$ and in the stationary reference frame $(i_{sD} + ji_{sQ})$ has been obtained from eqn (2.1-142) by the substitution of $\theta_g = \theta_r$. Equation (3.1-24) yields the following transformation between the stator current components of the special and rotor reference frames

$$i_{sx} = i_{sd} \cos \delta + i_{sq} \sin \delta \qquad (3.1\text{-}25)$$

$$i_{sy} = -i_{sd} \sin \delta + i_{sq} \cos \delta. \qquad (3.1\text{-}26)$$

Equation (3.1-23) will also be used for the development of the vector-control system of the IPM synchronous machine.

Implementation of the stator-flux-oriented control of the IPMSM in Cartesian coordinates

Similarly to that shown in Section 3.1.1, a current controlled PWM inverter provides the necessary control of the instantaneous currents of the IPMSM. The simplified schematic of the vector-controlled IPMSM drive is shown in Fig. 3.11.

In Fig. 3.11 a three phase uncontrolled rectifier produces a d.c. voltage across the filter capacitor C_F. The rectifier can be replaced by a battery bank too. The d.c. voltages supply power to the current controlled PWM inverter which supplies the IPM synchronous machine. Similarly to the PWM inverter discussed in Section 3.1.1, the inverter shown in Fig. 3.11 is a full-bridge three-phase inverter with six switching devices, which can be, for example, bipolar

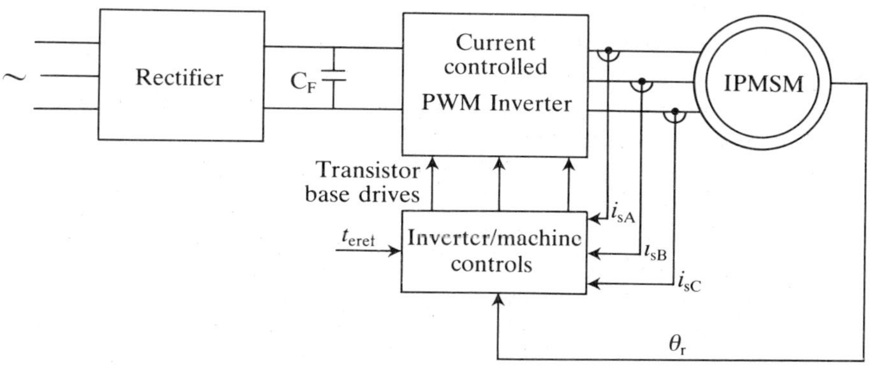

Fig. 3.11. Simplified schematic of the vector-controlled IPMSM drive system.

transistors. Each transistor is combined with a parallel freewheeling diode to provide a circulation path for the reactive stator currents of the machine. The drive system operates in all four quadrants.

The stator currents are monitored and because there are no zero-sequence currents, it is sufficient to monitor only two of these. The machine shaft is connected to a rotor position encoder, which supplies information on the rotor angle (θ_r) to the drive control system. The drive control electronics is also provided with information from the stator currents. The inverter/motor control also receives the reference value of the electromagnetic torque (t_{eref}) signal, and by utilizing the feedback signals of the currents and the rotor position, it generates the transistor base-drive signals. It should be noted that sinusoidal excitation of the IPMSM requires rotor-angle feedback with sufficient resolution to synchronize the sinusoidal references properly with the rotor position. This requirement is more demanding than for the comparable six-step square-wave current excitation configuration, for which the rotor angle information is necessary only in 60° increments.

For simplicity, in the first implementation of the drive system, it is assumed that the drive operates in the constant-torque mode, below base speed. In the constant-torque region, the stator flux is constant with the speed and the inverter is current controlled in the PWM mode to maintain the required flux–torque relationship. Thus the inverter has the characteristics of a current-source inverter. The stator flux can be controlled by the reactive stator currents, which can be magnetizing or demagnetizing.

If it is assumed that the drive can enter the constant-power operation (overspeed) region, then there are some special aspects which have to be considered. At high speeds the d.c. supply voltage is responsible for limiting the torque-speed envelope of the drive. This limit can be explained by considering that for any given direct- and quadrature-axis currents, the amplitude of the stator-voltage space phasor is dependent on nearly proportional to, the rotor speed. When the speed increases the resulting terminal line-to-line voltages approach the d.c. voltage and the voltage which is necessary to force the stator currents to their reference values decreases to zero. Under these conditions, at higher speed, the inverter saturates, the pulses in the phase-voltage waveforms drop out, the current control is lost and the inverter will generate constant-amplitude, six-step square-wave voltages. As the rotor speed increases in the constant-power region, the stator flux decreases inversely (with the speed) and the motor demands demagnetizing reactive currents to reduce the magnet flux (ψ_F). The saturation of the inverter requires the IPMSM control to change from current to voltage control. The transition between the PWM and square-wave modes must be fast and smooth under all conditions of operation of the drive. Such a transition involves some degradation in the torque control characteristics, since only the phase-angle of the voltage phasor can be adjusted during six-step voltage excitation; its magnitude cannot be adjusted.

From eqn (3.1-23) it follows that the torque control of the interior permanent-magnet synchronous machine can be performed similarly to the torque control of the separately excited d.c. machine. In the constant-torque region (below base speed) the torque can be controlled by controlling the quadrature-axis stator current i_{sy}. It is possible to control the stator flux by controlling the direct-axis stator current i_{sx}.

In Fig. 3.12 there the block diagram is shown of the torque control of the IPM synchronous machine in the constant-torque region, utilizing stator flux-oriented control. The inverter is current controlled, using the principles described in Section 3.1.1. The drive system is designed with an outer torque control loop, but it is possible to add speed and position control loops. In the form which is discussed, it is directly applicable to electrical vehicles.

In Fig. 3.12 the difference between the reference torque (t_{eref}) and the actual torque (t_e) serves as the input to the torque controller, which is a PI controller. Its output is the reference value of the torque-producing stator current components (i_{syref}). The reference value of the flux-producing stator current component (i_{sxref}) is generated from i_{syref} through the function generator FG2 as follows. The reference value of the modulus of the stator flux linkage $|\bar{\psi}_{sref}|$ is obtained by utilizing the function generator FG1. This serves to optimize the core losses in order to improve overall drive efficiency. Fig. 3.13 shows the non-linear characteristics implemented by function generator FG1. Point 1 corresponds to zero electromagnetic torque, since at this point the stator flux is equal to the magnet flux (ψ_F) and thus the load angle is zero. Point 2 corresponds to rated stator flux at rated torque.

The drive system incorporates the $|\bar{\psi}_s|$ flux control loop to prevent flux drift due to the variation of machine parameters. The difference between the reference stator flux $|\bar{\psi}_{sref}|$ and the actual value of the modulus of the stator flux $|\bar{\psi}_s|$ appears as the input of the stator flux controller, which is a PI controller, the output of which is Δi_{sxref}. The modulus of the actual stator flux is estimated as described below. The output signal of the flux controller is added to the output signal of function generator FG2 whose input is i_{syref}.

The role of FG2 can be understood by considering the phasor diagram shown in Fig. 3.10. In the constant-torque region, the electromagnetic torque can be controlled by i_{sy}, and $|\bar{\psi}_{sref}|$ can be maintained at a value determined by FG1. In Fig. 3.10, the locus of the stator flux-linkage space phasor is also shown when the electromagnetic torque is increased from zero to its rated value in the constant-torque region. At point 1, $t_e=0$, $i_{sx}=i_{sy}=0$ and $|\bar{\psi}_s|=\psi_F$. When the electromagnetic torque is increased, $|\bar{\psi}_s|$ will also increase, but i_{sx} and i_{sy} also have to increase until t_e is equal to its rated value. The function generator FG2 determines the relationship between i_{sxref} and i_{syref}, shown in Fig. 3.14.

In an electrically excited synchronous machine, $|\bar{\psi}_s|$ is controlled by the field current and thus the machine can always operate at unity power factor. However, in the IPMSM, $|\bar{\psi}_s|$, is controlled by the lagging stator current

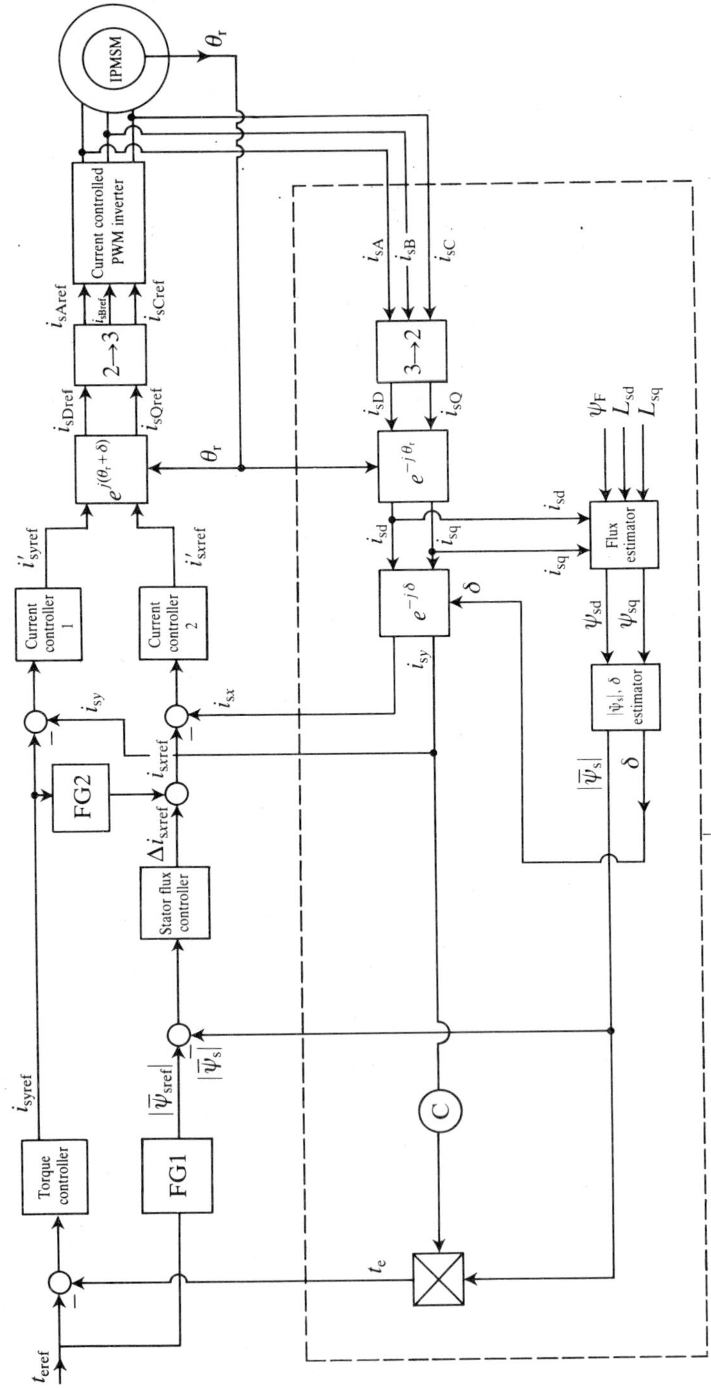

Fig. 3.12. Control block diagram of the stator-flux-oriented control of the IPMSM.

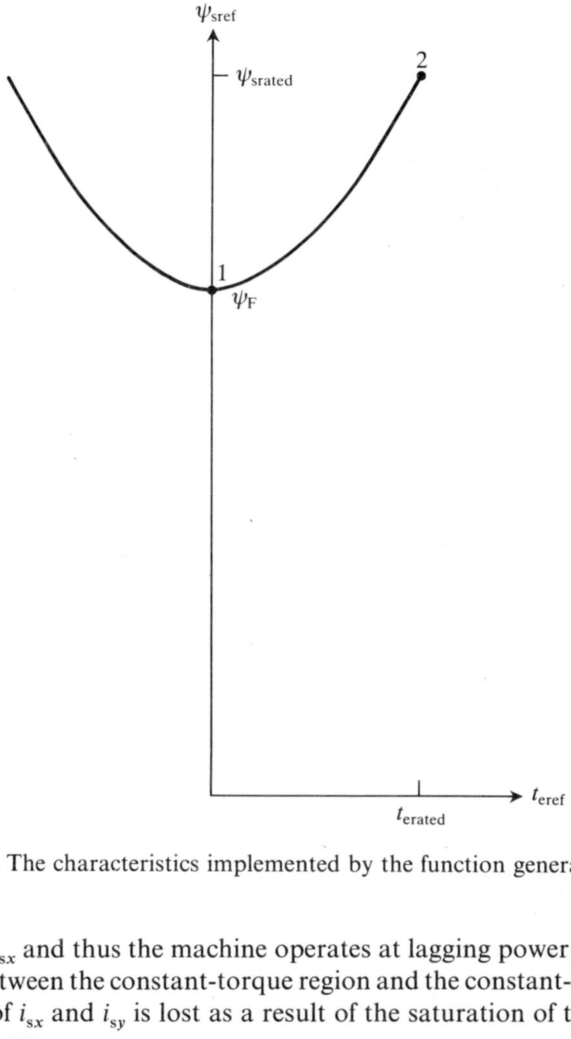

Fig. 3.13. The characteristics implemented by the function generator FG1.

component i_{sx} and thus the machine operates at lagging power factor. At the boundary between the constant-torque region and the constant-power region, the control of i_{sx} and i_{sy} is lost as a result of the saturation of the inverter as described above.

The reference signals i_{sxref}, i_{syref} are processed by current control loops, which allow vector control to be effective in partial saturation of the inverter, and if there is also a constant power region, help the smooth transition between the PWM and square-wave modes. The inputs to the two current controllers, which are PI controllers, are $i_{syref} - i_{sy}$ and $i_{sxref} - i_{sx}$ respectively. Below base speed, the operation of these current loops is redundant. The output currents of these controllers (i'_{sxref}, i'_{syref}) are fed into the block containing the complex transformation $e^{j(\theta_r + \delta)}$. This transformation is used to align i'_{sxref} with $\bar{\psi}_s$ and i'_{syref} with $\bar{U}_s$, as shown in Fig. 3.10, since this ensures the high dynamic performance of the system. Mathematically, it can be seen that it

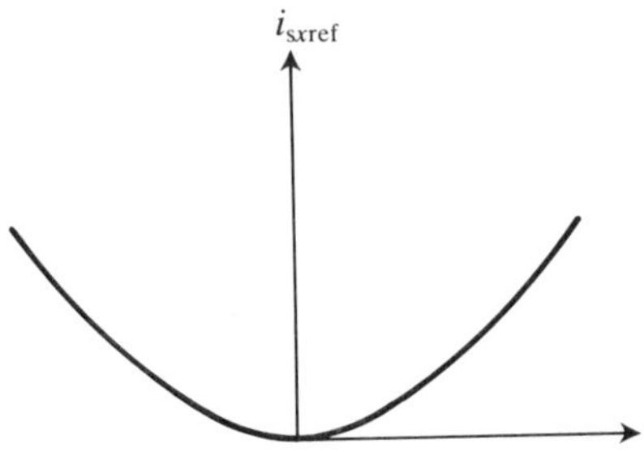

Fig. 3.14. The characteristic described by FG2.

is this transformation which is required, since it follows from eqn (3.1-24) that

$$i_{sD}+ji_{sQ}=(i_{sx}+ji_{sy})e^{j(\theta_r+\delta)}. \qquad (3.1\text{-}27)$$

The resolution of eqn (3.1-27) into its real- and imaginary-axis components gives

$$\begin{aligned} i_{sD}&=i_{sx}\cos(\theta_r+\delta)-i_{sy}\sin(\theta_r+\delta)\\ i_{sQ}&=i_{sx}\sin(\theta_r+\delta)+i_{sy}\cos(\theta_r+\delta) \end{aligned} \qquad (3.1\text{-}28)$$

and the unit-vector signals necessary for the transformation can be obtained as

$$\begin{aligned} \cos(\theta_r+\delta)&=\cos\theta_r\sin\delta-\sin\theta_r\sin\delta\\ \sin(\theta_r+\delta)&=\sin\theta_r\cos\delta+\cos\theta_r\sin\delta. \end{aligned} \qquad (3.1\text{-}29)$$

The three-phase reference stator currents are obtained from the stationary-axis stator reference currents i_{sDref}, i_{sQref} by the application of the two-to-three-phase transformation. It should be noted that in the absence of zero-sequence quantities, it is only necessary to obtain two current references. The reference currents are inputs to the inverter as described earlier. The actual three-phase stator currents (i_{sA}, i_{sB} i_{sC}) are monitored, but it is necessary to monitor only two of these in the absence of the zero-sequence currents.

The monitored values of the stator currents are used to obtain the electromagnetic torque (t_e), load angle (δ), modulus of the stator flux ($|\bar{\psi}_s|$), and the two-axis stator current components i_{sx} and i_{sy}, but it is also necessary to know the magnet flux (ψ_F), which can be determined by preliminary tests or can be calculated from the dimensions of the magnets and knowledge of the magnet material, as described in Section 3.1.

It is possible to obtain a fully digital implementation of the feedback signal processing, but to achieve high performance of the drive, it is necessary to compute all feedback signals (i_{sx}, i_{sy}, δ, $|\bar{\psi}_s|$, t_e) with high precision in real time. Thus a large-wordsize microcomputer with small sampling intervals is required. This could be achieved, for example, by the application of the Texas 32010 digital signal processor.

In this case the two-axis currents i_{sx} and i_{sy} are obtained from the monitored currents i_{sA}, i_{sB}, i_{sC} by utilizing the transformation $e^{-j(\theta_r+\delta)}$. Furthermore, the flux modulus $|\bar{\psi}_s|$ can be obtained from the two-axis stator flux-linkage components ψ_{sd} and ψ_{sq} as $|\bar{\psi}_s|=(\psi_{sd}^2+\psi_{sq}^2)^{1/2}$, where the flux-linkage components are obtained by considering eqn (3.1-20) and the torque can be obtained by considering eqn (3.1-19), according to which it is proportional to the product of $|\bar{\psi}_s|$ and i_{sy}.

Figure 3.12 also shows how these signals are generated. Thus the monitored stator currents are inputs to the three-phase to two-phase transformation block, the outputs of which are the stationary-axis currents i_{sD} and i_{sQ}. These are transformed into the two-axis components of the rotor reference frame (i_{sd}, i_{sq}) by the application of the transformation $e^{-j\theta_r}$, and finally i_{sd}, i_{sq} are transformed into the stator current components of the stator flux-oriented reference frame by the aid of transformation $e^{-j\delta}$. The inputs to the flux computation block are the current components i_{sd}, i_{sq}, ψ_F, L_{sd}, L_{sq} and its outputs ψ_{sd} and ψ_{sq}. These are inputs to the block which computes $|\bar{\psi}_s|$ and δ, as described above.

It is possible to have many other implementations of the vector control of the IPMSM, and rotor-oriented control of the IPMSM will be discussed in the next section.

Implementation of the rotor-oriented control of the IPMSM

Here rotor-oriented control of the interior permanent-magnet synchronous machine is described in both constant-torque and flux-weakening regions.

Implementation in the constant-torque region: It is assumed that the direct-axis of the special reference frame is aligned with the magnet flux (rotor-oriented control) and the stator current components expressed in the rotor reference frame (i_{sd} and i_{sq}) are controlled to control the torque. Figure 3.15 shows the block diagram of an implementation for the constant-torque region. The stator currents i_{sd} and i_{sq} are controlled in an open loop in such a way that the torque per stator ampere is maximum. All the other assumptions are the same as used in the previous section and it also follows that there are no damper circuits.

In Fig. 3.15, the reference torque serves as the input to the two function generators FG1 and FG2. Function generator FG1 gives the relationship between the torque and the direct-axis stator current component i_{sd}, and function generator FG2 gives the relationship between the torque and the quadrature-axis stator current i_{sq} for maximum torque per stator current ampere, if the effects of magnetic saturation are neglected. In a permanent magnet machine, which operates at a given speed and torque, optimal efficiency can be obtained by the application of an optimal voltage that minimizes the electrical losses. This optimum will not coincide with the condition of maximum torque per stator ampere owing to the presence of core

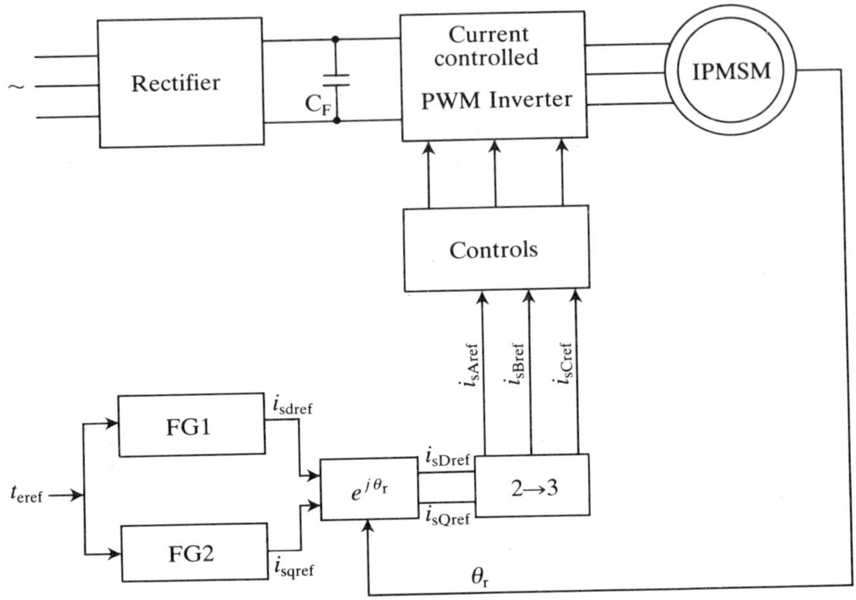

Fig. 3.15. Torque control of the IPMSM.

losses and the difference is more significant in the case of high-speed operation. In a permanent-magnet machine with no core losses, the machine which operates at maximum torque per stator ampere is optimally efficient. Such operation also leads to minimal copper, inverter, and rectifier losses. Furthermore, minimization of the stator currents for a given torque leads to higher current rating of the inverter and rectifier, and thus the overall cost of the system is reduced.

In the synchronous machine with surface-mounted magnets, the electromagnetic torque does not contain a reluctance torque term, since there is only a very small difference between the reluctances of the direct and quadrature axes. It has been shown that in this case, the torque is produced by the interaction of the stator currents with the air-gap flux and maximum torque is obtained when the space phasor of the stator currents $\bar{i}_s$ is in space quadrature to the magnet flux, and thus when the direct-axis stator current component (i_{sd}) is zero. However, in the synchronous machine with interior permanent magnets, there is a significant variation of the reluctance between the direct and quadrature axes, and as discussed before, $L_{sd} < L_{sq}$. As a consequence, there also exists a reluctance torque component, whose sign is negative compared with the reluctance torque of an electrically excited (wound rotor) salient-pole synchronous machine. Thus when this component is added to the torque component produced by the stator currents and the air-gap flux, maximum torque will be produced at the stator current angle $\beta > 0$, where this angle is the phase angle of the stator-current space phasor relative to the direct-axis of the rotor reference frame, as shown in Fig. 3.10. Thus if $\beta=0$ is chosen for the IPMSM, maximal torque per stator ampere cannot be obtained. In an IPMSM higher torque per ampere can be produced than in the PMSM with surface mounted magnets. The improved torque capability of the IPMSM is an advantage, but this must also be compared with the higher manufacturing costs.

To obtain generally applicable functions for the description of function generators FG1 and FG2, for any IPMSM, a normalized torque–current relationship is obtained and used in the function generators shown in Fig. 3.15. By selecting the following base value of the electromagnetic torque

$$t_{eb}=\tfrac{3}{2}P\psi_F i_b, \tag{3.1-30}$$

where ψ_F is the magnet flux and i_b is the base value of the currents

$$i_b=\frac{\psi_F}{L_{sq}-L_{sd}}, \tag{3.1-31}$$

the normalized value of the torque can be expressed as

$$t_{en}=\frac{t_e}{t_{eb}}. \tag{3.1-32}$$

Substitution of eqns (3.1-18), (3.1-30), and (3.1-31) into eqns (3.1-32) yields

$$t_{en}=\left(1-\frac{i_{sd}}{i_b}\right)\frac{i_{sq}}{i_b}=i_{sqn}(1-i_{sdn}), \tag{3.1-33}$$

where i_{sdn} and i_{sqn} are normalized values of the two-axis currents in the rotor reference frame,

$$i_{sdn}=\frac{i_{sd}}{i_b};\quad i_{sqn}=\frac{i_{sq}}{i_b}. \tag{3.1-34}$$

Thus it is possible to plot the constant-torque loci of the IPM synchronous machine in terms of the normalized values of the two-axis stator currents i_{sdn}, i_{sqn}. This is shown in Fig. 3.16, where on every locus the point which is nearest to the (0, 0) point (e.g. point A) corresponds to the minimum stator current, so that the maximum torque per stator ampere curve is obtained by connecting points CBAA′B′C′.

The maximum torque per stator ampere curve shown in Fig. 3.16 can be used to obtain the normalized stator current components as a function of the normalized torque for maximum torque per stator ampere. Thus the functions $i_{sdn}=f_1(t_{en})$ and $i_{sqn}=f_2(t_{en})$ shown in Fig. 3.17 are obtained and these are implemented by function gnerators FG1 and FG2 respectively.

In Fig. 3.15, the outputs of function generators FG1 and FG2 are the reference signals i_{sdref} and i_{sqref} respectively. These are rotated by utilizing the

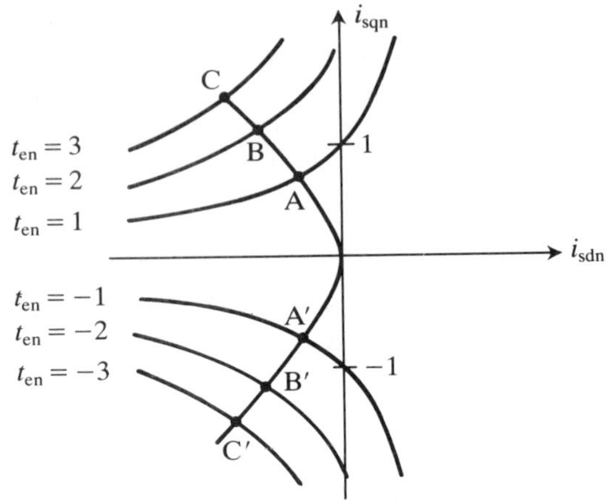

Fig. 3.16. Constant-torque loci for the IPMSM.

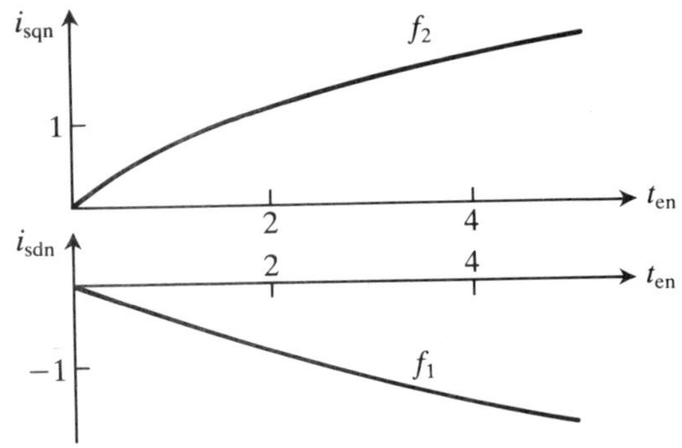

Fig. 3.17. The functions f_1 and f_2.

complex transformation $e^{j\theta_r}$ to yield the two-axis stator references in the stationary reference frame. Finally the two-to-three-phase transformation is used to obtain the three-phase reference currents which are used to control the current-controlled PWM inverter which supplies the IPMSM. An absolute rotor-position transducer is used to obtain the rotor angle and this angle is utilized during the $e^{j\theta_r}$ transformation. The combination of the current-controlled PWM inverter and rotor transducer yields a current-source type of control, where both the amplitude and the phase angle of the stator current are regulated.

It is only possible to obtain satisfactory operating characteristics of the drive described above at low frequencies. However, it is possible to achieve torque control at high speeds, in the flux-weakening mode, by commanding the demagnetizing component of the stator currents to keep the PWM controllers from saturating completely. This operating mode is different from the maximum torque per stator current condition, but is required because of the voltage limitations of the inverter. A possible implementation will now be described.

Implementation in the flux-weakening region: As discussed above, saturation of the current controllers of the current-controlled PWM inverter occurs at higher rotor speeds when the terminal voltages of the motor approach the ceiling voltage of the inverter. The rotor speed is an important factor which determines the saturation of the current controllers, since both the motor reactances and the induced stator e.m.f.s are proportional to the excitation frequency, which is synchronized with the rotor speed in order to yield smooth responsive torque control.

The maximum available per-phase fundamental stator voltage ($|\bar{u}_s|$) is determined by the d.c. link voltage, and can be expressed as

$$|\bar{u}_s| = (u_{sd}^2 + u_{sq}^2)^{1/2}, \tag{3.1-35}$$

where u_{sd} and u_{sq} are the direct- and quadrature-axis components of the stator voltages in the rotor reference frame and can be obtained as follows.

In the synchronously rotating reference frame the direct- and quadrature-axis stator voltage equations of a salient-pole synchronous machine are

$$u_{sd} = R_s i_{sd} + \frac{d\psi_{sd}}{dt} - \omega_1 \psi_{sq} \tag{3.1-36}$$

and

$$u_{sq} = R_s i_{sq} + \frac{d\psi_{sq}}{dt} + \omega_1 \psi_{sd}, \tag{3.1-37}$$

where R_s is the stator resistance and for the IPMSM the stator flux linkages ψ_{sd} and ψ_{sq} are defined by eqn (3.1-20). Thus in the steady state

$$U_{sd} = R_s I_{sd} - X_{sq} I_{sq} \tag{3.1-38}$$

$$U_{sq} = R_s I_{sq} + X_{sd} I_{sd} + \omega_1 \psi_F, \tag{3.1-39}$$

where $X_{sd} = \omega_1 L_{sd}$ and $X_{sq} = \omega_1 L_{sq}$ are the direct- and quadrature-axis stator reactances respectively and ω_1 is the excitation frequency (electrical rad/sec).

The maximum limits for the steady-state current components i_{sd} and i_{sq}, which correspond to maximal stator voltage $|\bar{u}_s|$ can thus be obtained by considering eqns (3.1-35), (3.1-38), and (3.1-39),

$$|\bar{u}_s|^2 = (R_s i_{sd} - X_{sq} i_{sq})^2 + (R_s i_{sq} + X_{sd} i_{sd} + \omega_1 \psi_F)^2. \tag{3.1-40}$$

In the complex d–q plane, specific values of the rotor speed, d.c. voltage and machine parameters, eqn (3.1-40) defines the locus of the stator current phasor $\bar{i}'_s = i_{sd} + ji_{sq}$. If, for simplicity, the ohmic voltage drops are neglected ($R_s = 0$), eqn (3.1-40) takes the recognizable form of an ellipse,

$$|\bar{u}_s|^2 = (X_{sq} i_{sq})^2 + \left[X_{sd}\left(i_{sd} + \frac{\omega_1 \psi_F}{X_{sd}}\right)\right]^2 \qquad (3.1\text{-}41)$$

or

$$\left(\frac{|\bar{u}_s|}{X_{sq}}\right)^2 = i_{sq}^2 + \left(\frac{X_{sd}}{X_{sq}}\right)^2 \left(i_{sd} + \frac{\omega_1 \psi_F}{X_{sd}}\right)^2. \qquad (3.1\text{-}42)$$

For any value of the rotor speed the stator current phasor $\bar{i}'_s = i_{sd} + ji_{sq}$ is directed away from the origin of the complex plane to a point on the ellipse, but because there is a maximal voltage, it must be always located inside the ellipse and not outside it. Such an ellipse will be referred to as a voltage-limit ellipse. For increasing rotor speeds, there is a group of ellipses, where the axes of the ellipses are inversely proportional to the rotor speed.

In comparison to the drive control shown in Fig. 3.15, in Fig. 3.18 a stator current feedback is shown for the identification of current-regulator saturation. In Fig. 3.18, the actual currents are first transformed into their direct- and quadrature-axis components (i_{sD}, i_{sQ}) in the stationary reference frame by the application of the three-phase to two-phase transformation and these components are transformed into the stator currents i_{sd}, i_{sq} formulated in the rotor reference frame, by utilizing the monitored rotor angle (θ_r). The direct-axis stator current reference i_{sdref} is obtained from the torque reference by utilizing function generator FG1, described above. The difference between the d-axis stator current i_{sd} and the reference current i_{sdref} is Δi_{sd} and this is used to identify the saturation of the current controllers of the inverter. When these are not saturated, this error is very small, since their gains are set high. The current difference Δi_{sd} serves as an input to the flux-weakening control.

Of course it is possible to perform the three-to-two-phase transformation and the $e^{-j\theta_r}$ transformation in one step and, similarly to the d-axis component in eqn (2.1-78),

$$i_{sd} = \tfrac{2}{3}[i_{sA} \cos \theta_r + i_{sB} \cos(\theta_r - 2\pi/3) + i_{sC} \cos(\theta_r - 4\pi/3)]. \qquad (3.1\text{-}43)$$

In Fig. 3.18 the quadrature-axis stator current reference i'_{sqref} is obtained from the torque reference by the application of function generator FG2 described above and i'_{sqref} is reduced in response to the increasing difference Δi_{sd}, which corresponds to the saturation of the current controllers of the inverter. Owing to the reduction of i'_{sqref}, at a given rotor speed, the reference-current space phasor $\bar{i}'_{sref} = i_{sdref} + ji_{sqref}$ will move from outside the so-called voltage limit ellipse—which represents the locus of the stator-current space phasor—to inside this ellipse and therefore a correspondence is forced between the reference and resultant current space phasors and therefore the current controllers of the inverter will work again.

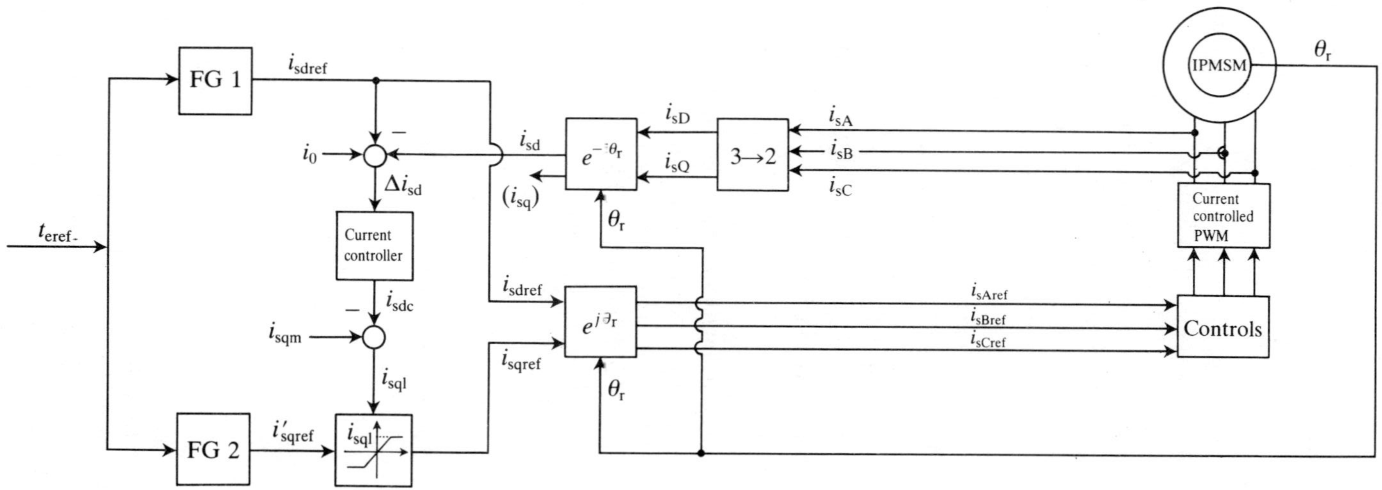

Fig. 3.18. IPMSM drive incorporating flux weakening.

The difference Δi_{sd} is fed into a PI current controller, the output of which is the current i_{sdc}. The quadrature-axis current reference i'_{sqref} is limited and the adjustable limiter shown in Fig. 3.18 ensures that if $i'_{sqref} < i_{sql}$, where i_{sql} is the limiting value of i_{sq} and is obtained as $i_{sqm} - i_{sdc}$, the output of the limiter is $i_{sqref} = i'_{sqref}$; otherwise the output signal is limited to $\pm i_{sql}$. For all speed and load conditions this provides the desired coupling of the flux-weakening operation from the normal current-controlled constant-torque operation until saturation of the current controllers of the inverter is reached. In Fig. 3.18 the values i_{sqm} and i_0 are adjusted to ensure adequate decoupling of the current and flux-weakening modes, while yielding high dynamic performance during the transition from one mode of operation to the other.

When the flux-weakening operation is inactive, the dynamic response of the drive is limited by the response characteristics of the current controllers of the inverter. However, during flux-weakening the dynamic response depends on the feedback loop of the direct-axis stator current (i_{sd}), and on the saturating current controllers of the inverter.

With the flux-weakening control described above, the upper speed limit is approximately equal to the overexcitation threshold speed of the interior permanent-magnet synchronous machine, which is the speed at which the induced stator e.m.f. $\omega_1 \psi_F$ is equal to the maximum available stator phase voltage $|\bar{u}_s|$.

It should be noted that in contrast to the synchronous machine with surface-mounted magnet, the synchronous machine with interior permanent magnets will move into the flux-weakening mode in the underexcitation speed mode, when $\omega_1 \psi_F < |\bar{u}_s|$. This is an advantage, since problems are caused if the motor back e.m.f. is allowed to exceed the d.c. source voltage. For example, d.c. link-voltage and current surges can arise if the inverter gating is suddenly removed at high speeds.

3.2 VECTOR CONTROL OF SALIENT-POLE SYNCHRONOUS MACHINES WITH ELECTRICALLY EXCITED ROTOR

3.2.1 Magnetizing-flux-oriented control

Basis for vector control

It is possible to design high-performance drive systems employing salient-pole synchronous machines with electrically excited rotor. In this section, the principle of the drive control will be based on field-oriented or more precisely, magnetizing-flux-oriented control. The control method can be applied to voltage-source inverter-fed synchronous machines, cycloconverter-fed synchronous machines and force-commutated current-source inverter-fed synchronous machines.

Vector control is not suitable for the load-commutated inverter-fed synchronous machine. In this drive, there is a phase-controlled rectifier on the supply side of the d.c. current link, which has a current control loop and operates as a controlled current source. The controlled d.c. current is passed through a d.c. link inductor to the thyristor-commutated three-phase inverter which supplies the synchronous machine with quasi-square-wave line currents. The role of the d.c. link inductor is to smooth the d.c. link current and to allow the rectifier and inverter to operate independently. Load commutation is ensured by overexcitation of the synchronous machine and thus the machine operates at leading power factor. Since the inverter is commutated by the induced voltages of the synchronous machine, load commutation will not function at a very low rotor speed, since in this case the induced voltage is low. The torque and the speed of the machine are controlled by the d.c. link current and the frequency of the inverter respectively. Vector control is not suitable for this drive because the stator currents of the synchronous machine cannot be freely controlled as, for example, in the case of the current-controlled PWM inverter, since the firing angle of the machine-side inverter must be formed to ensure safe commutation of the inverter and to yield a maximum link voltage and minimal link current for a given torque to achieve a higher power factor of the machine-side inverter. The control of the firing must be performed to give a minimal value of the turn-off angle (which is equal to the inverter advance angle minus the overlap angle), to give maximal motor power factor and efficiency, but it should be large enough not to produce commutation failure. Thus at any speed the turn-off angle must have a value, such that it will allow a safe recovery of the outgoing thyristor. Furthermore, as a result of operation at leading power factor, decoupling of the torque- and flux-producing components of the stator currents is not possible. Since this drive requires a smaller number of switching devices (e.g. thyristors) than those found in cycloconverter drives, and since higher frequencies and speeds can be obtained than with the drive using a cycloconverter, the load-commutated inverter-fed synchronous machine has found extensive high-power, high-speed applications. At such high power levels (several megawatts) there are no d.c. machines available. However, in these load-commutated inverter-fed synchronous drives high dynamic performance is usually not required, and thus the fact that vector control cannot be employed does not represent a great disadvantage.

Similarly to that described for smooth-air-gap machines, it is a main feature of the magnetizing-flux-oriented control of salient-pole synchronous machines that the stator current is divided into two components, a direct-axis component i_{sx} and a quadrature-axis component i_{sy} which determine the magnetization and the electromagnetic torque independently of each other. The quadrature-axis stator current component is in space quadrature to the magnetizing flux-linkage space phasor $\bar{\psi}_m$, and the electromagnetic torque is proportional to the product of the modulus of the magnetizing flux-linkage space phasor $|\bar{\psi}_m|$ and the quadrature-axis stator current i_{sy}. Fig. 3.19 shows the steady-state phasor diagram for the salient-pole synchronous machine.

In Fig. 3.19 ψ_{md} and ψ_{mq} are the magnetizing flux-linkage components in the reference frame fixed to the rotor and the space phasor of the stator currents $\bar{i}_s$ is also shown for the case of unity power factor. By using eqns (2.2-10) and (2.2-11), these flux components can be expressed in terms of the stator and rotor currents as

$$\psi_{md} = L_{md}(i_{sd} + i_{r\alpha} + i_{rF}) = L_{md} i_{md} \tag{3.2-1}$$

$$\psi_{mq} = L_{mq}(i_{sq} + i_{r\beta}) = L_{mq} i_{mq}, \tag{3.2-2}$$

where L_{md} and L_{mq} are the magnetizing inductances along the direct (d) and quadrature (q) axes fixed to the rotor, i_{sd} and i_{sq} are the direct- and quadrature-axis stator currents in the rotor reference frame, $i_{r\alpha}$ and $i_{r\beta}$ are the damper currents along the d and q axes respectively, which are zero in the steady state, i_{rF} is the field current and i_{md}, i_{mq} are the magnetizing current components in the rotor reference frame. The magnetizing flux-linkage space phasor is defined as $\bar{\psi}_m = \psi_{md} + j\psi_{mq}$ and is also shown in Fig. 3.19. As previously, a special (x, y) reference frame is chosen, where the direct axis (x) is aligned with the magnetizing flux-linkage space phasor and the special reference frame rotates at the speed of the magnetizing flux-linkage space phasor. In Fig. 3.19 the magnetizing-current space phasor is also shown and is defined as $\bar{i}_m = i_{md} + ji_{mq}$. As a result of saliency this is not coaxial with the magnetizing flux-linkage space phasor.

Implementation for a cycloconverter-fed salient-pole synchronous machine

Here the control scheme for the magnetizing-flux-oriented control of a salient-pole synchronous machine with electrically excited rotor winding is discussed.

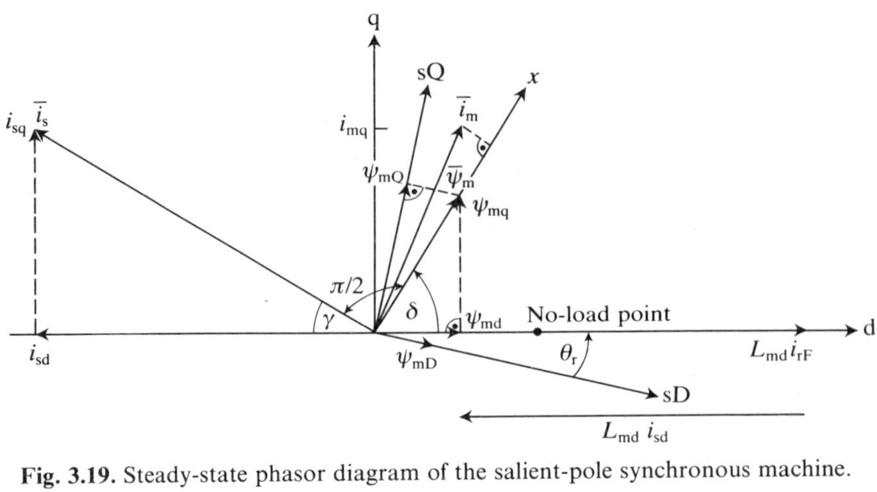

Fig. 3.19. Steady-state phasor diagram of the salient-pole synchronous machine.

The excitation winding on the rotor of the synchronous machine is supplied by a controllable three-phase rectifier. The stator windings are supplied by a cycloconverter, which utilizes the current control described in Section 3.1.1.

The cycloconverter is a frequency converter which converts power directly (without an intermediate d.c. link) from a fixed frequency to a lower frequency. On one specific implementation, the cycloconverter consists of three sets of antiparallel six-pulse bridge rectifiers without circulating currents. Figure 3.20 shows the schematic of the cycloconverter-fed synchronous machine drive.

Each of the motor phases is supplied through a three-phase transformer and an antiparallel thyristor bridge and the field winding is supplied by another three-phase transformer and a three-phase-rectifier using the bridge connection. The synchronous machine is equipped with rotor-position (θ_r) and rotor-speed (ω_r) transmitters and the stator currents (i_{sA}, i_{sB}, i_{sC}) and the excitation current (i_{rF}) are also monitored. The control circuits utilize the rotor speed, rotor position and monitored currents to control the relationship between the magnetizing flux and currents of the machine by modifying the voltages of the cycloconverter and the excitation current in the field winding.

Since each of the antiparallel converters carries only one direction of the output current, and the output voltage of each converter can have positive or

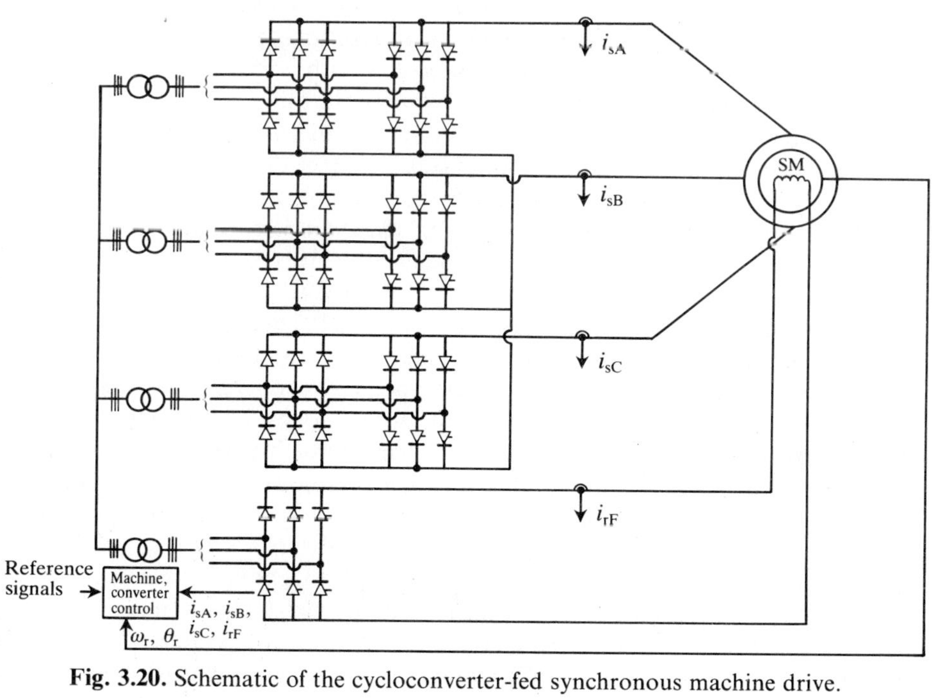

Fig. 3.20. Schematic of the cycloconverter-fed synchronous machine drive.

negative polarity, the cycloconverter is suitable for four-quadrant operation and can supply inductively or capacitively excited synchronous machines without the need for any extra equipment. The output currents of the cycloconverter have low harmonic content. The excellent current waveforms enable the synchronous machine to produce a uniform torque at standstill and during starting. In practice there are several power-circuit configurations to connect the cycloconverter to the supply. In one of the applications there is an isolation transformer (or three separate transformers) between the supply network and the cycloconverter phases as shown in Fig. 3.20. This is necessary to ensure that the three sets of antiparallel bridge rectifiers can operate independently.

It should be noted that since the bridge rectifiers are line-commutated, the frequency range is limited to approximately half the supply frequency and thus with a line frequency of 50 Hz, the output frequency of a 6-pulse cycloconverter can vary between approximately 0 and 25 Hz. However, this limited output frequency is of no importance in low-speed drives.

At higher voltages and frequencies the utilization of the cycloconverter can be improved by changing the output voltages of the cycloconverter from sinusoidal to trapezoidal waveforms since the amplitude of the fundamental voltage component exceeds the maximum instantaneous value by 15%. However, it should be ensured that air-gap harmonics, which are uneven multiples of 3, should be as low as possible, to give the best utilization of the drive system. This can be achieved by appropriate machine design.

In the past, for large power and low speed applications, d.c. drives have been used. However, owing to the problems associated with d.c. machines discussed earlier and also because in large drives, owing to the limited output power of d.c. machines, it would be necessary to employ a multimotor d.c. drive, it is more economical to use a single-motor a.c. drive. Since the cycloconverter has practically unlimited output power, it is very suitable for driving low-speed and large-horsepower a.c. motors, e.g. mine-shaft winders, rolling mill drives used in the steel industry, or gearless cement-mill drives. Conventionally low speeds could be obtained by, for example, the use of a four-pole machine together with a gearset having a large gear ratio. However, the maintenance and cost of these large gears is a disadvantage which can be avoided by using the cycloconverter-fed synchronous machine. Cycloconverter-fed synchronous machines are also used in ship propulsion drives, large compressors, large excavators and large belt conveyor systems. In such drives it is preferable to use synchronous machines rather than induction machines, because of their high power factor and large torque capability at low speeds. Furthermore, a synchronous machine draws less reactive power from the supply. In comparison with converter-fed d.c. drives, in the cycloconverter-fed synchronous machine drive, commutation problems do not exist, the drive has higher efficiency and higher overload capacity and there are lower maintenance and ventilation requirements. The rotor diameter of the synchronous machine can be kept smaller in comparison with an equivalent d.c. machine and thus the inertia is reduced. The physical size of the synchronous machine fed by a cycloconverter is smaller than that of the converter-fed d.c. machine of equal power rating.

The control scheme described in the present section is suitable for, say, a rolling mill application. For the control system of a rolling mill, the following characteristics are required:

- high torque and power control response;
- quick acceleration/deceleration;
- frequent start/stop operation;
- four-quadrant operation including field-weakening mode;
- high overload capacity from zero to maximum speed;
- production of constant torque at zero speed;
- small torque oscillations.

It should be noted that similar characteristics are required for propulsion systems, where until recently, d.c. drives have played a dominant role. However, in the cycloconverter-fed propulsion drive it can be a great disadvantage to employ isolation transformers which are bulky and heavy. Therefore these can be eliminated by isolating the three stator phases of the synchronous machine and by supplying all the stator phases separately from a separate phase of the cycloconverter (cycloconverter with isolated output phases).

All the characteristics described above are almost perfectly combined in the drive containing a salient-pole synchronous machine fed by the line-commutated cycloconverter and subjected to field-oriented control. With the implementation shown in Fig. 3.21, operation is possible in both constant-flux and flux-weakened modes.

Derivation of the reference signals: In Fig. 3.21 the reference rotor speed (ω_{rref}) and the actual monitored value of the rotor speed (ω_{r}) are compared and their difference is fed into the speed controller, which is a PI controller. The output voltage of the speed controller is proportional to the developed electromagnetic torque (t_{e}) of the synchronous machine, and thus the reference torque (t_{eref}) is obtained. This is divided by the modulus of the magnetizing flux-linkage space phasor ($|\bar{\psi}_{\text{m}}|$) to yield the reference value of the torque-producing stator current component i_{syref}.

The monitored rotor speed serves as input to function generator FG1, which below base speed yields a constant value of the magnetizing flux reference $|\bar{\psi}_{\text{mref}}|$; above base speed this flux is reduced. The flux-linkage reference $|\bar{\psi}_{\text{mref}}|$ is compared with the actual value of the magnetizing flux

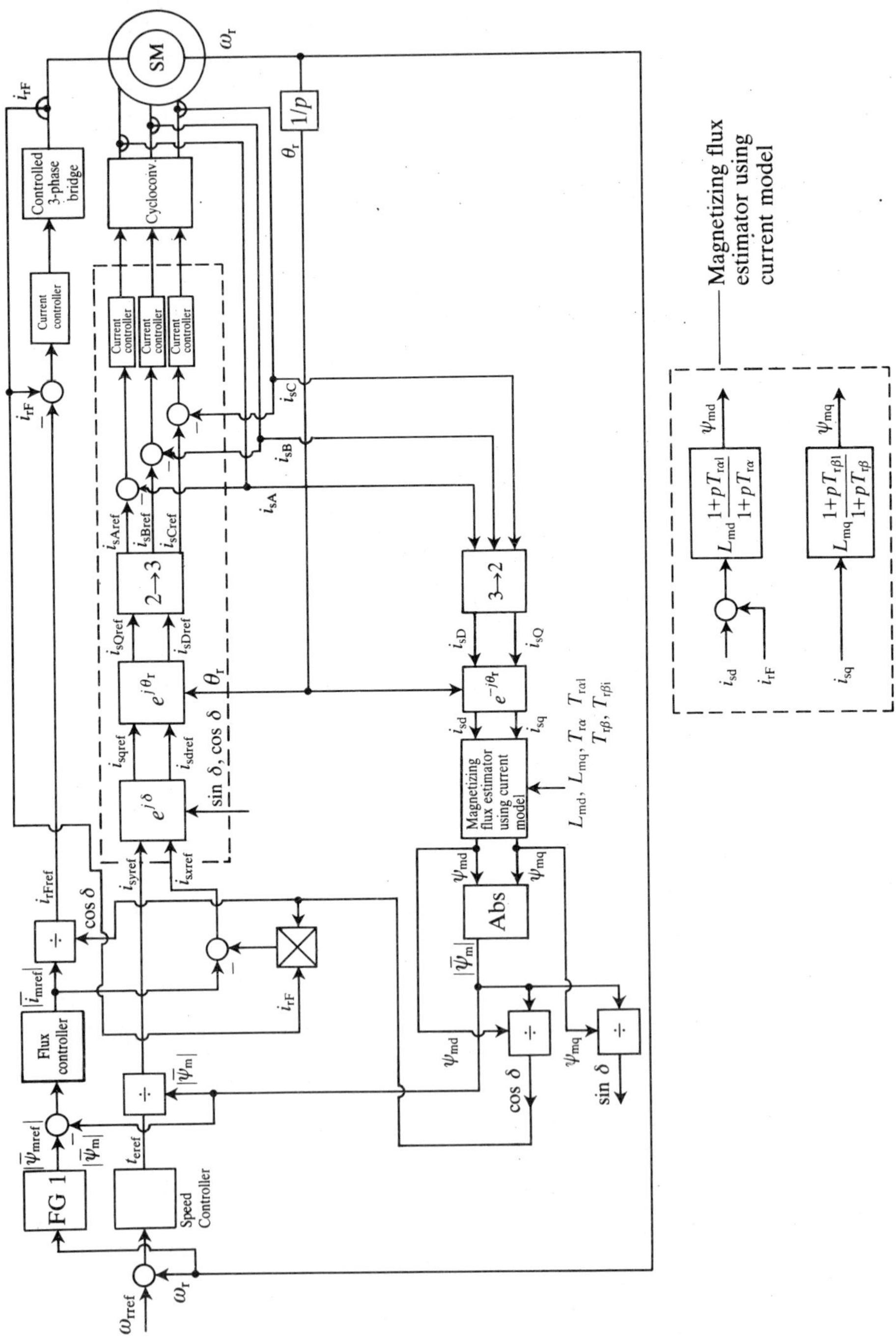

Fig. 3.21. Block diagram of field-oriented control of a cyclocoverter-fed salient-pole synchronous machine.

linkage $|\bar{\psi}_m|$ and their difference is fed into the flux controller, which is also a PI-controller. The flux controller maintains the magnetizing flux linkage in the machine at a preset value, independent of the load. To ensure optimal utilization of the motor, the magnetizing flux-linkage reference $|\bar{\psi}_{mref}|$ is raised to a very high value during starting, to yield a high breakdown torque, but for normal operating conditions it is set to the rated value. The output of the flux controller is the reference value of the magnetizing current $|\bar{i}_{mref}|$.

In Fig. 3.21 the reference value of the field current (i_{rFref}) and the reference values of the magnetizing and torque-producing stator current components (i_{sxref}, i_{syref}) are obtained as follows. The general expression for the magnetizing-current space phasor in the rotor reference frame is obtained from eqns (3.2-1) and (3.2-2), as

$$\bar{i}_m = i_{md} + ji_{mq} = \frac{\psi_{md}}{L_{md}} + \frac{j\psi_{mq}}{L_{mq}} = i_{sd} + i_{r\alpha} + i_{rF} + j(i_{sq} + i_{r\beta}) \tag{3.2-3}$$

and in accordance with Fig. 3.19, it follows from this equation that because of magnetic saliency, the magnetizing flux-linkage space phasor and the magnetizing-current space phasor are not coaxial. In the steady state the damper currents $i_{r\alpha}$ and $i_{r\beta}$ are zero.

The control strategy is defined in such a way that in the steady state there is no reactive power drawn from the stator. In this case the power factor is maximum ($\cos\phi_s = 1$, where ϕ_s is the phase angle between the stator-current space phasor and the stator-voltage space phasor) and the stator currents are optimal, i.e. at the minimal level. The zero-reactive-power condition can be fulfilled by controlling the stator component i_{sx} to be zero and thus from Fig. 3.19 and eqn (3.2-3), the following expression is obtained for the reference value of the magnetizing current (reference of the modulus of the magnetizing current vector), if it is assumed that the space angle of the magnetizing current space phasor and the space angle of the magnetizing flux linkage space phasor with respect to the direct-axis of the rotor are approximately equal,

$$|\bar{i}_{mref}| = i_{rFref}\cos\delta = (i_{rFref}^2 - i_{syref}^2)^{1/2}, \tag{3.2-4}$$

where i_{rFref} and i_{syref} are the reference values of the field current and quadrature-axis stator current respectively, and the latter current component is formulated in the reference frame fixed to the special reference frame. Thus it follows from eqn (3.2-4) that the reference value of the field current can be obtained as

$$i_{rFref} = \frac{|\bar{i}_{mref}|}{\cos\delta} \tag{3.2-5}$$

or as

$$i_{rFref} = (|\bar{i}_{mref}^2|^2 + i_{syref}^2)^{1/2}. \tag{3.2-6}$$

Because the field winding has a large inductance, it is sometimes necessary to keep the magnetizing flux linkage constant by keeping the field voltage transiently high. It is therefore effective to provide an additional magnetizing current component i_{sx} from the stator in the transient state. It should be noted that i_{sx} is the direct-axis stator current component in the special reference frame, and thus it is collinear with the magnetizing flux-linkage space phasor. Thus in the transient state the modulus of the magnetizing-current space phasor is not equal to the magnetizing current component produced by the field current in the real axis of the special reference frame ($i_{rF}\cos\delta$), but is the sum of the current components $i_{rF}\cos\delta$ and i_{sx}. It follows that the reference value of the stator current component in the special reference frame can be obtained as

$$i_{sxref} = |\bar{\imath}_{mref}| - i_{rF}\cos\delta. \tag{3.2-7}$$

This reference value becomes zero in the steady-state when the required magnetizing current is produced only by the field winding.

The stator current components i_{sxref}, i_{syref} are first transformed into the stator current components i_{sdref}, i_{sqref} established in the rotor reference frame by utilizing eqn (3.1-24):

$$i_{sdref} = \cos\delta\, i_{sxref} - \sin\delta i_{syref} \tag{3.2-8}$$

$$i_{sqref} = \sin\delta\, i_{sxref} + \cos\delta i_{syref}. \tag{3.2-9}$$

These components are transformed into the stationary-axes stator current components i_{sDref}, i_{sQref} by a similar transformation, but taking into account that the phase displacement between the stator direct axis (sD) and the direct axis of the rotor (d) is θ_r in accordance with Fig. 3.19:

$$i_{sDref} = \cos\theta_r\, i_{sdref} - \sin\theta_r i_{sqref} \tag{3.2-10}$$

$$i_{sQref} = \sin\theta_r\, i_{sdref} + \cos\theta_r i_{sqref}. \tag{3.2-11}$$

Of course it is possible to combine these two transformations into a compound transformation and thus the following reference values of the currents are obtained:

$$i_{sDref} = \cos(\theta_r + \delta) i_{sxref} - \sin(\theta_r + \delta) i_{syref} \tag{3.2-12}$$
$$i_{sQref} = \sin(\theta_r + \delta) i_{sxref} + \cos(\theta_r + \delta) i_{syref}. \tag{3.2-13}$$

These equations are similar to the eqns (3.1-28). It should be noted that in eqns (3.2-12) and (3.2-13) the angle $\theta_r + \delta$ appears; this is the space angle between the magnetizing flux-linkage space phasor and the real-axis of the stationary reference frame fixed to the stator. By utilizing the inverse of the transformation matrix defined in eqn (2.1-69), the obtained two-axis stator current references are transformed into the three-phase stator current references by the application of the three-phase to two-phase transformation indicated by the block containing the symbol '2→3'. The reference stator currents are compared with their respective monitored values, and their differences are fed into the respective stator current controllers, which are PI-controllers. The output signals from these current controllers are used to generate the firing pulses of the cycloconverter which supplies the salient-pole synchronous machine.

Similarly, the reference value of the field current and the actual value of the field current are compared and their difference serves as input to the field current controller. This supplies the necessary signal to a controlled three-phase bridge rectifier.

To obtain the reference values of the three-phase stator currents and the reference value of the field current, it is necessary to utilize the rotor speed and the modulus and phase angle of the magnetizing flux-linkage space phasor in various reference frames. Thus it is necessary to obtain $|\bar{\psi}_m|$, the load-angle δ, the rotor speed ω_r and the rotor angle θ_r. For this purpose the rotor speed is monitored, e.g. by a d.c. tachogenerator. In Fig. 3.21 the rotor angle is obtained by using the integration block. In practice this can be obtained by using a resolver, or by other techniques described in earlier sections. The load angle, the components and the modulus of the magnetizing-flux space phasor can be obtained as described in the next section.

Determination of the components of the magnetizing flux-linkage space phasor using a current model: The two-axis components of the magnetizing flux-linkage space phasor ψ_{md}, ψ_{mq} can be determined in many ways. It is possible to use a method similar to the one described in the case of the interior permanent-magnet machine. Thus these flux linkages can be determined from the monitored values of the machine currents and some machine parameters. However, in contrast to the technique described in Section 3.1.2 for the IPMSM, there are now damper circuits as well and it is not possible to monitor the damper currents directly. Therefore it is not possible to use directly the flux linkage equations, eqns (3.2-1) and (3.2-2), but when the voltage equations of the machine are considered and the damper currents are eliminated from the rotor voltage equations, it is possible to obtain expressions for the required magnetizing flux components in terms of the stator currents and the field current. Such a current model is now described.

It follows from eqns (3.1-36) and (3.1-37) that the stator voltage equations of the salient-pole synchronous machine in the reference frame fixed to the rotor can be put into the form,

$$u_{sd} = R_s i_{sd} + \frac{d\psi_{sd}}{dt} - \omega_r \psi_{sq} \tag{3.2-14}$$

$$u_{sq} = R_s i_{sq} + \frac{d\psi_{sq}}{dt} + \omega_r \psi_{sd}, \tag{3.2-15}$$

where the two-axis components of the stator flux linkages are

$$\psi_{sd}=L_{sl}i_{sd}+\psi_{md} \quad (3.2\text{-}16)$$

and

$$\psi_{sq}=L_{sl}i_{sq}+\psi_{mq}. \quad (3.2\text{-}17)$$

R_s and L_{sl} are the resistance and leakage inductance respectively of a stator winding and the magnetizing flux-linkage components are defined by eqns (3.2-1) and (3.2-2). The voltage equation for the field winding can be put into the following form in the reference frame fixed to the rotor:

$$u_{rF}=R_{rF}i_{rF}+\frac{d\psi_{rF}}{dt}, \quad (3.2\text{-}18)$$

where R_{rF} is the resistance of the field winding. ψ_{rF} is the flux linking the field winding and can be expressed as:

$$\psi_{rF}=L_{Fl}i_{rF}+\psi_{md}, \quad (3.2\text{-}19)$$

where L_{Fl} is the leakage inductance of the field winding. Similar equations hold for the damper circuits:

$$0=R_{r\alpha}i_{r\alpha}+\frac{d\psi_{r\alpha}}{dt} \quad (3.2\text{-}20)$$

$$0=R_{r\beta}i_{r\beta}+\frac{d\psi_{r\beta}}{dt}, \quad (3.2\text{-}21)$$

where $R_{r\alpha}$ and $R_{r\beta}$ are the resistances of the damper windings along the direct and quadrature axes respectively, $i_{r\alpha}$ and $i_{r\beta}$ are the currents in the two damper windings and $\psi_{r\alpha}$ and $\psi_{r\beta}$ are the flux components linking the damper windings. These can be expressed as

$$\psi_{r\alpha}=L_{r\alpha l}i_{r\alpha}+\psi_{md} \quad (3.2\text{-}22)$$

$$\psi_{r\beta}=L_{r\beta l}i_{r\beta}+\psi_{mq}, \quad (3.2\text{-}23)$$

where $L_{r\alpha l}$ and $L_{r\beta l}$ are the leakage inductances of the damper windings in the direct and quadrature axes of the rotor respectively.

Expressions for the magnetizing flux linkages which contain the stator and field current components and not the damper currents can be obtained from the damper voltage equations by eliminating the damper currents. Thus by considering eqns (3.2-1), (3.2-20), and (3.2-22), the voltage equation for the α-axis damper winding can be put into the following form

$$T_{r\alpha}\frac{d\psi_{md}}{dt}+\psi_{md}=L_{md}(i_{sd}+i_{rF})+L_{md}T_{r\alpha l}\frac{d(i_{rF}+i_{sd})}{dt}, \quad (3.2\text{-}24)$$

where $T_{r\alpha}$ is the time constant of the damper winding along the α-axis, $T_{r\alpha}=(L_{md}+L_{r\alpha l})/R_{\alpha}$, and $T_{r\alpha l}$ is the leakage time constant of the same damper winding, $T_{r\alpha l}=L_{r\alpha l}/R_{\alpha}$.

Similarly by considering eqns (3.2.-2), (3.2-21) and (3.2-23), the voltage equation of the damper winding along the β axis can be obtained as

$$T_{r\beta}\frac{d\psi_{mq}}{dt}+\psi_{mq}=L_{mq}i_{sq}+L_{mq}T_{r\beta l}\frac{di_{sq}}{dt}, \tag{3.2-25}$$

where $T_{r\beta}$ is the time constant of the β-axis damper winding, $T_{r\beta}=(L_{mq}+L_{r\beta l})/R_{\beta}$, and $T_{r\beta l}$ is the leakage time constant of the damper winding in the β-axis, $T_{r\beta l}=L_{r\beta l}/R_{\beta}$.

Equations (3.2-24) and (3.2-25) can be arranged as

$$\psi_{md}=L_{md}(i_{sd}+i_{rF})\frac{1+pT_{r\alpha l}}{1+pT_{r\alpha}} \tag{3.2-26}$$

$$\psi_{mq}=L_{mq}i_{sq}\frac{1+pT_{r\beta l}}{1+pT_{r\beta}} \tag{3.2-27}$$

where $p=d/dt$. When the magnetizing inductances and the time constants used in these equations are known (e.g. by utilizing the results of measurements performed prior to starting up the drive system) and the currents i_{sd}, i_{sq}, i_{rF} are obtained from monitored currents, eqns (3.2-26) and (3.2-27) can be used to obtain the magnetizing flux linkages ψ_{md}, ψ_{mq} in the direct and quadrature axes of the rotor. For this purpose it is necessary to obtain i_{sd} and i_{sq} from the monitored stator currents i_{sA}, i_{sB}, i_{sC}, and therefore to use the inverse of the above transformations. This is also shown in Fig. 3.21, where the monitored stator currents are first transformed into their stationary two-axis components i_{sD} and i_{sQ} by the application of the three-phase to two-phase transformation and these components are then transformed into the i_{sd} and i_{sq} currents by the application of the block containing the $e^{-j\theta_r}$ transformation. The currents i_{sd}, i_{sq}, and i_{rF} serve as inputs to the block which contains the magnetizing flux linkage estimator, which is based on eqns (3.2-26) and (3.2-27).

To achieve successful field-oriented control, it is extremely important to obtain accurate values of the flux-linkage components. However, this requires accurate values of the machine parameters in the current model described by eqns (3.2-26) and (3.2-27).

In contrast to the current model, it is also possible to use a voltage model, where the magnetizing flux-linkage components in the reference frame fixed to the stator (ψ_{mD}, ψ_{mQ}) are directly obtained from the terminal voltages. In this case the monitored terminal voltages of the salient-pole synchronous machine are reduced by their respective ohmic and leakage voltage drops and the resulting magnetizing voltages are integrated. However, when such a scheme is implemented, at low speeds problems can arise concerning the accuracy of the determined flux-linkage components. It is possible to have a hybrid implementation, where at lower speeds (say below 10% of the rated speed), the current model is used for the estimation of the flux-linkage components and at higher speed the voltage model is used.

Derivation of the load angle and modulus of the magnetizing flux-linkage space phasor: By considering the vector diagram shown in Fig. 3.19, the load angle can be obtained from

$$\cos \delta = \frac{\psi_{md}}{|\bar{\psi}_m|} \tag{3.2-28}$$

or

$$\sin \delta = \frac{\psi_{mq}}{|\bar{\psi}_m|} \tag{3.2-29}$$

and the modulus of the magnetizing flux-linkage space phasor is obtained as

$$|\bar{\psi}_m| = (\psi_{md}^2 + \psi_{mq}^2)^{1/2}. \tag{3.2-30}$$

Equations (3.2-28) and (3.2-29) are also utilized in Fig. 3.21, and the block labelled 'Abs' yields the absolute value of the magnetizing flux-linkage space phasor in accordance with eqn (3.2-30).

In the control scheme described in Fig. 3.21 it has been necessary to use the monitored rotor angle in the blocks containing the transformation $e^{j\theta_r}$ and its inverse. However, when the space phasor of the magnetizing flux linkages is directly determined in terms of its stationary-axes components, ψ_{mD} and ψ_{mQ}, which are obtained from the monitored stator voltages and currents in the way described above, it is not necessary to use the monitored value of the rotor angle for the determination of the two-axis stator current references i_{sDref}, i_{sQref} since it follows from Fig. 3.19 that

$$\cos(\delta + \theta_r) = \frac{\psi_{mD}}{|\bar{\psi}_m|} \tag{3.2-31}$$

or

$$\sin(\delta + \theta_r) = \frac{\psi_{mQ}}{|\bar{\psi}_m|} \tag{3.2-32}$$

and thus the required angle in the compound transformation $e^{j(\theta_r + \delta)}$ is obtained from ψ_{mD} and $|\bar{\psi}_m|$ or ψ_{mQ} and $|\bar{\psi}_m|$ by using eqn (3.2-31) or eqn (3.2-32).

It should be noted that by using high-speed microprocessors or signal processors it is possible to perform the computation of the direct- and quadrature-axis magnetizing flux linkages, the modulus of the magnetizing flux linkage space phasor and the load angle, as well as all the required transformations.

Current control with e.m.f. compensation: The control system described in Fig. 3.21 is similar to the control of a d.c. drive with a bridge converter. However, at higher output frequencies of the cycloconverter, even in the steady state, owing to a.c. induced stator voltages, which disturb the current control system shown in Fig. 3.21, there will be a phase error and also an amplitude (gain) error which the current controllers cannot eliminate. There are several ways of eliminating these errors and in the present section a current control scheme with e.m.f. compensation is described.

In the current control system shown in Fig. 3.22, which replaces the components surrounded with a broken line in Fig. 3.21, feedforward voltages are added to the outputs of the current controllers and the cycloconverter is mainly controlled by the feedforward stator voltage references u_{isAref}, u_{isBref} and u_{isCref}. These voltages are obtained by utilizing the rotor speed and the direct- and quadrature-axis components of the stator flux linkages. It follows from eqns (3.2-14) and (3.2-15) that the direct- and quadrature-axis components of the feedforward e.m.f.s in the reference frame fixed to the rotor are

$$u_{isd} = -\omega_r \psi_{sq} \tag{3.2-33}$$

and

$$u_{isq} = \omega_r \psi_{sd}, \tag{3.2-34}$$

where the direct- and quadrature-axis stator flux linkages have been defined by eqns (3.2-16) and (3.2-17) respectively. Thus to obtain the corresponding two-axis components u_{isD} and u_{isQ} in the stationary reference frame, by considering eqn (2.1-143), it is necessary to perform the transformation $u_{isD}+ju_{isQ}=(u_{isd}+ju_{isq})e^{j\theta_r}$, and the transformation $e^{j\theta_r}$ is also shown in Fig. 3.22. This transformation is followed by the two-phase to three-phase transformation and thus the induced voltages u_{isA}, u_{isB} and u_{isC} are obtained.

The induced stator voltages can be obtained from eqns (3.2-33) and (3.2-34), together with eqns (3.2-16) and (3.2-17). Thus the block containing the induced voltage estimator uses as its inputs the monitored rotor speed and the stator currents i_{sd}, i_{sq}. These stator current components are obtained from the monitored three-phase stator currents by utilizing the three-phase transformation and the complex transformation $e^{-j\theta_r}$, as shown in Fig. 3.21.

Current control with no e.m.f. compensation, but utilizing controllers operating in the magnetizing-flux-oriented reference frame: It is possible to have an implementation of the current control loops where there is no e.m.f. compensation but the drive will still have excellent control characteristics. In such a control system integrating current controllers operating in the

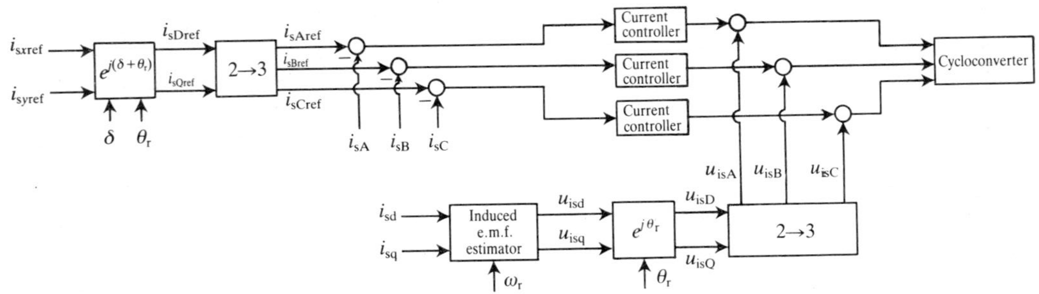

Fig. 3.22. Block diagram of field-oriented control of the salient-pole synchronous machine incorporating stator current control with e.m.f. compensation.

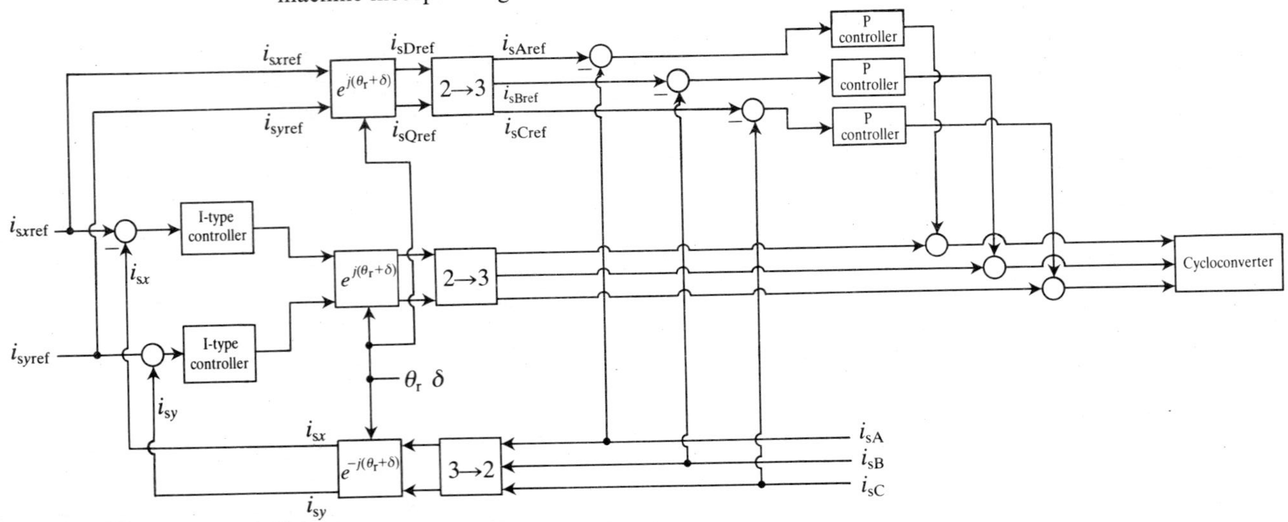

magnetizing-flux-oriented reference frame are used, since the deviations in the steady state between the reference stator currents and the actual values of the stator currents will be zero when i_{sx} and i_{sy} are controlled by integrating (I) controllers. Fig. 3.23 shows a possible implementation and this should replace the components surrounded with a broken line in Fig. 3.21.

In Fig. 3.23 the differences $i_{sxref}-i_{sx}$ and $i_{syref}-i_{sy}$ serve as inputs to the respective I controllers. The output signals of these controllers are transformed into the three-phase components of the stationary reference frame by the successive application of the transformation $e^{j(\theta_r+\delta)}$ and the three-phase to two-phase transformation, since the first transformation transforms the real- and imaginary-axis components of the magnetizing-flux-oriented reference frame into the real- and imaginary-axis components of the stationary reference frame. The stator currents i_{sx} and i_{sy} are obtained from the monitored values of the three-phase stator currents by utilizing the three-phase to two-phase transformation and also the transformation $e^{-j(\theta_r+\delta)}$.

However, when only the currents i_{sx} and i_{sy} are controlled, any imbalance and nonlinearity of the gain of the $e^{j(\theta_r+\delta)}$ and 2→3 coordinate converters, gate pulse generators of the cycloconverter, etc. are amplified by the gain of the stator circuit of the synchronous machine under consideration. It is possible to eliminate this deficiency by controlling the three-phase stator currents, which are a.c. quantities using proportional (P) controllers. For this purpose it is necessary to obtain the signals $i_{sAref}-i_{sA}$, $i_{sBref}-i_{sB}$, and $i_{sCref}-i_{sC}$ respectively, and in Fig. 3.23 the reference stator currents i_{sAref}, i_{sBref}, and i_{sCref} are obtained from i_{sxref} and i_{syref} by the application of the transformation $e^{j(\theta_r+\delta)}$ and the two-phase to three-phase transformation. The output signals of the P controllers are added to the output signals of the transformed signals of the I controllers and the signals obtained are the commands for the gate control of the cycloconverter. With the implementation shown in Fig. 3.23 a high-response current control is achieved.

3.2.2 Rotor-oriented control of reluctance machines

As discussed in Section 3.1, in a synchronous machine with three-phase stator windings and salient-pole rotor configuration, as a result of saliency, a reluctance torque component exists since the rotor tries to align itself in the minimum-reluctance position with respect to the synchronously rotating air-gap flux wave. This principle is utilized in reluctance machines. A reluctance machine can be considered as a salient-pole synchronous machine with three-phase stator windings and no rotor excitation.

In earlier constructions, rotor saliency was achieved by removing certain teeth from the rotors of conventional squirrel cages. Such reluctance machines with low output power have been used for a long time. Their inferior performance combined with their relatively high price have resulted in their limited use. However, as a result of recent developments, more reliable and robust rotor constructions exist. In the new type of reluctance machines there is a segmental rotor, a rotor with flux barriers, or an axially laminated rotor. These machines have higher output power which is comparable to that of corresponding induction machines. However, it is a distinct advantage over induction machines that reluctance machines run at exactly the synchronous speed—which is solely determined by the stator excitation frequency and the number of poles—and there are no rotor currents. These are important aspects in many industrial applications.

It is possible to operate reluctance motors without a rotor cage. Conventionally, it is necessary to have a cage winding in a relucatance machine, since the machine does not have a starting torque when excited from a constant frequency supply. By using the cage winding, the machine runs up to synchronous speed by induction motor action where the rotor locks into synchronism with the field produced by the stator. For sinusoidal stator excitation, at synchronous speed the only role of the cage winding is to damp the oscillations in the rotor speed. However, if variable stator excitation is used, it is not necessary to have a rotor cage for starting purposes, since the excitation can be controlled in such a way that the motor is always kept in synchronism. Reluctance machines with cageless rotors lead to a reduction of the rotor losses, improved efficiency, higher power factor and higher torque/weight ratio.

It follows from eqns (3.1-18) and (3.1-19) that the electromagnetic torque produced by a reluctance machine with no damper windings is

$$t_e = \tfrac{3}{2}P(\psi_{sd}i_{sq} - \psi_{sq}i_{sd}) = \tfrac{3}{2}P(L_{sd} - L_{sq})i_{sd}i_{sq}, \tag{3.2-35}$$

where $\psi_{sd} = L_{sd}i_{sd}$ and $\psi_{sq} = L_{sq}i_{sq}$ are the direct- and quadrature-axis stator flux linkages respectively in the reference frame fixed to the rotor, and L_{sd}, L_{sq} are the stator inductances along the direct and quadrature axes respectively. Thus it can be seen that under linear magnetic conditions the electromagnetic torque is the product of a constant times the direct-axis stator current i_{sd} and the quadrature-axis stator current i_{sq}. There are many types of possible rotor configurations which influence the stator inductances L_{sd} and L_{sq} and thus have a direct effect on the torque. It is possible to improve the performance of the reluctance machine significantly by increasing the ratio L_{sd}/L_{sq}. In reluctance machines with axially laminated rotors, it is possible to achieve values up to approximately 5.5.

In this section the rotor-oriented control of a reluctance machine is discussed. The machine is supplied by a PWM voltage source inverter and is current controlled along the direct (d) and quadrature (q) axes of the rotating reference frame fixed to the rotor. It is assumed that the drive operates in both constant-torque and constant-power regions. Figure 3.24 shows the block diagram of the drive system.

In Fig. 3.24 the rotor speed (ω_r) is monitored and the rotor angle (θ_r) is also obtained. However, it should be noted that the integration block is only

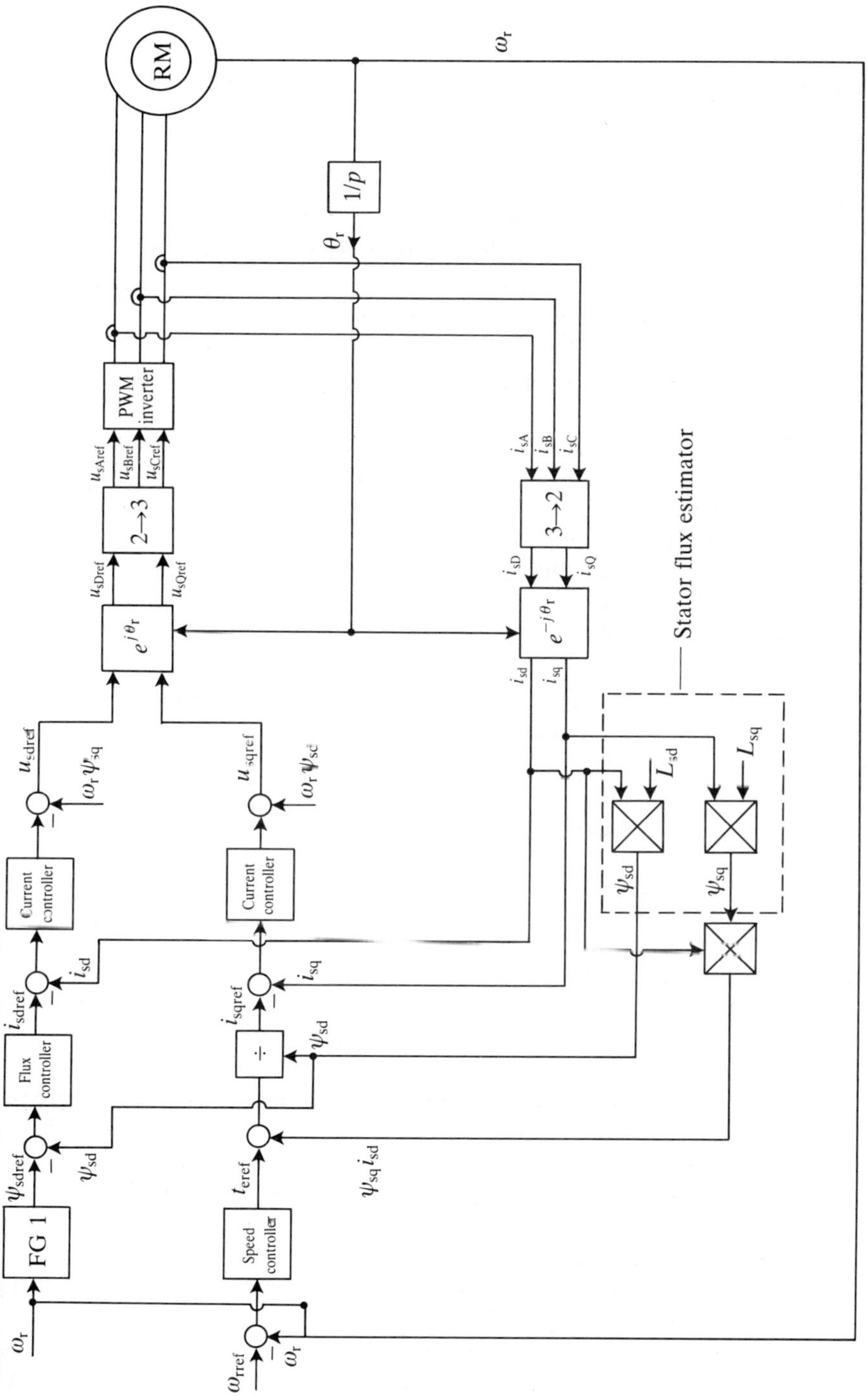

Fig. 3.24. Vector control of the reluctance machine.

symbolic; there are many ways of obtaining the variation of the rotor angle in real time, by the use of a resolver. It is also possible to obtain the rotor speed from the rotor angle from the expression,

$$\omega_r = \cos\theta_r \frac{d(\sin\theta_r)}{d\theta_r} - \sin\theta_r \frac{d(\cos\theta_r)}{d\theta_r}. \qquad (3.2\text{-}36)$$

In Section 6.3 a high-resolution rotor speed estimator is described. The monitored rotor speed is fed into the function generator FG1, the output of which is the direct-axis stator flux reference (ψ_{sdref}). During field weakening, this reference signal is a function of the rotor speed.

The direct-axis stator flux linkage ψ_{sd} is obtained with the stator flux-estimation circuit surrounded with the dashed lines. Below base speed, the direct-axis stator flux loop is current controlled. The difference between ψ_{sdref} and ψ_{sd} is fed into a flux controller, a PI controller, and the output is the direct-axis stator current reference i_{sdref}. The error $\Delta i_{sd} = i_{sdref} - i_{sd}$ is fed into a current controller, also a PI controller, and the output signal of this controller is added to the rotational voltage component, $\omega_r \psi_{sq}$, where ψ_{sq} is the quadrature-axis stator flux linkage expressed in the reference frame fixed to the rotor. Thus d-axis stator voltage reference u_{sdref} is obtained and it should be noted that it is expressed in the rotor reference frame.

The speed error $\omega_{rref} - \omega_r$ is fed into a speed controller to produce the reference signal for the electromagnetic torque (t_{eref}). From eqn (3.2-35), it is possible to obtain the reference torque as

$$t_{eref} = \psi_{sd} i_{sqref} - \psi_{sq} i_{sd} \qquad (3.2\text{-}37)$$

and thus the quadrature-axis stator current reference i_{sqref} is obtained as

$$i_{sqref} = \frac{t_{eref} + \psi_{sq} i_{sd}}{\psi_{sd}}. \qquad (3.2\text{-}38)$$

The error $\Delta i_{sq} = i_{sqref} - i_{sq}$ is fed into the quadrature-axis current controller (a PI controller) and the output signal together with the rotational voltage component $\omega_r \psi_{sd}$ is added to yield the q-axis stator voltage reference u_{sqref}, which is expressed in the reference frame fixed to the rotor.

With eqn (2.1-143), the voltage components u_{sdref} and u_{sqref} are transformed into their stationary-axes components u_{sDref}, u_{sQref} via the complex transformation $e^{j\theta_r}$, and these components are finally transformed into the three-phase stator voltage references u_{sAref}, u_{sBref}, and u_{sCref}, by the application of the two-phase to three-phase transformation shown in the block labelled '2→3'. These are the reference voltages for the inverter.

It has been necessary to use the stator current components i_{sd}, i_{sq} which correspond to the rotor reference frame. In Fig. 3.24 these are obtained from the monitored three-phase stator currents (i_{sA}, i_{sB}, i_{sC}) by the application of the three-phase to two-phase and $e^{-j\theta_r}$ complex transformations. Because of the absence of the zero-sequence currents, it is possible to obtain i_{sd} and i_{sq} by utilizing only two monitored stator currents.

The stator flux components are obtained by the stator flux estimator circuit shown in Fig. 3.24, which uses $\psi_{sd}=L_{sd}i_{sd}$ and $\psi_{sq}=L_{sq}i_{sq}$ and assumes L_{sd} and L_{sq} to be known; they can even vary with i_{sd} and i_{sq}. It is also possible to implement other flux estimators, e.g. those which involve monitored machine terminal voltages and currents.

BIBLIOGRAPHY

Adkins, B. and Harley, R. G. (1978). *The general theory of alternating current machines.* Chapman and Hall, London.

Alaküla, M. and Vas, P. (1989). Influence of saturation on the dynamic performance of vector-controlled salient-pole synchronous machines. *4th International Conference on Electrical Machines and Drives*, IEE, London, 298–302.

Arkadan, A. A., Demerdash, N. A., Vaidya, J. G., and Shah, M. J. (1988). Impact of load on winding inductances of permanent magnet generators with multiple damping circuits using energy perturbation. *IEEE Transactions on Energy Conversion* **3**, 880–9.

Bausch, H. (1978). Large power variable speed a.c. machines with permanent magnet excitation. *Electric Energy Conference, Adelaide*, pp. 265–71.

Bayer, K. H., Waldmann, H., and Weibelzahl, M. (1972). Field-oriented control of a synchronous machine with the new Transvektor control system. *Siemens Review* **39**, 220–3

Binns, K. J. and Chaaban, F. B. (1988). The relative merits of rare earth permanent magnet materials for use in the excitation of permanent magnet machines. In *Proceedings of the International Conference on Electrical Machines, Pisa*, 463–6.

Bolognani, S. and Buja, G. S. (1987). Dynamic characteristics of a P.W.M. voltage source inverter-fed P.M. brushless motor drive under field orientation. *Electric Energy Drive Conference, Adelaide*, pp. 91–5.

Bose, B. K. (1986). *Power electronics and drives.* Prentice Hall, Englewood Cliffs, New Jersey.

Bose, B. K. (1987). A high performance inverter-fed system of an interior permanent magnet synchronous machine, *IEEE IAS Annual Meeting, Atlanta*, pp. 269–76.

Bose, B. K. and Szczesny, P. M. (1988). A microcomputer-based control and simulation of an advanced IPM synchronous machine drive system for electrical vehicle propulsion. *IEEE Transactions on Industrial Electronics* **35**, 547–59.

Chalmers, B. J. (1988). *Electric motor handbook.* Butterworth, London.

Colby, R. S. and Novotny, D. W. (1986). Efficient operation of surface mounted P.M. synchronous motors, *IEEE IAS Annual Meeting, Denver*, pp. 806–13.

Consoli, A. and Abela, A. (1986). Transient performance of permanent magnet a.c. motor drives, *IEEE Transactions on Industry Applications*, **IA-22**, 32–41.

Fratta, A. and Vagati, A. (1987). A reluctance motor drive for high dynamic performance applications, *IEEE IAS Annual Meeting, Atlanta*, pp. 295–302.

Honsinger, V. B. (1982). The fields and parameters of interior type permanent magnet a.c. machines, *IEEE Transactions on PAS*, **PAS-101**, 867–76.

Hussels, P. (1984). Flux and torque control of synchronous machine with compensation of inaccuracy due to variations of motor parameters. In IEE Conference Publication 234, *Power Electronics and Variable Speed Drives*, pp. 377–80, Institution of Electrical Engineers, London.

Jahns, T. M. (1987). Flux-weakening regime operation of an interior permanent-magnet synchronous motor drive. *IEEE Transactions on Industry Applications* **IA-23**, 681–9.

Jahns, T. M., Kliman, G. B., and Neumann, T. W. (1986). Interior permanent-magnet synchronous motors for adjustable-speed drives. *IEEE Transactions on Industry Applications*, **IA-22**, 738–47.

Krause, P. C. (1986). *Analysis of electrical machinery*. McGraw-Hill, New York.

Lawrenson, P. J. and Agu, L. A. (1964). Theory and performance of polyphase reluctance machines. *Proceedings of the IEEE* **111**, 1435–45.

Lawrenson, P. J. and Gupta, S. K. (1967). Development in the performance and theory of segmental reluctance motors, *Proceedings of the IEE* **114**, 645–50.

Leonhard, W. (1985). *Control of electrical drives*. Springer-Verlag, Berlin.

Leonhard, W. (1986). Microcomputer control of high dynamic performance a.c. drives—A survey, *Automatica* **22**, 1–19.

Leonhard, W. (1988). Adjustable-speed a.c. drives. *Proceedings of the IEEE* **76**, 455–70.

Leonhard, W. (1988). Field-orientation for controlling a.c. machines—Principle and application. *3rd International Conference on Power Electronics and Variable Speed Drives*, pp. 277–82. IEE, London.

Lessmeier, R., Schumacher, W., and Leonhard, W. (1986). Microprocessor-controlled a.c. servo drives with synchronous or induction motors: which is preferable? *IEEE Transactions on Industry Applications* **IA-22**, 812–19.

Letas, H. H. and Leonhard, W. (1983). Dual-axis servo-drive in cylindrical coordinates using permanent magnet synchronous motors with microprocessor control. *Conference Proceedings of Comunel 83*, pp. II.23–II.30, Toulouse.

Miller, T. J. E. (1981). Methods for testing permanent magnet polyphase motors. *IEEE IAS Annual Meeting*, pp. 494–9.

Miller, T. J. E. (1981). Transient performance of permanent magnet a.c. machines. *IEEE IAS Annual Meeting*, pp. 500–3.

Murphy, J. M. D. and Turnbull, F. G. (1987). Power Electronic Control of A.C. Motors. Pergamon Press, Oxford.

Novotny, D. W. and Lorenz, R. D. (1985). Introduction to field orientation and high performance a.c. drives. *Tutorial Course, IEEE IAS Annual Meeting, Toronto.*

Nakano, T., Ohsawa, H., and Endoh, K. (1984). A high-performance cycloconverter-fed synchronous machine drive system, *IEEE Transactions on Industry Applications* **IA-20**, 1278–84.

Ogasawara, S., Nishimura, M., Akagi, H., Nabae, A., and Nakanishi, Y. (1986). A high performance a.c. servo system with permanent magnet synchronous motors. *IEEE Transactions on Industrial Electronics* **IE-33**, 87–91.

Pillay, P. and Krishnan, R. (1987). Modelling, analysis and simulation of a high performance, vector controlled, permanent magnet synchronous motor drive, *IEEE IAS Annual Meeting, Atlanta*, pp. 254–61.

Rahman, M. A. and Little, T. A. (1984). Dynamic performance analysis of permanent magnet synchronous motors. *IEEE Transactions on PAS*, **PAS-103**, 1277–82.

Richter, E. and Neumann, T. W. (1984). Saturation effects in salient-pole synchronous machines with permanent magnet excitation. In *Proceedings of International Conference on Electrical Machines, Lausanne*, pp. 603–6.

Sebastian, T. and Slemon, G. R. (1987). Transient modelling and performance of variable speed permanent magnet motors. *IEEE IAS Annual Meeting*, pp. 35–9.

Schiferl, R. and Lipo, T. A. (1986). Power capability of salient-pole permanent magnet synchronous motors in variable-speed drive applications. *IEEE IAS Annual Meeting*, pp. 23–31.

Sneyers, B., Novotny, D. W., and Lipo, T. A. (1985). Field-weakening in buried permanent magnet a.c. motor drives, *IEEE Transactions on Industry Applications* **IA-21**, 398–404.

Takeda, Y., Morimoto, S., Hirasa, T., and Fuchi, K. (1988). Most suitable control method for permanent magnet synchronous motors. In *Proceedings of the International Conference on Electrical Machines, Pisa*, pp. 53–8.

Tso, S. K. and Wai, L. K. (1987). A novel implementation of high-performance synchronous motor drives. *BICEM*, pp. 667–70.

Vas, P., Alaküla, M., Brown, J. E., and Hallenius, K. E. (1988). Field-oriented control of saturated a.c. machines. *3rd International Conference on Power Electronics and Variable-speed drives*, pp. 283–6. IEE, London.

Vauhkonen, V. and Mård, M. (1987). Compensated field-oriented control of cycloconverter-fed synchronous motors in rolling-mill drives. In *Proceedings of Beijing International Conference on Electrical Machines, Beijing*, pp. 679–82.

4 Vector control of induction machines

Induction machines have been used for over a hundred years. Because of its simplicity, ruggedness, reliability, efficiency, low cost, compactness, and economical and volume manufacturing advantages, the induction machine with a squirrel-cage rotor is the most widely used machine at fixed speed. This is especially the case at lower power levels, where on the rotor there is a squirrel-cage winding, which is manufactured by die-casting. However, only recent developments in the field of variable-speed drives have made possible the large scale application of variable-speed induction motor drives.

Although the induction machine is superior to the d.c. machine with respect to size, weight, rotor inertia, efficiency, maximum speed, reliability, cost, etc., because of its highly non-linear dynamic structure with strong dynamic interactions, it requires more complex control schemes than, say, a separately excited d.c. machine. The general dynamic model of the induction machine can be represented by a sixth-order state-space equation, where the inputs to the stator are voltage and frequency and the outputs can be rotor speed, rotor position, electromagnetic torque, stator or rotor flux linkages, magnetizing flux linkage, stator or rotor currents, magnetizing current or a combination of these. Furthermore, the cost of a.c. power converters is higher for the variable-speed induction motor drive than for the converters which can supply d.c. machines. Until recently the cost of the introduction of the variable-speed induction motor has been prohibitive and the complexity of control has made its development difficult and acceptance reluctant. However, the rapid developments in the field of power electronics, whereby better and more powerful semiconductor devices are available (with higher switching speeds, high conducting currents and very high blocking voltages, which can be turned on and off etc.) and where the power devices and circuits are packaged into modular form, and the existence of powerful and inexpensive microprocessors, which allow the complex control functions of the a.c. drive to be performed by utilizing software instead of expensive hardware, mean that a.c. drives employing induction machines can be considered as economical alternatives to adjustable-speed d.c. drives.

Some of the other functional advantages of the application of microprocessors or digital techniques are:

- Cost reduction in control electronics;
- Improved reliability, due to the reduction of the number of components;
- Standard universal hardware is required and the only changes are to the software, which is very flexible and can be easily modified;

- Digital transmission requires a minimal amount of cabling and is very tolerant to noise; it eliminates drift and electromagnetic interference problems;
- Very high accuracy, excellent repeatability, linearity, and stability with different setting ranges;
- Centralized operator communications, monitoring, and diagnostics. The diagnostics programs monitor the operation of the system. Some parts of the programs monitor the various semiconductor devices (this is continuously performed), current/voltage transducers, supply voltages, speed/rotor angle detectors, serial data communication, etc.;
- Complex, high-speed arithmetic and capability of decision making.

Some of the service and diagnostic features of the application of microprocessors or digital circuits are:

- Only a few standard modules are required without any special adjustments;
- There is the possibility of plug-in memory modules for user programs, parameters and modifications to functions, etc.;
- Powerful system software for on-line measurements, control parameter setting (e.g. current control parameter setting, speed control parameter setting, etc.) and testing;
- Automatic location of hardware faults with the help of system and user software.

It is expected that the present trends in technology will result in cost reduction and performance improvement of a.c. drives and will thus lead to an even more widespread application of these drives. The MOS bipolar power devices, which include the Insulated Gate Transistor (IGT), which is also known as the GEMFET or COMFET, and the MOS-Controlled Thyristor (MCT), can be turned on and off with a MOS gate and have such excellent characteristics that it is expected that in future they will revolutionize power converters and will profoundly affect the design of future controllers and their use.

The method of vector control described in earlier sections for synchronous machines can be extended to induction machines. However, whilst in the electrically excited synchronous machine and the permanent-magnet synchronous machine, the space angle between the field winding and the direct axis of the stator can be directly measured, for the induction machine the space angle of the rotor flux-linkage space phasor with respect to the direct axis of the stator is not a directly measurable quantity. Furthermore, in the case of the converter-fed induction machine, as well as supplying active power, the converter must supply reactive power for magnetization, since there is no external excitation as in the case of the synchronous machine. Thus both excitation (reactive) and torque producing (active) currents must simultaneously exist in the stator windings of the induction machine.

The methods discussed in this chapter can be applied to induction machines with quadrature-phase stator windings ('two-phase' machine) or to induction machines with three-phase stator windings. The induction machine can have either a wound rotor (slip-ring machine) or a rotor with squirrel-cage winding and it is possible to implement vector control on both types of induction machines. The squirrel cage can be of the double-cage or the single-cage type.

The mechanism of electromagnetic torque production in smooth-air-gap machines has been described in Section 2.1.6 and various forms of the expressions for the electromagnetic torque have been given in Section 2.1.8. It has been shown that in special reference frames fixed to the magnetizing flux-linkage, the stator flux-linkage or the rotor flux-linkage space phasor, the expression for the electromagnetic torque of the smooth-air-gap machine is similar to the expression for the torque of the separately excited d.c. machine. This suggests that torque control of the induction machine can be performed by the decoupled control of the flux- and torque-producing components of the stator currents, which is similar to controlling the field and armature currents in the separately excited d.c. machine. However, it should be noted that in the squirrel-cage induction machine it is not possible to monitor the rotor currents directly. The stator currents of the induction machine can be separated into the flux- and torque-producing components by utilizing the transformations described in Section 2.1.8, whilst in the d.c. machine, as a result of the decoupled orthogonal field and armature axes, it is straightforward to achieve independent control of the flux and torque. It follows from Section 2.1.8 that the implementation of vector control (stator flux-oriented control, rotor flux-oriented control or magnetizing flux-oriented control) requires information on the modulus and space angle (position) of the stator-flux, rotor-flux, or magnetizing-flux space phasors respectively. The control can be performed in a reference frame fixed to the stator-flux, rotor-flux, or magnetizing-flux space phasor respectively, and the direct- and quadrature-axis stator currents are obtained in the corresponding reference frame. These stator currents are similar to the field and armature currents of the separately excited d.c. machine.

In the case of induction machines rotor-flux-oriented control is usually employed, although it is possible to implement stator-flux- and also magnetizing-flux-oriented control, as described below. With rotor-flux-oriented control there are two main implementations to obtain the modulus and space angle of the rotor flux-linkage space phasor. When the so-called direct rotor-flux-oriented control is used (flux-feedback control), these quantities are directly measured (by using Hall-effect sensors, search coils, or tapped stator windings of the machine etc.) or they are calculated from a so-called flux model. However, in one specific form of the so-called indirect rotor-flux-oriented control (flux-feedforward control), the modulus and space angle of the rotor flux-linkage space phasor are obtained by utilizing the monitored stator currents and the rotor speed. The space angle of the rotor flux-linkage space phasor is then obtained as the sum of the monitored rotor angle (θ_r) and the computed reference value of the slip angle (θ_{sl}), where the latter quantity gives the position of the rotor flux-linkage space phasor relative to the direct axis of the rotor, as shown below. However, it will also be shown in this chapter that the slip-angle (θ_{sl}) can be calculated from the reference values of the torque- and flux-producing stator currents and strongly depends on the rotor parameters (rotor time constant, which is equal to the ratio of the rotor inductance to the rotor resistance) of the machine under consideration, and when inaccurate parameters are used it is not possible to achieve correct field orientation. The rotor time constant can change as a result of saturation, variation of the temperature and effects of current displacement. Thus it is preferable to implement schemes which incorporate some form of on-line parameter adaption. In practice, despite its parameter sensitivity and the fact that it requires the use of shaft encoders, which are expensive, the indirect method has gained more widespread application, since it does not require flux sensors or a flux model.

It should be noted that with stator-flux-oriented control or magnetizing-flux-oriented control there can also be two main different implementations, direct and indirect, and they will be discussed later in this section.

Induction machines represent an alternative to synchronous machines in high-performance servo-drive applications. Although, both the permanent-magnet synchronous machine and the induction machine are suitable for this type of application, for the selection of the appropriate machine, the following points have to be considered:

- Induction machines are more difficult to control;
- If field weakening is excluded, the drive containing the induction machine requires an inverter with higher rating (in the PM synchronous machine there are no rotor losses, and if there is field weakening, the machine is overexcited and draws large stator currents resulting in large stator losses);
- Because of the rotor losses, an induction machine normally requires forced cooling (a PM synchronous machine has no rotor losses and natural cooling can be sufficient);
- For the same torque produced, the efficiency (which has direct effects on the size of the machine) for the induction machine is lower (a PM synchronous machine has high efficiency because there are no rotor losses);
- An induction machine can be designed for higher flux densities (in a PM synchronous machine this is restricted by the magnets);

- An induction machine is cheaper (in a PM synchronous machine the price of the required magnets is still high);
- In case of the drive containing the induction machine, field-weakening (constant-power mode) is easily achieved over a wide speed range (in a PM synchronous machine this is not the case);
- The drive with the induction machine requires the application of a more complicated microcomputer. (The microcomputer can be simpler for the permanent-magnet synchronous machine, since it does not require the application of a signal processor; or the machine can even be designed without any microprocessor.)

Vector control can be applied to an induction machine supplied by a voltage source inverter or by a current source inverter or by a cycloconverter. It is simpler to implement the control when controlled-current operation is achieved. The vector-controlled induction machine can achieve four-quadrant operation with high dynamic response.

In the following section, first the rotor-flux-oriented control of induction machines is discussed in detail and, as they are very similar, this will be followed by shorter discussions of the stator-flux-oriented and magnetizing-flux-oriented control techniques.

4.1 ROTOR-FLUX-ORIENTED CONTROL OF INDUCTION MACHINES

In this part of the chapter the rotor-flux-oriented control of induction machines is discussed for the case when the machine is supplied by impressed stator voltages, impressed stator currents, or impressed rotor currents. Both direct and indirect methods are discussed.

4.1.1 Control of an induction machine supplied by a voltage source inverter

It is assumed that the induction machine is supplied by a pulse-width modulated (PWM) voltage source thyristor inverter, whose switching frequency is low, usually in the range of 100 Hz–1 kHz. Such converters are used up to ratings of several 100 kW and one of the main applications is the high-dynamic-performance position-controlled servo drive. At lower power levels, inverters containing transistors are used where the switching frequency is high. At higher power levels, converters employing thyristors and gate turn-off thyristors are used.

Because of the low switching frequency, it is not possible to achieve fast closed-loop current control of the stator currents with, for example, sinusoidal stator current references in the steady state. In this drive the stator currents can contain time harmonics with large amplitudes. Since the stator currents cannot be assumed to be impressed by fast control loops, it is necessary to utilize the stator voltage equations as well, once the equation of the stator reference voltages are established. However, it will be shown in Section 4.1.2 that a significant simplification arises in the equations and thus in the drive control system if the induction machine can be considered to be supplied by current sources. This is the case for an induction machine supplied by a voltage-source transistor inverter with fast current control (used at lower power levels at a high switching frequency, usually above 15 kHz), for a cycloconverter-fed induction machine (used for high-power, low-speed applications), where the stator currents are approximately sinusoidal and where the stator currents can be controlled by individual current control loops, or for an induction machine supplied by, say, a current-source thyristor inverter.

Stator voltage equations in the rotor-flux-oriented reference frame

In this section the stator voltage equations are derived and formulated in the reference frame fixed to the rotor-flux linkage space phasor. There are many ways to obtain the stator voltage equations in this reference frame. However, a straightforward method is followed here, whereby the space-phasor forms of the voltage equations formulated in the general reference frame (see Section 2.1.7) are used directly. The rotor-current space phasor is deliberately expressed in terms of the so-called rotor magnetizing-current space phasor and thus the resulting voltage equations will contain the modulus and space angle of the rotor magnetizing-current space phasor (or rotor flux-linkage space phasor), which are necessary to implement vector control.

It has been shown in Section 2.1.8, eqn (2.1-195), that the so-called rotor magnetizing-current space phasor in the rotor-flux-oriented reference frame ($\bar{\imath}_{mr}$) is obtained by dividing the rotor flux-linkage space phasor established in this reference frame ($\bar{\psi}_{r\psi r}$) by the magnetizing inductance (L_m). Thus under linear magnetic conditions (where the magnetizing inductance is constant), $\bar{\imath}_{mr}$ and $\bar{\psi}_{r\psi r}$ are proportional. For convenience, eqn (2.1-195) is repeated here, and this also gives the relationship between the rotor magnetizing space phasor to the stator and rotor current space phasors:

$$\bar{\imath}_{mr}=\frac{\bar{\psi}_{r\psi r}}{L_m}=\bar{\imath}_{s\psi r}+(1+\sigma_r)\bar{\imath}_{r\psi r} \tag{4.1-1}$$

where σ_r is the rotor leakage factor (it is the ratio of the rotor leakage inductance to the magnetizing inductance, as shown in Section 2.1.8).

By considering eqns (2.1-148) and (2.1-150), which give the stator space phasor voltage and stator flux-linkage space phasor equations in the general reference frame respectively, and by assuming linear magnetic conditions (i.e. L_m = constant and the leakage inductances are also constant), the following stator voltage equation is obtained in the reference frame fixed to the rotor flux-linkage space phasor, which rotates at the speed ω_{mr} (this speed is defined in eqn (2.1-188) as the first time derivative of the space angle ρ_r, which is the space angle of the rotor magnetizing-current space phasor with respect to the direct axis of the stationary reference frame):

$$\bar{u}_{s\psi r} = R_s \bar{i}_{s\psi r} + L_s \frac{d\bar{i}_{s\psi r}}{dt} + L_m \frac{d\bar{i}_{r\psi r}}{dt} + j\omega_{mr} L_s \bar{i}_{s\psi r} + j\omega_{mr} L_m \bar{i}_{r\psi r}. \quad (4.1\text{-}2)$$

It follows from eqn (2.1-189) that in the special rotor-flux-oriented (x, y) reference frame, the space phasor of the stator currents can be expressed in terms of the space phasor of the stator currents established in the stationary reference frame ($\bar{i}_s$) as

$$\bar{i}_{s\psi r} = i_{sx} + j i_{sy} = \bar{i}_s e^{-j\rho_r} = (i_{sD} + j i_{sQ}) e^{-j\rho_r} \quad (4.1\text{-}3)$$

and a similar expression can be obtained for the stator-voltage space phasor, since the stator quantities in the same reference frame must be transformed by the same transformation,

$$\bar{u}_{s\psi r} = \bar{u}_s e^{-j\rho_r} = u_{sx} + j u_{sy} = (u_{sD} + j u_{sQ}) e^{-j\rho_r}. \quad (4.1\text{-}4)$$

From eqn (4.1-1) and the fact that as a result of the special selection of the reference frame the rotor magnetizing current space phasor is coaxial with the direct-axis and thus $\bar{i}_{mr} = |\bar{i}_{mr}|$, the rotor-current space phasor in the special reference frame is obtained in terms of the stator-current and rotor magnetizing-current space phasors as

$$\bar{i}_{r\psi r} = \frac{|\bar{i}_{mr}| - i_{s\psi r}}{1 + \sigma_r}. \quad (4.1\text{-}5)$$

Substitution of eqn (4.1-5) into eqn (4.1-2) yields the following differential equation for the stator currents, if both sides of the equation are divided by the stator resistance R_s and if it is expressed in the form required by a time-delay element:

$$T_s' \frac{d\bar{i}_{s\psi r}}{dt} + \bar{i}_{s\psi r} = \frac{\bar{u}_{s\psi r}}{R_s} - j\omega_{mr} T_s' \bar{i}_{s\psi r} - (T_s - T_s')\left(j\omega_{mr}|\bar{i}_{mr}| + \frac{d|\bar{i}_{mr}|}{dt}\right). \quad (4.1\text{-}6)$$

T_s' is the stator transient time constant of the machine, $T_s' = L_s'/R_s$, where L_s' is the stator transient inductance, $L_s' = (L_s - L_m^2/L_r)$, and can be expressed in terms of the total leakage factor σ and the stator inductance as $L_s' = \sigma L_s$. Thus it is possible to define σ as $1 - L_m^2/(L_s L_r)$. T_s is the stator time constant, $T_s = L_s/R_s$, and thus it is also possible to express T_s' as σT_s. It follows that the term $(T_s - T_s')$ can be expressed in terms of the total leakage factor and the stator time constant as $(1-\sigma)T_s$.

By resolving eqn (4.1-6) into its real (x) and imaginary axis (y) components, the following two-axis differential equations are obtained for the stator currents:

$$T_s' \frac{\mathrm{d}i_{sx}}{\mathrm{d}t} + i_{sx} = \frac{u_{sx}}{R_s} + \omega_{mr} T_s' i_{sy} - (T_s - T_s') \frac{\mathrm{d}|\bar{\imath}_{mr}|}{\mathrm{d}t} \tag{4.1-7}$$

$$T_s' \frac{\mathrm{d}i_{sy}}{\mathrm{d}t} + i_{sy} = \frac{u_{sy}}{R_s} - \omega_{mr} T_s' i_{sx} - (T_s - T_s')\omega_{mr}|\bar{\imath}_{mr}|. \tag{4.1-8}$$

The relationship between the stationary-axis voltage (u_{sD}, u_{sQ}) and stator current components (i_{sD}, i_{sQ}) and the corresponding voltage (u_{sx}, u_{sy}) and current (i_{sx} i_{sy}) components can be obtained by utilizing the transformations defined by eqns (4.1-3) and (4.1-4) respectively. It follows from eqns (4.1-7) and (4.1-8) that with respect to the stator currents i_{sx} and i_{sy}, the induction machine behaves as a first-order time-delay element, whose time constant is equal to the stator transient time constant of the machine, and its gain is equal to the inverse of the stator resistance. However, it can be seen that there is an unwanted coupling between the stator circuits on the two axes. For the purposes of rotor-flux-oriented-control, it is the direct-axis stator current i_{sx} (rotor flux-producing component) and the quadrature-axis stator current i_{sy} (torque-producing component) which must be independently controlled. However, since the voltage equations are coupled, and the coupling term in u_{sx} also depends on i_{sy} and the coupling term in u_{sy} also depends on i_{sx}, u_{sx}, and u_{sy} cannot be considered as decoupled control variables for the rotor flux and electromagnetic torque. The stator currents i_{sx} and i_{sy} can only be independently controlled (decoupled control) if the stator voltage equations [eqns (4.1-7) and (4.1-8)] are decoupled and the stator current components i_{sx} and i_{sy} are indirectly controlled by controlling the terminal voltages of the induction machine. Various implementations of the necessary decoupling circuit will be described in the next section.

Decoupling circuits in the rotor-flux-oriented reference frame

In this section the decoupling circuits valid for the case of an ideal drive are derived first and then the derivation is given of the necessary decoupling circuits if there is an inverter time delay.

Decoupling circuits of an ideal drive: From eqns (4.1-7) and (4.1-8) there is in the direct-axis voltage equation a rotational voltage coupling term $\omega_{mr} L_s' i_{sy}$, and thus the quadrature-axis stator current i_{sy} affects the direct-axis stator voltage u_{sx}. Similarly, in the quadrature-axis voltage equation there is a rotational voltage coupling term $-\omega_{mr} L_s' i_{sx} - (L_s - L_s')\omega_{mr}|\bar{\imath}_{mr}|$ and thus the direct-axis stator current i_{sx} affects the quadrature-axis voltage u_{sy}. However, by assuming an ideal drive (which has no extra time delays apart from the natural ones, i.e. there is no inverter dead time, no delay due to signal processing etc., and for which the parameters are those used in the equations) and by assuming constant rotor-flux operation ($|\bar{\imath}_{mr}| = \text{const.}$), it follows from eqns (4.1-7) and (4.1-8) that the stator current components can be independently controlled if the decoupling rotational voltage components,

$$u_{dx} = -\omega_{mr} L_s' i_{sy} \tag{4.1-9}$$

$$u_{dy} = \omega_{mr} L_s' i_{sx} + (L_s - L_s')\omega_{mr}|\bar{i}_{mr}|, \tag{4.1-10}$$

are added to the outputs ($\hat{u}_{sx}$, $\hat{u}_{sy}$) of the current controllers which control i_{sx} and i_{sy} respectively. This can be proved by considering that $\hat{u}_{sx} + u_{dx}$ yields the direct-axis terminal voltage component, and the voltage on the output of the direct-axis current controller is

$$\hat{u}_{sx} = R_s i_{sx} + L_s' \frac{di_{sx}}{dt}. \tag{4.1-11}$$

Similarly, $\hat{u}_{sy} + u_{dy}$ gives the quadrature-axis stator voltage component and the voltage on the output of the quadrature-axis current controller is

$$\hat{u}_{sy} = R_s i_{sy} + L_s' \frac{di_{sy}}{dt}. \tag{4.1-12}$$

Thus $\hat{u}_{sx}$ and $\hat{u}_{sy}$ directly control the stator currents i_{sx} and i_{sy} through a simple time delay element with the stator transient time constant T_s'. The required decoupling circuit is shown in Fig. 4.1, where the decoupling voltages u_{dx} and u_{dy} are obtained from i_{sx}, i_{sy}, $|\bar{i}_{mr}|$, and ω_{mr} by using eqns (4.1-9) and (4.1-10) respectively.

In the decoupling circuit described above the inputs are the stator currents i_{sx} and i_{sy}, the rotor magnetizing current $|\bar{i}_{mr}|$ and the angular speed of the rotor flux-linkage space phasor, ω_{mr}. The current components i_{sx} and i_{sy} can be obtained from the measured three-phase stator currents by utilizing the transformations described by eqn (4.1-3) and $|\bar{i}_{mr}|$ and ω_{mr} can be obtained by utilizing a so-called flux-model, which can be obtained by considering the

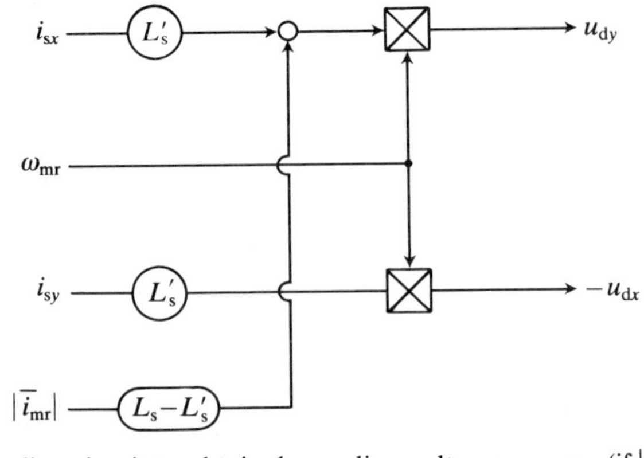

Fig. 4.1. Decoupling circuit to obtain decoupling voltages u_{dx}, u_{dy} (if $|\bar{i}_{mr}|$ = constant). Inputs are i_{sx}, i_{sy}, ω_{mr}, $|\bar{i}_{mr}|$.

rotor voltage equations, as shown in the following section. However, it is also possible to have an implementation of the decoupling circuit where the stator currents are not utilized as input quantities, but instead, the output voltages of the stator current controllers ($\hat{u}_{sx}$, $\hat{u}_{sy}$) are used. For this purpose eqns (4.1-11) and (4.1-12) are utilized and the stator currents i_{sx}, i_{sy} are expressed in terms of the output voltages of the current controllers. When these are substituted into eqns (4.1-9) and (4.1-10), the following expressions are obtained for the decoupling voltage components:

$$u_{dx} = -\omega_{mr}\frac{T_s'}{1+T_s'p}\hat{u}_{sy} \tag{4.1-13}$$

$$u_{dy} = \omega_{mr}\left[(L_s - L_s')|\bar{i}_{mr}| + \frac{T_s'}{1+T_s'p}\hat{u}_{sx}\right], \tag{4.1-14}$$

where $p = d/dt$ and it should be noted that these equations are valid if the rotor flux is constant. The decoupling circuit corresponding to eqns (4.1-13) and (4.1-14) is shown in Fig. 4.2, where the two blocks contain a lag element with time constant T_s' and gain T_s'.

It can be seen that the decoupling circuit shown in Fig. 4.2 also utilizes the stator resistance of the induction machine in addition to the machine parameters which have been used in the implementation shown in Fig. 4.1. The direct-axis reference value of the inverter voltage (in the rotor-flux-

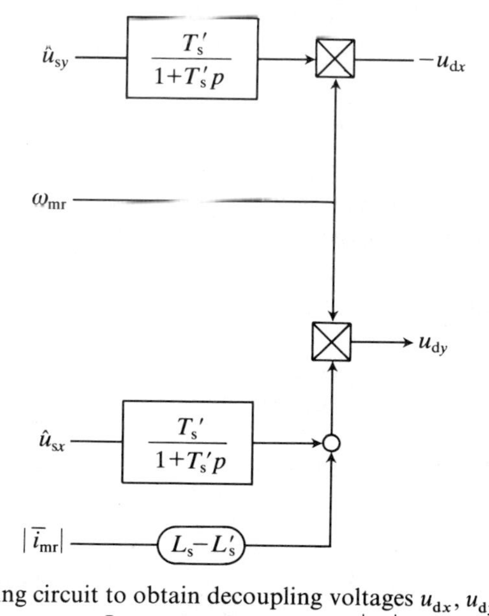

Fig. 4.2. Decoupling circuit to obtain decoupling voltages u_{dx}, u_{dy} (if $|\bar{i}_{mr}| = \text{constant}$). Inputs are $\hat{u}_{sx}$, $\hat{u}_{sy}$, ω_{mr}, $|\bar{i}_{mr}|$.

oriented reference frame) is again obtained as the sum of the direct-axis voltage obtained from the decoupling circuit and the direct-axis voltage which is present on the output of the direct-axis stator current controller. Similarly, the quadrature-axis reference value of the inverter voltage (in the rotor-flux-oriented reference frame) is obtained as the sum of the quadrature-axis voltage obtained from the decoupling circuit and the quadrature-axis voltage which is present on the output of the quadrature-axis stator current controller. To obtain the three-phase stator voltage references of the inverter, these have to be transformed into the two-axis stator voltage components by the application of the transformation $e^{j\rho_r}$ [see eqn (4.1-4)] and this transformation is followed by the two-phase to three-phase transformation.

Decoupling circuits considering the effect of the inverter time delay: It has been assumed above that the drive is ideal. However, the main difference between a real drive and an idealized drive is that in the real drive there are dead-time components due to the delay of the inverter, signal processing (if a microcomputer or microprocessor is employed) etc. These have to be taken account of, since they cause unwanted coupling terms and it is therefore not possible to perform the required decoupling of the stator circuits using the circuit described above. The unwanted coupling can cause instabilities etc. and make the vector control scheme totally inoperative. It is now shown how the decoupling circuits have to be modified in order to take account of these effects.

For simplicity it is assumed that the total dead time T is concentrated at the end of the signal process, between the control circuits and the machine, i.e. in the inverter. The lag effect of the inverter is described by a first-order delay element, thus in the stationary reference frame

$$\frac{d\bar{u}_s}{dt} + \bar{u}_s = \bar{u}_{sref} \qquad (4.1\text{-}15)$$

holds, where $\bar{u}_s$ is the space phasor of the stator voltages in the reference frame fixed to the stator and $\bar{u}_{sref}$ is its reference value. By using eqn (4.1-4), it is possible to transform eqn (4.1-15) into the rotor-flux-oriented reference frame. Thus the inverter can be described by the following first-order differential equation in the rotor-flux-oriented reference frame:

$$T\frac{d\bar{u}_{s\psi r}}{dt} + j\omega_{mr}T\bar{u}_{s\psi r} + \bar{u}_{s\psi r} = \bar{u}_{s\psi rref}, \qquad (4.1\text{-}16)$$

resolution of which into direct quadrature axes components yields

$$T\frac{du_{sx}}{dt} + u_{sx} = \omega_{mr}Tu_{sy} + u_{sxref} \qquad (4.1\text{-}17)$$

$$T\frac{du_{sy}}{dt} + u_{sy} = -\omega_{mr}Tu_{sx} + u_{syref}. \qquad (4.1\text{-}18)$$

Thus it can be seen that the two equations are coupled; in the direct-axis equation there is an unwanted coupling $\omega_{mr}Tu_{sy}$ and in the quadrature-axis $-\omega_{mr}Tu_{sx}$. When the delay time is zero, the coupling disappears as expected. If the machine is operated in the field-weakening range (above base speed), it follows from eqns (4.1-17) and (4.1-18) that the undesirable decoupling terms will be the largest during field weakening. In the constant-flux range (below base speed), the effect of the delay time can be neglected. Equations (4.1-17) and (4.1-18) can be implemented by the circuit shown in Fig. 4.3.

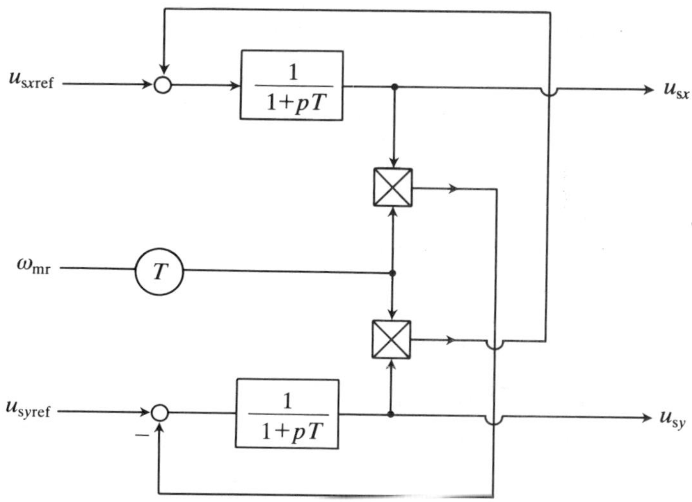

Fig. 4.3. Circuit for inverter in rotor-flux-oriented reference frame.

Thus it is possible to obtain a combined decoupling circuit of the machine and inverter system, which is a series connection of a decoupling circuit of the machine (as described in Fig. 4.1 or Fig. 4.2) and the decoupling circuit of the inverter as shown in Fig. 4.3. If, for example, the decoupling circuit of the machine used is the one shown in Fig. 4.1, the combined decoupling circuit takes the form shown in Fig. 4.4.

However, it is not possible to decouple the machine and the drive system completely when this method is used, since the machine and the inverter represent a total system, which cannot be decoupled by considering the machine and the inverter individually. However, by combining the equations of the machine with the equations of the inverter, it is possible to obtain a full decoupling circuit for the machine–inverter drive system as a whole. This will now be discussed.

For simplicity $|\bar{i}_{mr}|=$constant is assumed, but similar considerations hold when this assumption is not made. Thus substitution of the expression for u_{sx}, which can be obtained by considering eqn (4.1-7), into eqn (4.1-17) yields the

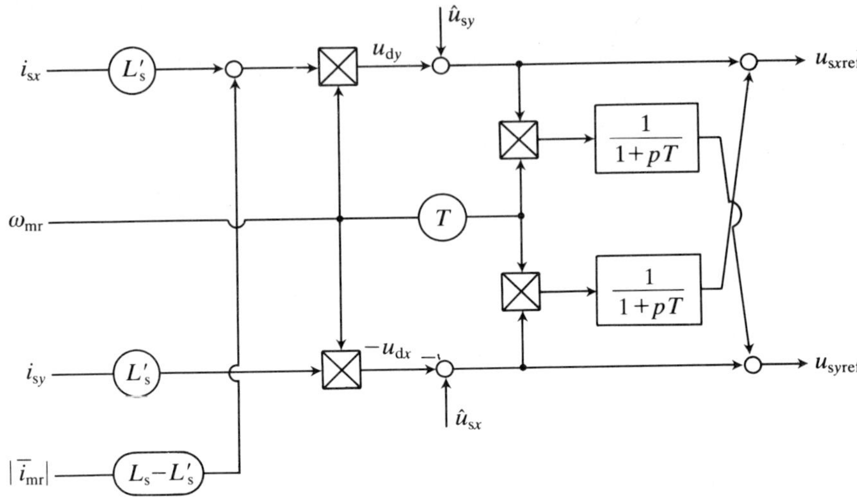

Fig. 4.4. Series connection of the decoupling circuits of the machine and inverter.

following expression for the direct-axis stator voltage reference:

$$u_{sxref} = [R_s + (L'_s + TR_s)p + TL'_s p^2]i_{sx} - \omega_{mr}(L'_s + TR_s + 2TL'_s p)i_{sy} - \omega^2_{mr}[TL'_s i_{sx} + T(L_s - L'_s)|\bar{i}_{mr}|] - TL'_s i_{sy}(p\omega_{mr})$$
$$= \hat{u}_{sx} + u_{dx}, \tag{4.1-19}$$

where $p = d/dt$. Similarly, substitution for u_{sy} [given by eqn (4.1-8) into eqn (4.1-18)] gives

$$u_{syref} = [R_s + (L'_s + TR_s)p + TL'_s p^2]i_{sy} + \omega_{mr}(L'_s + TR_s + 2TL'_s p)i_{sx} + \omega^2_{mr} TL'_s i_{sy} + [TL'_s i_{sx} + T(L_s - L'_s)|\bar{i}_{mr}|]\,(p\omega_{mr})$$
$$= \hat{u}_{sy} + u_{dy}, \tag{4.1-20}$$

for the quadrature-axis stator voltage reference. If $T = 0$, u_{dx}, and u_{dy} in eqns (4.1-19) and (4.1-20) take the forms of eqns (4.1-9) and (4.1-10) respectively, which is an expected result. However, when T is not zero, the required full decoupling circuit corresponding to the full inverter–machine system is shown in Fig. 4.5 and this has been obtained by considering eqns (4.1-19) and (4.1-20).

Although the circuit shown in Fig. 4.5 is much more complex then the one shown in Fig. 4.4, when this is used in the implementation of the vector-controlled drive, the drive will operate correctly for any value of the total dead time T.

In this and following sections various flux models are described, which yield

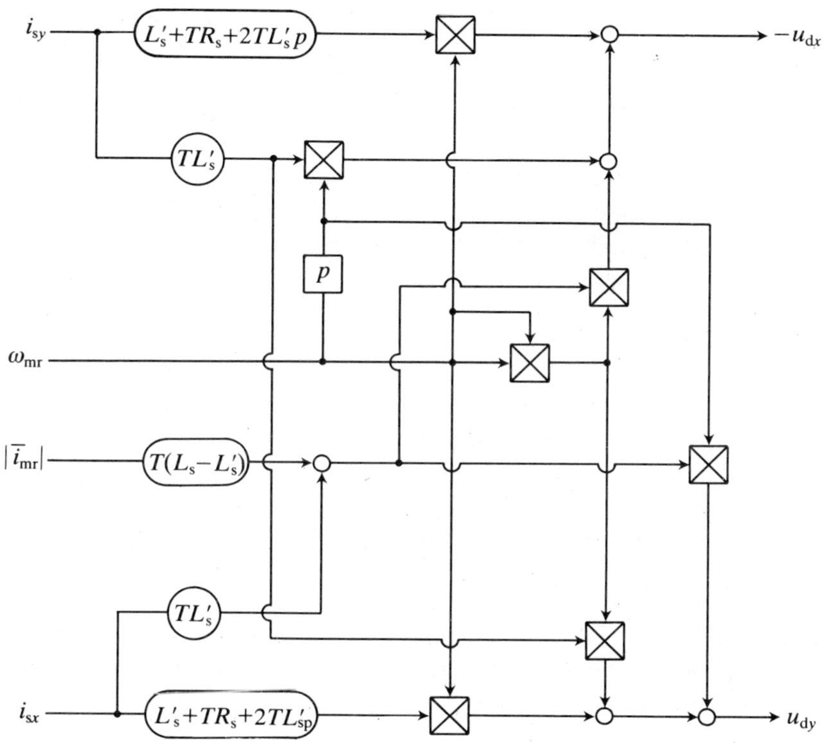

Fig. 4.5. Full decoupling circuit considering the inverter machine system.

the modulus and speed (or phase angle) of the rotor flux (or rotor magnetizing-current) space phasor.

Rotor voltage equations for the rotor flux model in the rotor-flux-oriented reference frame

The rotor voltage equations expressed in the rotor-flux-oriented reference frame can be used to obtain the modulus and phase angle of the rotor-flux space phasor or they can be used to obtain the modulus of the so-called rotor magnetizing current $|\bar{i}_{mr}|$ and its speed ω_{mr}. This flux model is derived in the present section.

By considering eqn (2.1-149), which gives the rotor voltage equation in the space-phasor form established in the general reference frame, the rotor voltage equation of the induction machine, in the special rotor-flux-oriented reference frame will take the following form:

$$0 = R_r \bar{i}_{r\psi r} + \frac{d\bar{\psi}_{r\psi r}}{dt} + j(\omega_{mr} - \omega_r)\bar{\psi}_{r\psi r}, \qquad (4.1\text{-}21)$$

where $\bar{\psi}_{r\psi r}$ is the rotor flux-linkage space phasor in the rotor-flux-oriented reference frame and has been defined in eqns (2.1-190) and (2.1-193). Alternatively, it follows from eqn (4.1-1) that because in the special reference frame $\bar{\imath}_{mr} = |\bar{\imath}_{mr}|$,

$$\psi_{r\psi r} = L_m |\bar{\imath}_{mr}|, \tag{4.1-22}$$

which gives a linear relationship if the magnetizing inductance is assumed to be constant. Substitution of eqn (4.1-22) into eqn (4.1-21) yields the following rotor voltage differential equation if the effects of main flux saturation are neglected (L_m = constant):

$$0 = R_r \bar{\imath}_{r\psi r} + L_m \frac{d|\bar{\imath}_{mr}|}{dt} + j(\omega_{mr} - \omega_r) L_m |\bar{\imath}_{mr}|. \tag{4.1-23}$$

By substituting eqn (4.1-5) into eqn (4.1-23) and dividing by R_r, which is the rotor resistance, finally the following equation is obtained, which is deliberately put into the form which is similar to the differential equation describing a time delay element,

$$T_r \frac{d|\bar{\imath}_{mr}|}{dt} + |\bar{\imath}_{mr}| = \bar{\imath}_{s\psi r} - j(\omega_{mr} - \omega_r) T_r |\bar{\imath}_{mr}|. \tag{4.1-24}$$

By resolving into real- and imaginary-axis components, the following extremely simple equations are obtained which describe the flux model in the rotor-flux-oriented reference frame:

$$T_r \frac{d|\bar{\imath}_{mr}|}{dt} + |\bar{\imath}_{mr}| = i_{sx} \tag{4.1-25}$$

$$\omega_{mr} = \omega_r + \frac{i_{sy}}{T_r |i_{mr}|}. \tag{4.1-26}$$

In eqn (4.1-26) the term $i_{sy}/(T_r|\bar{\imath}_{mr}|)$ represents the angular rotor frequency (angular slip frequency of the rotor flux) ω_{sl}, and it follows that the angular speed of the rotor flux is equal to the sum of the angular rotor speed and the angular slip frequency of the rotor flux. If $|\bar{\imath}_{mr}|$ is constant, it follows from eqn (4.1-25) that $|\bar{\imath}_{mr}| = i_{sx}$, which is in accordance with Section 2.1.8. The modulus of the rotor flux-linkage space phasor can be kept at a desired level by controlling the direct-axis stator current i_{sx}, as seen from eqn (4.1-25), but if there is no field weakening (below base speed) the electromagnetic torque is determined by the quadrature-axis stator current i_{sy}, in accordance with eqn (2.1-197).

Figure 4.6 shows the flux models of the induction machine in the rotor-oriented reference frame, based on eqns (4.1-25) and (4.1-26). The implementation shown in Fig. 4.6(a) utilizes the monitored values of the stator currents (i_{sA}, i_{sB}, i_{sC}), the monitored value of the rotor speed (ω_r) and the rotor time constant (T_r). The three stator currents are monitored and transformed into

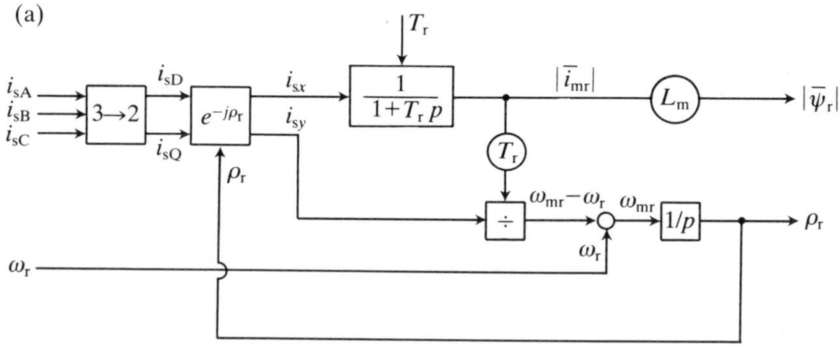

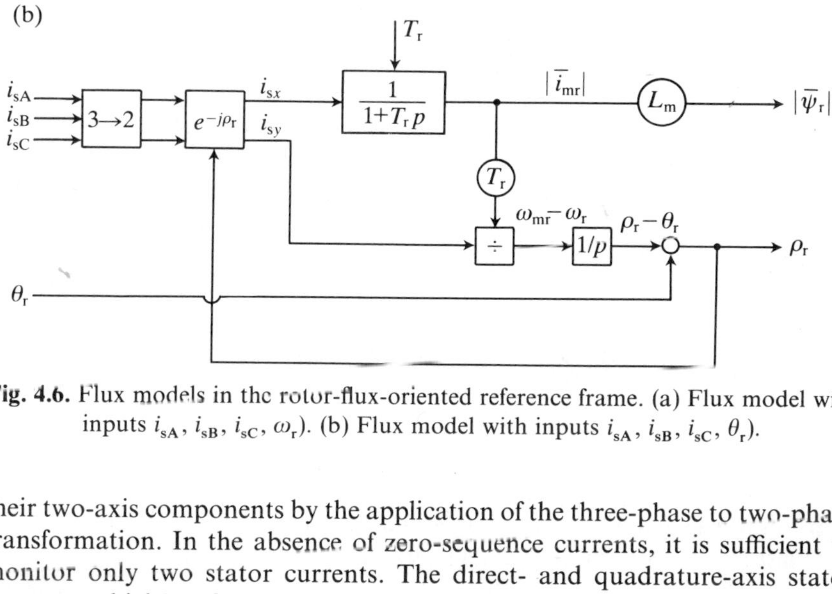

Fig. 4.6. Flux models in the rotor-flux-oriented reference frame. (a) Flux model with inputs i_{sA}, i_{sB}, i_{sC}, ω_r). (b) Flux model with inputs i_{sA}, i_{sB}, i_{sC}, θ_r).

their two-axis components by the application of the three-phase to two-phase transformation. In the absence of zero-sequence currents, it is sufficient to monitor only two stator currents. The direct- and quadrature-axis stator currents, which are formulated in the stationary reference frame fixed to the stator (i_{sD}, i_{sQ}), are then transformed into the two-axis stator current components in the rotor-flux-oriented reference frame (i_{sx}, i_{sy}), by utilizing the transformation given in eqn (4.1-3). The current component i_{sx} serves as an input to a first-order time-delay element with gain 1 and time constant T_r, the output of which is the modulus of the rotor magnetizing current. This is proportional to the modulus of the rotor magnetizing flux-linkage space phasor and the proportionality factor is equal to the magnetizing inductance of the machine (which has been assumed to be constant). The quadrature-axis stator current i_{sy} is divided by $T_r|\bar{i}_{mr}|$, thus yielding the angular slip frequency of the rotor flux, and when the rotor speed is added to this, finally ω_{mr} is obtained. Integration of ω_{mr} yields the angle ρ_r, which defines the position of the rotor-flux space phasor with respect to the real axis of the stationary reference frame. This angle is used in the transformation block $e^{-j\rho_r}$.

However, it is possible to have a similar flux model where, instead of using the monitored values of the rotor speed, the monitored values of the rotor angle (θ_r) are utilized. This is shown in Fig. 4.6(b). It can be seen that the outputs of this flux-model are again $|\bar{i}_{mr}|$, ρ_r and $|\bar{\psi}_r|$, but not the speed of the rotor flux-linkage space phasor ω_{mr}.

It is an important feature of eqns (4.1-25) and (4.1-26) that there is a strong dependency on the rotor time constant. If there is an inaccurate value of T_r in the flux models described above, it could lead to an unwanted coupling between the x and y axes, and therefore to a deteriorated dynamic performance of the drive with unwanted instabilities. This problem can be avoided by, for example, the application of on-line parameter-adaption schemes which yield accurate values of the rotor time constant. However, without parameter adaption more acceptable performance can be obtained at lower power levels than at higher power levels. The technique of Model Reference Adaptive Control (MRAC) will be used below to obtain an identification algorithm for the rotor resistance of a squirrel-cage induction machine. In Section 6.2 other models are described to obtain the rotor parameters of induction machines. It should be noted that in contrast to the large sensitivity to the rotor parameters, the implemented drive is less sensitive to variations in the stator parameters (which are, however, needed in the decoupling circuit).

Rotor voltage equations for flux models in the stationary reference frame

There are several ways to obtain the modulus and phase angle of the rotor-flux space phasor by utilizing certain machine parameters and various monitored quantities, such as the actual stator currents, the rotor speed and the stator voltages. Three different implementations will now be discussed. Since it is the intention to use actual stator voltages and/or currents, the equations will be formulated in the stationary reference frame fixed to the stator.

Flux model utilizing the monitored rotor speed and stator currents: It is possible to obtain the modulus and phase angle of the rotor-flux space phasor from a circuit, which in addition to the monitored rotor speed or rotor angle, also utilizes the monitored stator currents. However, in contrast to the implementation shown in Fig. 4.6, the stator currents do not have to be transformed into their values in the rotor-flux-oriented reference frame. For this purpose again the rotor voltage space-phasor equation is used, but in order to have an equation which directly contains the stator currents expressed in the reference frame fixed to the stator, the rotor voltage equation formulated in the stationary reference frame must be used. It should be noted that the effects of magnetic saturation are again neglected.

It follows from eqn (2.1-125) that in the stationary reference frame the rotor voltage equation takes the form,

$$0 = R_r \bar{i}'_r + \frac{d\bar{\psi}'_r}{dt} - j\omega_r \bar{\psi}'_r, \tag{4.1-27}$$

where $\bar{i}'_r$ and $\bar{\psi}'_r$ are the rotor-current and rotor-flux space phasors respectively, but expressed in the stationary reference frame. Similarly to the definition used in eqn (4.1-1), the rotor magnetizing-current space phasor expressed in the stationary reference frame is obtained by dividing the rotor flux-linkage space phasor expressed in the stationary reference frame by the magnetizing inductance,

$$\bar{i}_{mr} = \frac{\bar{\psi}'_r}{L_m} = \bar{i}_s + (1+\sigma_r)\bar{i}'_r, \tag{4.1-28}$$

where $\bar{i}_{mr}$ and $\bar{i}_s$ are the rotor magnetizing current and stator phasor currents respectively in the stationary reference frame. By considering eqn (4.1-28), the rotor current space phasor can be expressed in terms of the rotor magnetizing space phasor, and when this expression is substituted into eqn (4.1-27), the following equation is obtained,

$$T_r \frac{d\bar{i}_{mr}}{dt} = \bar{i}_s - \bar{i}_{mr} + j\omega_r T_r \bar{i}_{mr}. \tag{4.1-29}$$

In the stationary reference frame, the rotor magnetizing-current space phasor and the stator current space phasor can be expressed in terms of their direct- and quadrature-axis components as $\bar{i}_{mr} = i_{mrD} + ji_{mrQ}$ and $\bar{i}_s = i_{sD} + ji_{sQ}$, and therefore resolution of eqn (4.1-29) into real and imaginary axis components gives the following two differential equations,

$$T_r \frac{di_{mrD}}{dt} = i_{sD} - i_{mrD} - \omega_r T_r i_{mrQ} \tag{4.1-30}$$

$$T_r \frac{di_{mrQ}}{dt} = i_{sQ} - i_{mrQ} + \omega_r T_r i_{mrD}. \tag{4.1-31}$$

An implementation of eqns (4.1-30) and (4.1-31) is shown in Fig. 4.7, where first the three-phase stator currents are transformed into their two-axis components by the application of the three-phase to two-phase transformation. According to eqns (4.1-30) and (4.1-31), first the signals $i_{sD} - i_{mrD} - \omega_r T_r i_{mrQ}$ and $i_{sQ} - i_{mrQ} + \omega_r T_r i_{mrD}$ are obtained. These are then divided by the rotor time constant (T_r) and are integrated to yield the direct- and quadrature-axis rotor magnetizing current components (i_{mrD}, i_{mrQ}). A rectangular-to-polar converter is used to obtain the modulus ($|\bar{i}_{mr}|$) and the phase angle (ρ_r) of the rotor magnetizing flux-linkage space phasor. If required, $|\bar{i}_{mr}|$ can be multiplied by the magnetizing inductance (L_m) to yield the modulus of

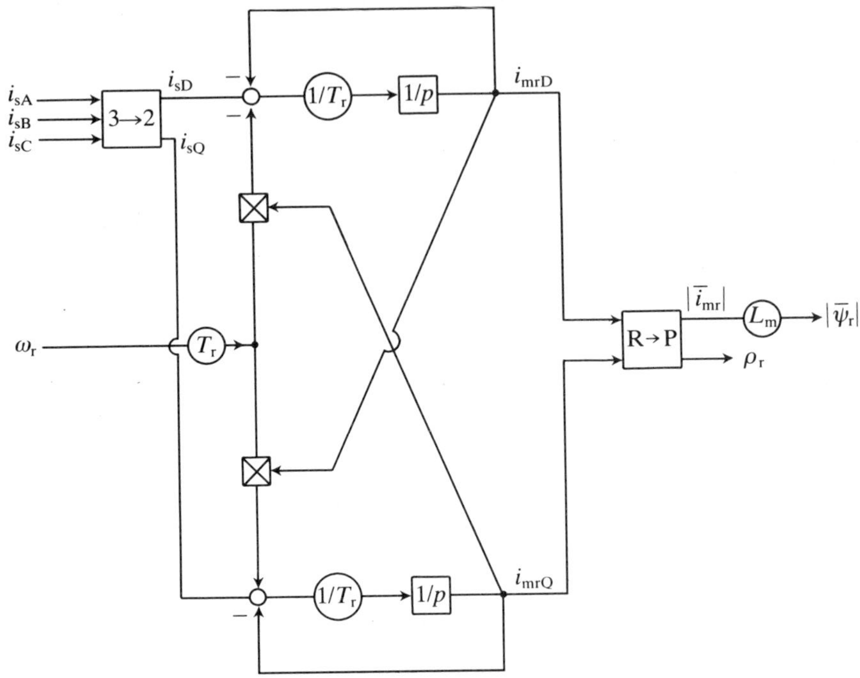

Fig. 4.7. Flux model in the stationary reference frame (the inputs are i_{sA}, i_{sB}, i_{sC}, ω_r).

the rotor flux-linkage space phasor. This scheme is also dependent on the rotor time constant of the machine and can be used over the entire speed range, including standstill. When this model is used, the angle ρ_r at the output of the model has to be differentiated to obtain ω_{mr} which is required in the decoupling circuits described in Figs 4.1, 4.2, 4.3, and 4.4. When compared with the flux models shown in Fig. 4.6, the flux model described in Fig. 4.7 yields less accurate values of the modulus and position of the rotor flux-linkage space phasor.

From eqn (4.1-29), it can be shown that in the steady state, where $p=\mathrm{d}/\mathrm{d}t=j\omega_1$, the rotor magnetizing current can be expressed in terms of the stator current phasor as

$$\bar{I}_{mr}=\frac{\bar{I}_s}{1+j(\omega_1-\omega_r)T_r}, \tag{4.1-32}$$

from which it follows that at high speeds, where the difference $(\omega_1-\omega_r)=s\omega_1$ (where s is the slip) is small, a small error in the monitored value of the rotor speed will result in a large error in the rotor magnetizing current (in its modulus and space angle), and this is especially pronounced in its phase angle. The other source of errors is the rotor time constant, which is also temperature dependent, although it follows from eqn (4.1-32) that only at no load ($s=0$) is the rotor magnetizing current (or the rotor flux) not influenced by the rotor time constant. This is of course a physically expected result, since at no load there are no currents in the rotor. However, it will be shown in the next section that it is possible to utilize a linear combination of the stator and rotor voltage equations, and these equations are not so sensitive to the rotor speed. Thus an implementation based on these equations requires less accurate rotor speed monitoring. Furthermore, it will also be shown that the system to be considered is affected by changes in the temperature only in the low speed region.

Flux model utilizing monitored rotor speed and stator voltages and currents: From eqns (2.1-148) and (2.1-149), which describe the stator voltage and flux linkage equations respectively in the space-phasor form, it follows that in the stationary reference frame the stator voltage equations can be put into the following form, if again the effects of magnetic saturation are neglected,

$$\bar{u}_s = R_s \bar{\imath}_s + L_s \frac{d\bar{\imath}_s}{dt} + L_m \frac{d\bar{\imath}_r'}{dt}, \qquad (4.1\text{-}33)$$

where R_s and L_s are the stator resistance and self-inductance L_m is the magnetizing inductance, and $\bar{\imath}_s$ and $\bar{\imath}_r'$ are the space phasors of the stator and rotor currents respectively in the stationary reference frame.

If in eqn (4.1-33) the rotor-current space phasor is expressed in terms of the rotor magnetizing current ($\bar{\imath}_{mr}$) defined by eqn (4.1-28), the following space-phasor voltage equation is obtained for the stator

$$\bar{u}_s = R_s \bar{\imath}_s + \sigma L_s \frac{d\bar{\imath}_s}{dt} + (1-\sigma) L_s \frac{d\bar{\imath}_{mr}}{dt}, \qquad (4.1\text{-}34)$$

where σ is the resultant leakage constant $\sigma = (1 - L_m^2/L_s L_r)$. Thus it follows from eqn (4.1-34) that

$$(1-\sigma) T_s \frac{d\bar{\imath}_{mr}}{dt} = \frac{\bar{u}_s}{R_s} - \bar{\imath}_s - T_s' \frac{d\bar{\imath}_s}{dt}, \qquad (4.1\text{-}35)$$

where $T_s = L_s/R_s$ and $T_s' = L_s'/R_s$ are the stator time constant and stator transient time constant respectively. When this equation is added to eqn (4.1-29), the following differential equation is obtained:

$$\frac{d\bar{\imath}_{mr}}{dt}[T_r + T_s(1-\sigma)] = \frac{\bar{u}_s}{R_s} + (j\omega_r T_r - 1)\bar{\imath}_{mr} - T_s' \frac{d\bar{\imath}_s}{dt}. \qquad (4.1\text{-}36)$$

It follows that if certain machine parameters and the stator voltages and currents (in the stationary reference frame) are known, it is possible to use eqn (4.1-36) directly to obtain the rotor magnetizing-current space phasor. However, for this purpose, eqn (4.1-36) must be resolved into its real- and imaginary-axis components. Since in the stationary reference frame $\bar{i}_{mr}=i_{mrD}+ji_{mrQ}$, $\bar{i}_s=i_{sD}+ji_{sQ}$, and $\bar{u}_s=u_{sD}+ju_{sQ}$, it follows from eqn (4.1-35) that its real- and imaginary-axis forms are

$$\frac{di_{mrD}}{dt}[T_r+T_s(1-\sigma)]+i_{mrD}=\frac{u_{sD}}{R_s}-\omega_r T_r i_{mrQ}-T_s'\frac{di_{sD}}{dt} \qquad (4.1\text{-}37)$$

$$\frac{di_{mrQ}}{dt}[T_r+T_s(1-\sigma)]+i_{mrQ}=\frac{u_{sQ}}{R_s}+\omega_r T_r i_{mrD}-T_s'\frac{di_{sQ}}{dt}. \qquad (4.1\text{-}38)$$

An implementation of eqns (4.1-37) and (4.1-38) is shown in Fig. 4.8, where the input quantities are the monitored value of the rotor speed ω_r, and the monitored values of the three-phase stator voltages and currents. These are then transformed into the two-axis components of the stator currents (i_{sD}, i_{sQ}) and stator voltages (u_{sD}, u_{sQ}) by the application of the three-phase to two-phase transformation. There are two integrators, labelled in the block '$1/p$' ($p=d/dt$). The two-axis components of the rotor magnetizing currents i_{mrD}, i_{mrQ} are then converted into the modulus $|\bar{i}_{mr}|$ and phase angle ρ_r of the

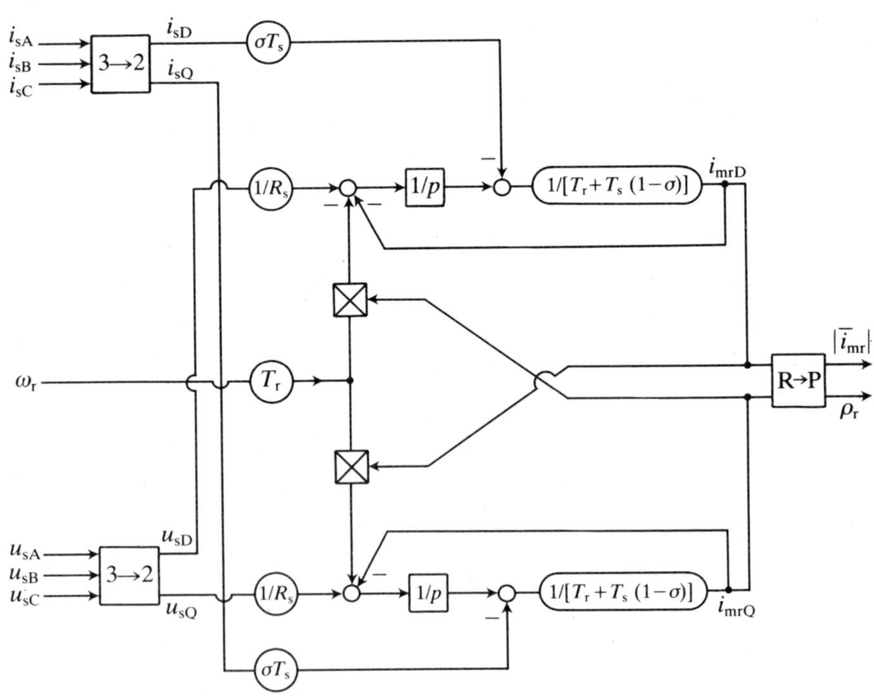

Fig. 4.8. Flux model in the stationary reference frame (the inputs are u_{sA}, u_{sB}, u_{sC}, i_{sA}, i_{sB}, i_{sC}, ω_r).

rotor magnetizing current space phasor with a rectangular-to-polar converter. It follows that the machine parameters to be used in this circuit are the stator resistance, the total leakage constant, and the stator and rotor time constants. When off-line parameter identification is used, these can be obtained by the application of conventional tests. Other techniques are described in Section 6.2.

It can be shown by considering eqn (4.1-36), that in the steady state the following expression is obtained for the rotor magnetizing current phasor (by utilizing $d/dt = j\omega_1$):

$$\bar{I}_{mr} = \frac{\bar{U}_s/R_s - j\omega_1 T_s' \bar{I}_s}{1 + j[(\omega_1 - \omega_r)T_r + \omega_1 T_s(1-\sigma)]}. \tag{4.1-39}$$

from which it follows that at high speeds, where $\omega_1 - \omega_r$ is small, a small error in the measured value of the rotor speed will not influence $\bar{I}_{mr}$ as much as in eqn (4.1-32), since at high values of the rotor speed in the imaginary part of the denominator of eqn (4.1-39), the term $\omega_1 T_s(1-\sigma)$ will dominate and this is independent of the speed.

To see the influence of the changes in temperature, eqn (4.1-39) is now rearranged into a form where the temperature-dependent parameters are only present in the denominator. For this purpose both the numerator and the denominator of eqn (4.1-39) are multiplied by $R_s/L_s = 1/T_s$ and, with $T_s' = \sigma T_s$,

$$\bar{I}_{mr} = \frac{\bar{U}_s/L_s - j\omega_1 \sigma \bar{I}_s}{1/T_s + j[(\omega_1 - \omega_r)T_r/T_s + \omega_1(1-\sigma)]}. \tag{4.1-40}$$

When $\omega_1 = \omega_r$, the denominator becomes $1/T_s + j\omega_1(1-\sigma)$, and the only parameter influenced by the change in the stator temperature is T_s, since it depends on R_s, but the effect is negligible. However, at low speeds or at standstill, large errors can arise in $\bar{I}_{mr}$, due to the change of R_s caused by the variation of the temperature. Higher precision can be obtained if instead of using the 'cold' value of the stator resistance a 'hot' value is used, if the temperature is sensed and a simple compensation circuit is implemented, or if a thermal model is used to correct the stator resistance.

It should be noted that instead of eqn (4.1-36), it is possible to use another equation, obtained in a similar way but by using different linear combinations of eqns (4.1-29) and (4.1-35). By increasing the effect of eqn (4.1-29), some improvement can be achieved in the sensitivity of $\bar{I}_{mr}$ in the low-speed region, but this would be obtained at the expense of greater sensitivity to the errors (in the monitored rotor speed and due to temperature variation) at higher speeds.

Flux model utilizing monitored stator voltages and currents: It is possible to establish a flux model which does not use the monitored rotor speed, but only the monitored values of the stator voltages and stator currents for the determination of the modulus and the space angle of the rotor magnetizing-current space phasor. For this purpose eqn (4.1-34) can be used. In the stationary reference frame $\bar{u}_s = u_{sD} + ju_{sQ}$ and $\bar{i}_s = i_{sD} + ji_{sQ}$, and therefore resolution of eqn (4.1-34) into its real- and imaginary-axis components yields

$$(1-\sigma)T_s \frac{\mathrm{d}i_{mrD}}{\mathrm{d}t} = \frac{u_{sD}}{R_s} - i_{sD} - T_s' \frac{\mathrm{d}i_{sD}}{\mathrm{d}t} \tag{4.1-41}$$

$$(1-\sigma)T_s \frac{\mathrm{d}i_{mrQ}}{\mathrm{d}t} = \frac{u_{sQ}}{R_s} - i_{sQ} - T_s' \frac{\mathrm{d}i_{sQ}}{\mathrm{d}t}. \tag{4.1-42}$$

In Fig. 4.9 there is shown an implementation of eqns (4.1-41) and (4.1-42), where the input quantities are the monitored values of the three-phase stator voltages and currents respectively (u_{sA}, u_{sB}, u_{sC}, i_{sA}, i_{sB}, i_{sC}) and the required parameters are the stator resistance, the total leakage constant and the stator inductance. Similarly to Fig. 4.8, two integrators are used and the modulus ($|\bar{i}_{mr}|$) and the phase angle (ρ_r) of the rotor magnetizing-current space phasor are obtained from its two-axis components by the application of a rectangular-to-polar converter.

It follows from eqns (4.1-41) and (4.1-42) that when these equations are multiplied by the stator resistance, the direct- and quadrature-axis magnetiz-

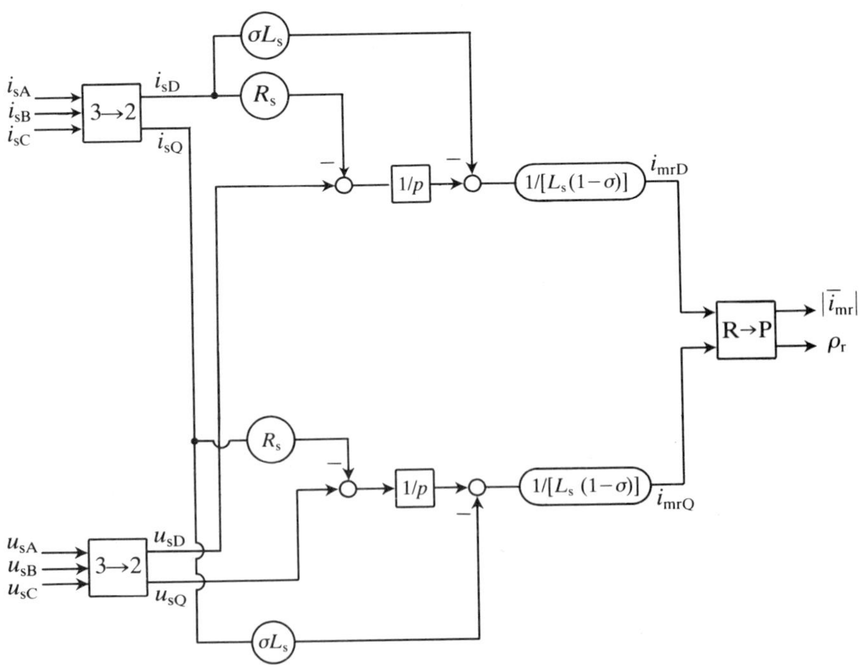

Fig. 4.9. Flux model in the stationary reference frame (the inputs are u_{sA}, u_{sB}, u_{sC}, i_{sA}, i_{sB}, i_{sC}).

ing-current components in the stator reference frame are essentially obtained by the integration of the direct- and the quadrature-axis magnetizing voltages respectively. These voltages are obtained by monitoring the two-axis components of the terminal voltages and by reducing them by the corresponding stator ohmic drops and by the corresponding voltage drops across the transient inductance of the stator (L'_s). When the magnetizing inductance is assumed to be very large (infinite), a voltage drop across the transient stator inductance is equal to the voltage drop across a fictitious resultant leakage inductance, which is obtained by connecting the leakage inductances of the stator and rotor in series. It should be noted that such an assumption leads to only a few percent error in the determination of the stator transient inductance. However, at low stator frequencies, the stator ohmic drops will dominate and accurate ohmic voltage drop compensation must be performed prior to the integration. However, due to the temperature dependency of the stator resistance, this is difficult to perform and with such an implementation a lower frequency limit for useful operation is approximately 3 Hz with a 50 Hz supply. It should also be noted that at low frequencies it is not possible to perform drift-free analogue integration.

Expression for the developed electromagnetic torque; utilization of the steady-state equivalent circuit

By considering the expression for the electromagnetic torque shown in eqn (2.1-197), which is repeated here for convenience,

$$t_e = \tfrac{3}{2}P\frac{L_m^2}{L_r}|\bar{i}_{mr}|i_{sy}, \qquad (4.1\text{-}43)$$

it follows that if the parameters of the machine are considered to be constant and $|\bar{i}_{mr}|$ is constant, then the electromagnetic torque is proportional to the quadrature-axis stator current expressed in the rotor-flux-oriented reference frame and thus the torque will respond instantaneously with i_{sy}. If there is a change in i_{sx}, it follows by considering eqns (4.1-25) and (4.1-26) that there will be a delayed response in the torque, and the delay is determined by the rotor time constant.

In the steady state, eqn (4.1-43) can also be derived by considering the steady-state equivalent circuit of the induction machine. This will now be discussed. The expression for the electromagnetic torque will be obtained by utilizing the relationship between the air-gap power and the torque and the expression for the air-gap power will be obtained directly from the steady-state equivalent circuit.

In the steady state, if the induction machine is supplied by a sinusoidal symmetrical three-phase supply voltage system, the stator voltages are $u_{sA}(t)=\sqrt{2}U_s\cos(\omega_1 t)$, $u_{sB}=\sqrt{2}U_s\cos(\omega_1 t-2\pi/3)$, and $u_{sC}=\sqrt{2}U_s\cos(\omega_1 t-4\pi/3)$, where U_s is the r.m.s. value of the line-to-neutral voltages. Thus it follows from the definition of the stator-voltage space phasor in the stationary reference frame given by eqn (2.1-61), that $\bar{u}_s = \sqrt{2}U_s e^{j\omega_1 t}$. The space phasor of the stator currents can be similarly defined. Thus by considering that in the steady state, the instantaneous values of the stator currents are $i_{sA}(t) = \sqrt{2}I_s \cos(\omega_1 t - \phi_s)$, $i_{sB} = \sqrt{2}I_s \cos(\omega_1 t - \phi_s - 2\pi/3)$, and $i_{sC} = \sqrt{2}I_s \cos(\omega_1 t - \phi_s - 4\pi/3)$, it follows, by the application of eqn (2.1-4), that the space phasor of the stator currents in the stationary reference frame is $\bar{i}_s = \sqrt{2}I_s e^{j(\omega_1 t - \phi_s)} = \sqrt{2}\bar{I}_s e^{j(\omega_1 t)}$. Thus by considering eqns (4.1-27), (4.1-28), and (4.1-34) and the steady-state expressions for $\bar{u}_s$ and $\bar{i}_s$, the resulting equations can be represented by the equivalent circuit shown in Fig. 4.10(a).

It is a main feature of the equivalent circuit shown in Fig. 4.10(a), that in contrast to the well-known steady-state equivalent circuit of the induction machine, now the rotor leakage inductance is not present in the rotor branch.

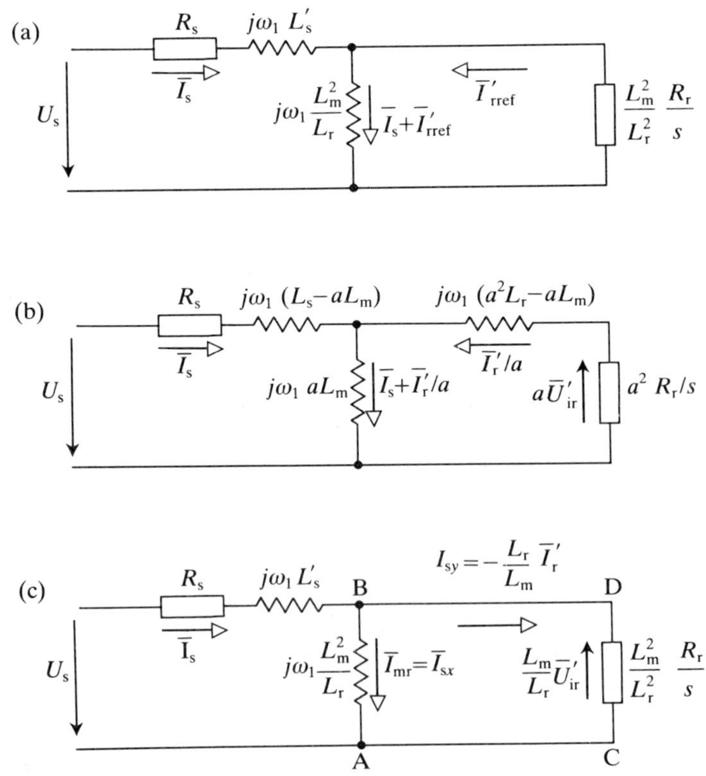

Fig 4.10. Steady-state equivalent circuits of the induction machine. (a) Equivalent circuit where the rotor leakage inductance is not present in the rotor branch. (b) Equivalent circuit incorporating the effects of the general turns ratio. (c) Equivalent circuit which contains the torque- and flux-producing stator current components.

It should be noted that the rotor resistance (R_r) is divided by the slip (s) and R_r/s is multiplied by a^2, where a is a specially selected turns ratio $a = L_m/L_r$. In Fig. 4.10(a) the referred value of the rotor current phasor in the steady-state is present; this is $\bar{I}'_{rref} = \bar{I}'_r/a$ and $\bar{I}_{mr} = \bar{I}_s + \bar{I}'_{rref}$ is the rotor magnetizing-current phasor in the steady state in accordance with that stated in Section 2.1.3 and also in agreement with eqn (4.1-28). Furthermore, in the equivalent circuit shown, the referred value of the magnetizing inductance is present, which is equal to $aL_m = L_m^2/L_r$ and owing to the special referring factor, instead of the stator leakage inductance, the stator transient inductance ($L'_s = \sigma L_s$) is present.

The equivalent circuit shown in Fig. 4.10(a) could have also been obtained from the space-phasor stator and rotor voltage equations in the stationary reference frame, but by considering a general value of the turns ratio (a). For this purpose, first eqns (2.1-35) and (2.1-51) are considered, which define the stator and rotor flux-linkage space phasors respectively ($\bar{\psi}_s$, $\bar{\psi}'_r$) in the stationary reference frame. By considering the steady state and adding the term $aL_m\bar{I}_s - aL_m\bar{I}_s$ to the expression for $\bar{\Psi}_s$,

$$\bar{\Psi}_s = (L_s - aL_m)\bar{I}_s + aL_m(\bar{I}_s + \bar{I}'_r/a)$$

is obtained. The referred value of the rotor flux-linkage space phasor is obtained by multiplying $\bar{\Psi}'_r$ by the general turns ratio (a), and by adding the term $aL_m\bar{I}'_r/a - aL_m\bar{I}'_r/a$. Thus

$$a\bar{\Psi}'_r = (a^2L_r - aL_m)\bar{I}'_r/a + aL_m(\bar{I}_s + \bar{I}'_r/a)$$

is obtained. When these expressions are combined with the stator and rotor voltage equations, the equivalent circuit shown in Fig. 4.10(b) is obtained. It follows from Fig. 4.10(b) that if the general turns ratio is selected as $a = L_m/L_r$, in the rotor branch the inductive part will vanish and the equivalent circuit shown in Fig. 4.10(a) is obtained.

It is possible to redraw the equivalent circuit shown in Fig. 4.10(a) as the equivalent circuit shown in Fig. 4.10(c), which contains the torque and rotor flux-producing stator current components ($\bar{I}_{sy}$, $\bar{I}_{sx}$). It follows from Fig. 4.10(c) that the stator current $\bar{I}_s$ is divided into $\bar{I}_{mr} = \bar{I}_{sx}$ and $\bar{I}_{sy}$ components. The current $\bar{I}_{mr} = \bar{I}_{sx}$ flows through the referred magnetizing reactance $\omega_1 L_m^2/L_r$ and the current $\bar{I}_{sy}$ flows through the referred rotor resistance $(L_m^2/L_r^2)R_r/s$. Furthermore, $(L_m/L_r)\bar{U}'_{ir}$ is the referred value of the voltage drop across the referred rotor resistance and thus $\bar{U}'_{ir} = -j\omega_1\bar{\Psi}'_r$. In Fig. 4.10(c) the current $\bar{I}_{sx}$ is the rotor flux-producing stator current component and this can be proved by considering that the voltage across points A and B ($\bar{U}_{AB}$) is equal to the voltage across points C and D ($\bar{U}_{CD}$) and thus $\bar{I}_{sx} j\omega_1 L_m^2/L_r = j\omega_1(L_m/L_r)\bar{\Psi}'_r$ and it follows that $\bar{\Psi}'_r = L_m\bar{I}_{sx}$, so $\bar{I}_{sx}$ is indeed the rotor flux-producing stator current.

That $\bar{I}_{sy}$ is the torque-producing stator current can be proved by considering that in the steady state, the electromagnetic torque can be obtained as $T_e = P_{ag}/\omega_1$, where P_{ag} is the air-gap power, i.e. the power that crosses the air-gap. By considering the equivalent circuit shown in Fig. 4.10(c), P_{ag} can be expressed as $P_{ag}=(3P/2)|\bar{U}_{AB}||\bar{I}_{sy}|$, where $|\bar{U}_{AB}|=\omega_1(L_m^2/L_r)|\bar{I}_{sx}|$. Thus the electromagnetic torque can be expressed as

$$T_e=(3P/2)(L_m^2/L_r)|\bar{I}_{sx}||\bar{I}_{sy}|=(3P/2)(L_m/L_r)|\bar{\Psi}_r'||\bar{I}_{sy}|.$$

As expected, this is similar to eqn (4.1-43).

The equivalent circuit shown in Fig. 4.10(c) can also be used to obtain the expression for the angular slip frequency in the steady-state. If the modulus of the voltage across points A and B $|\bar{I}_{sx}|\omega_1 L_m^2/L_r$ is equal to the modulus of the voltage across points C and D, which in terms of the referred rotor resistance can be expressed as $|\bar{I}_{sy}|(L_m^2/L_r^2)R_r/s$, it follows that $s\omega_1=|\bar{I}_{sy}|/(T_r|\bar{I}_{sx}|)$, where $s\omega_1=\omega_{sl}$ is the angular slip frequency, and T_r is the rotor time constant ($T_r=L_r/R_r$). The same result also follows from eqn (4.1-26) if the steady state is considered, and $\omega_{mr}-\omega_r=s\omega_1$.

Since in the steady state the torque-producing stator current varies linearly with the angular slip frequency, and thus there is no pull-out slip or pull-out torque (see also Section 4.3.4), static instability does not arise in the induction machine subjected to rotor-flux-oriented control. However, in contrast to this, when, say, stator-flux-oriented control of the induction machine is performed, as shown in Section 4.2, the torque-producing stator current does not vary linearly with the angular slip frequency and theoretically static instability can arise. The details of this will be discussed in Section 4.3.4, but it should be noted that from the general equivalent circuit shown in Fig. 4.10(b), if the general turns ratio is chosen to be $a=L_s/L_m$, the inductive term in the stator branch is eliminated. In this case the current $\bar{I}_s+(L_m/L_s)\bar{I}_r'$ flows across the referred value of the magnetizing inductance and this current is equal to (L_m/L_s) times the so-called stator magnetizing current, which in the steady state takes the form, $\bar{I}_{ms}=(L_s/L_m)\bar{I}_s+\bar{I}_r'$. A similar definition of this current will be used in Section 4.2. By choosing $a=L_s/L_m$, it is possible to use the equivalent circuit shown in Fig. 4.10(b) to obtain the expression for the electromagnetic torque in the steady state in terms of the stator flux-producing stator current component and the torque-producing stator current component or to obtain the expression for the angular slip frequency in terms of the same two current components.

Implementation of the PWM voltage-source inverter-fed induction machine drive utilizing a speed sensor and the actual values of i_{sx}, i_{sy}

This implementation of the PWM voltage-source inverter-fed induction machine drive uses the concepts discussed in previous sections. Figure 4.11 shows the schematic of the rotor-flux-oriented control of a voltage-source inverter-fed induction machine.

In this and the following sections three different implementations are described for the PWM voltage-source inverter-fed induction machine drive which have found widespread applications. The first one utilizes one of the

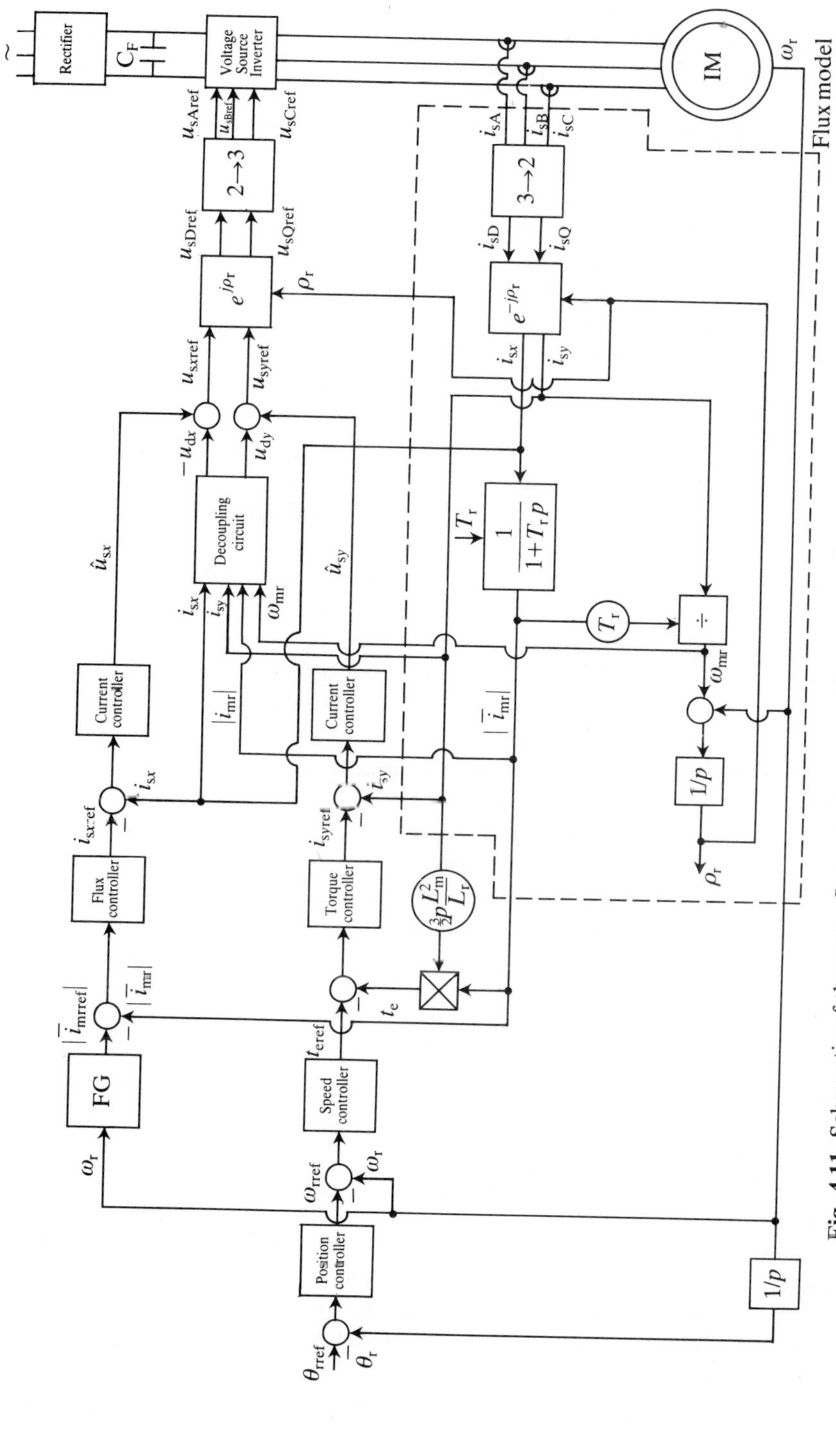

Fig. 4.11. Schematic of the rotor-flux-oriented control of a voltage-source inverter-fed induction machine.

decoupling circuits described above, which uses the actual values of the direct- and quadrature-axis stator currents in the rotor-flux-oriented reference frame (i_{sx}, i_{sy}) and contains a closed-loop control of the rotor position, rotor speed, electromagnetic torque and rotor flux. In the second, simpler implementation, decoupling is achieved by using the reference values of the same stator current components (i_{sxref}, i_{syref}); there is closed-loop control of the rotor speed, which is obtained by the application of a sensor. In the third implementation, there is closed-loop control of the flux- and torque-producing currents and of the rotor speed; a decoupling circuit is utilized which is similar to the one used in the first implementation, and uses the actual values of i_{sx} and i_{sy}. However, the rotor speed is not obtained by a sensor, but is estimated by utilizing Model Reference Adaptive Control.

In Fig. 4.11 the I.M. is supplied by a voltage-source PWM inverter, which is supplied by a diode bridge rectifier through a filter capacitor. The output voltages of the inverter are controlled by a pulse-width modulation technique. There are many types of modulation techniques which will not be discussed in detail here, but it should be noted that the so-called suboscillation technique has gained widespread application. When this technique is used, the output voltage of the inverter is generated by comparing a triangular carrier wave of frequency (switching frequency) f_s (in order to avoid beat effects, usually $f_s = Nf_1$, where f_1 is the fundamental frequency and N is the pulse number, an integer) with a sinusoidal modulating wave with fundamental frequency; the natural points of intersection determine the switching instants. The pulse-width modulated output voltages contain time harmonics and these lead to unwanted losses in the machine. There are a number of so-called harmonic elimination techniques, but when such a technique is used, although certain time harmonics (e.g. the fifth and the seventh) are almost eliminated, other harmonics will have increased magnitudes (e.g. the ninth and the eleventh). However, since the harmonic losses in the machine are determined by the r.m.s. value of the harmonic currents, it is these currents which have to be minimized in order to reduce these losses. This concept has lead to the development of the so-called minimum harmonic current technique. It is a common feature of these techniques that they are based on off-line computation of optimal pulse patterns for steady-state operation of the drive under consideration. Thus the optimum conditions can only be obtained in the steady state and not in the transient state. Furthermore, at low stator frequencies, off-line methods can be ineffective. To overcome this problem, on-line computation of the pulse patterns is required and by using on-line optimization algorithms it is possible to ensure, for example, minimum harmonic torque at minimum switching frequency or minimum torque pulsations. When the inverter switching frequency is reduced, the efficiency of the drive is increased as a result of the decrease in the switching losses, which can be especially important in thyristor inverters with forced commutation.

The flux model shown in Fig. 4.6 is used to obtain the angle ρ_r, which is used in the transformation blocks $e^{j\rho_r}$ and $e^{-j\rho_r}$. Furthermore, this flux model is used to obtain the angular speed of the rotor flux ω_{mr} and the modulus of the rotor magnetizing current $|\bar{i}_{mr}|$, since these are also used in the decoupling circuit, which is now assumed to be the one shown in Fig. 4.1. The modulus of the rotor magnetizing-current space phasor is also used to obtain the electromagnetic torque, in accordance with eqn (4.1-43).

A position controller, which can be a proportional controller (since the rotor speed ω_r is equal to the first time derivative of the rotor angle θ_r) provides as its output the reference value of the rotor speed (ω_{rref}). The speed controller, a PI controller, provides the reference torque (t_{eref}), and the torque controller, also a PI controller, gives the reference value of the quadrature-axis stator current in the rotor-flux-oriented reference frame (i_{syref}). Since in the steady state, the fundamental components of the stator voltages increase with stator frequency, but only a specific maximal output voltage of the inverter is available, it follows that with constant rotor magnetizing current ($|\bar{i}_{mr}|$), at a specific frequency, these voltages would exceed the maximally available value. Thus above a certain frequency, an increase of speed is only possible if the rotor magnetizing current is reduced. In Fig. 4.11 field weakening is achieved by the application of the function generator FG, the output of which is the reference value of the modulus of the rotor magnetizing current space phasor ($|\bar{i}_{mrref}|$), which is speed dependent. Below base speed a constant (maximal) value is obtained (this value is only limited by main flux saturation), and above base speed this is reduced, inversely proportional to the rotor speed. The reference signal $|\bar{i}_{mrref}|$ is compared with the actual value of the rotor magnetizing current and the error serves as input to the flux controller, also a PI controller. Its output is the direct-axis stator current reference expressed in the rotor-flux-oriented reference frame (i_{sxref}). In Fig. 4.11 it is possible to leave out the quadrature-axis stator current controller, since the torque controller performs a similar role, and in this case there is no need to feed back the current i_{sy}; there is only the torque feedback, which is, however, proportional to the quadrature-axis stator current if the magnetizing inductance is constant.

The error signals $i_{sxref}-i_{sx}$ and $i_{syref}-i_{sy}$ serve as inputs to the respective current controllers and the outputs from these are added to the corresponding outputs of the decoupling circuit as described above for the ideal drive. Thus the direct- and quadrature-axis reference stator voltages u_{sxref} and u_{syref} are obtained; these are established in the rotor-flux-oriented reference frame, and therefore they have to be transformed by $e^{j\rho_r}$ to obtain the two-axis stator voltages references in the stationary reference frame (u_{sDref}, u_{sQref}). This is followed by the application of the two-phase to three-phase transformation indicated by the '2→3' block and finally the reference values of the three-phase stator voltages are obtained. These signals are used to control the pulse-width modulator which transforms these reference signals into appropriate on–off switching signals to command the inverter phases.

With the given scheme it is possible to produce full torque even at standstill, since the stator currents flow even at standstill and supply the magnetizing currents. The implementation shown in Fig. 4.11 gives a high dynamic performance drive. In addition to the high dynamic response resulting from the decoupling control, there is no pull-out effect and if there is too quick a change of the speed reference or if the induction machine is overloaded, the electromagnetic torque cannot exceed the specified maximal level, since the speed error signal will saturate.

It is possible to obtain a fully digital implementation of the total control system, where all the control tasks and pulse-width modulation (which can be an on-line technique) are performed by a single 16-bit microprocessor, the currents are sensed in an analog manner (e.g. by the application of Hall sensors) and the rotor position or rotor speed is obtained by using a single optical encoder. Since the pulse-width modulation and all the control tasks are performed by the same microprocessor, this allows the application of optimization criteria to the modulation, since all the machine quantities (e.g. voltages and currents) are known. The possibilities for using various optimization schemes have been discussed above.

It should be noted that there are other ways to obtain the reference value of the rotor magnetizing current. For example, it is also possible to have an implementation, where the function generator in Fig. 4.11 is not present but the two-axis voltages u_{sxref}, u_{syref} are used to obtain the absolute value of the reference stator-voltage space phasor, $|\bar{u}_{\text{sref}}| = (u_{\text{sxref}}^2 + u_{\text{syref}}^2)^{1/2}$. Field weakening can be automatically performed by the application of an extra control loop which contains a limiting stator voltage controller (a PI controller) the input of which is the error $u - |\bar{u}_{\text{sref}}|$, where u is a constant voltage reference (the ceiling voltage of the inverter), and the output is $|\bar{\imath}_{\text{mref}}|$. The voltage controller attempts to remove the error $(u - |\bar{u}_{\text{sref}}|)$ by adjusting the rotor magnetizing current $|\bar{\imath}_{\text{mr}}|$. When the motor operates below base speed, $|u_{\text{sref}}| < u$ and the voltage controller will be saturated and the maximal $|\bar{\imath}_{\text{mr}}|$ is produced, so the rotor flux is kept at the saturation limit of the induction machine. However, when the speed increases above base speed, the inverter will approach its maximal output voltage and when the maximal output voltage of the inverter is reached ($|\bar{u}_{\text{sref}}|$ is equal to u), the voltage controller will give a smaller value of $|\bar{\imath}_{\text{mr}}|$ in order to limit the modulus of the stator voltages. Thus it follows that field weakening is automatically performed with this implementation. In contrast to permanent-magnet synchronous machines, where there is only a limited field weakening range (because the large air-gap causes a small synchronous reactance), in the induction machine, field weakening can be achieved over a wide speed range with constant power. This feature is extremely useful in spindle-drive applications but it can be utilized in position-controlled feed drives as well.

Implementation of the PWM voltage-source inverter-fed induction machine utilizing a speed sensor and the reference values of i_{sxref}, i_{syref}

Decoupling circuit: In this section an implementation of the rotor-flux-oriented control is presented for a voltage-source inverter-fed induction machine where the necessary decoupling circuit uses the reference values of the direct- and quadrature-axis components of the stator currents i_{sxref}, i_{syref}. Operation with constant rotor flux is assumed and the effects of magnetic saturation are neglected.

When the rotor flux ($\bar{\psi}_r$) is constant and since $\bar{\psi}_r = L_m \bar{i}_{mr}$ and thus under linear magnetic conditions the rotor-magnetizing current $|\bar{i}_{mr}|$ is also constant, it follows from eqn (4.1-25) that $|\bar{i}_{mr}| = i_{sx}$, where i_{sx} is the direct-axis component of the stator current in the rotor-flux-oriented reference frame. Under these conditions it follows from eqns (4.1-7) and (4.1-8) that if the term $L'_s \, di_{sy}/dt$ is neglected, the stator voltage components in the rotor-flux-oriented reference frame are:

$$u_{sx} = R_s i_{sx} - \omega_{mr} L'_s i_{sy} \tag{4.1-44}$$

$$u_{sy} = R_s i_{sy} + \omega_{mr} L_s i_{sx}. \tag{4.1-45}$$

It should be noted that in eqn (4.1-44), the stator transient inductance L'_s is present and in eqn (4.1-45) the stator inductance L_s is present in the direct- and quadrature-axis rotational voltage terms respectively. Furthermore, it follows from eqns (4.1-44) and (4.1-45) that in the direct-axis voltage equation, the rotational voltage is affected by the quadrature-axis stator current (i_{sy}) and in the quadrature-axis voltage equation, the rotational term is influenced by the direct-axis stator current (i_{sx}). Thus the rotor flux (or i_{sx}) is not solely determined by the direct-axis stator voltage expressed in the rotor-flux-oriented reference frame (u_{sx}), but is also influenced by the quadrature-axis stator current and similarly, the torque-producing stator current component is not solely determined by the quadrature-axis stator voltage expressed in the rotor-flux-oriented reference frame u_{sy}, but is also dependent on the direct-axis stator current. However, it is possible to have an independent control of the flux-producing stator current component (i_{sx}) and the torque-producing stator current component (i_{sy}), when the unwanted coupling terms are cancelled and thus the rotor flux is controlled by u_{sx} and the electromagnetic torque is controlled by u_{sy}, which are independent of each other. It follows from eqns (4.1-44) and (4.1-45) that in the decoupling circuit the actual values of i_{sx} and i_{sy} have to be used. However, by assuming that the lag time between i_{sy} and u_{sy} is small and thus can be neglected, it is possible to utilize the reference currents in the decoupling circuit and in this case $\hat{u}_{sx} = R_s i_{sxref}$ and $\hat{u}_{sy} = R_s i_{syref}$. This concept is utilized in the implementation of the rotor-flux-oriented control of the voltage-source inverter-fed induction machine drive described in the following section.

Drive implementation: The schematic of the rotor-flux-oriented control of a voltage-source inverter-fed induction machine drive utilizing the concepts described above is shown in Fig. 4.12. This is simpler than the drive described in Fig. 4.11, but it can yield a satisfactory dynamic response.

In Fig. 4.12, the reference value of the rotor flux is $|\bar{\psi}_{rref}|$ and when it is

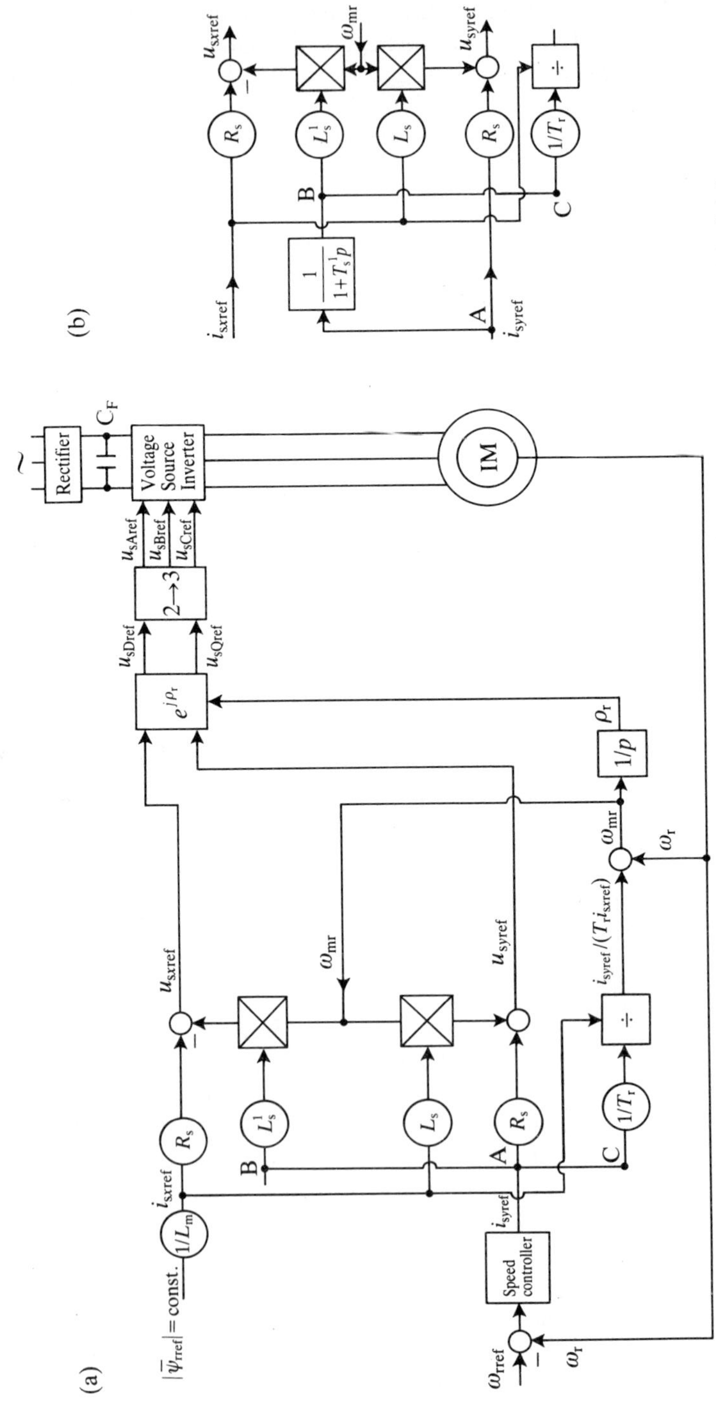

Fig. 4.12. Schematic of the rotor-flux-oriented control of a voltage-source induction machine drive. (a), (b) See text.

divided by the magnetizing inductance, the rotor magnetizing current is obtained, which is equal to the direct-axis stator current reference i_{sxref}. The reference value of the rotor speed (ω_{rref}) is compared with its actual value (ω_r) and the error serves as input to the speed controller (a PI controller). The output of the speed controller is the torque reference, which is, however, proportional to the quadrature-axis stator current reference (i_{syref}).

The direct- and quadrature-axis stator current references are used in the decoupling circuit, which utilizes the principles described above. Thus i_{sxref} is multiplied by the stator resistance (R_s) and the value obtained is reduced by $-\omega_{mr}L_s' i_{syref}$. Thus the direct-axis stator reference voltage component expressed in the rotor-flux-oriented reference frame (u_{sxref}) is obtained. Similarly, the quadrature-axis stator current reference is multiplied by the stator resistance and the rotational voltage component $\omega_{mr}L_s i_{sxref}$ is added to this voltage and thus the quadrature-axis stator voltage reference expressed in the rotor-flux-oriented reference frame (u_{syref}) is obtained. The voltage references u_{sxref} and u_{syref} are transformed into the two-axis voltage components of the stationary reference frame (u_{sDref}, u_{sQref}), by utilizing the transformation $e^{j\rho_r}$, where ρ_r is the space angle of the rotor flux-linkage space phasor with respect to the real axis of the stationary reference frame. For completeness the transformed values are also given in their expanded forms. Thus by considering eqn (4.1-4), but using reference values of the voltages,

$$u_{sDref}+ju_{sQref}=(u_{sxref}+ju_{syref})e^{j\rho_r} \tag{4.1-46}$$

or, from the resolution of eqn (4.1-46) into its real and imaginary components,

$$u_{sDref}=\cos\rho_r u_{sxref}-\sin\rho_r u_{syref} \tag{4.1-47}$$

$$u_{sQref}=\sin\rho_r u_{sxref}+\cos\rho_r u_{syref}. \tag{4.1-48}$$

These two-axis voltage references are then transformed into their three-phase reference values by the application of the two-phase to three-phase transformation indicated by the '2→3' block. It is possible to obtain the three-phase reference values by simply considering that in the absence of zero-sequence voltages, the projections of the voltage space phasor on the corresponding axes yield the instantaneous values of the phase voltages [see eqns (2.1-70), (2.1-71), and (2.1-72)],

$$u_{sAref}=\mathrm{Re}(\bar{u}_{sref})=\mathrm{Re}(u_{sDref}+ju_{sQref})=u_{sDref}, \tag{4.1-49}$$

where $\bar{u}_{sref}$ is the reference value of the stator-voltage space phasor in the stationary reference frame,

$$u_{sBref}=\mathrm{Re}(a^2\bar{u}_{sref})=\frac{-u_{sDref}}{2}+\sqrt{3}\,\frac{u_{sQref}}{2} \tag{4.1-50}$$

and

$$u_{sCref} = \mathrm{Re}(a^2\bar{u}_{sref}) = \frac{-u_{sDref}}{2} - \sqrt{3}\,\frac{u_{sQref}}{2}. \tag{4.1-51}$$

Of course it is also possible to use the Euler forms of the space-phasor equations. Thus the space phasor of the stator voltage references in the rotor-flux-oriented reference frame can be expressed as

$$\bar{u}'_{sref} = u_{sxref} + ju_{syref} = (u^2_{sxref} + u^2_{syref})^{1/2} e^{j\phi_u} \tag{4.1-52}$$

where $\phi_u = \tan^{-1}(u_{syref}/u_{sxref})$.

From eqn (4.1-46), in the stationary reference frame the space phasor of the stator reference voltages is

$$\bar{u}_{sref} = \bar{u}'_{sref} e^{j\rho_r} = (u^2_{sxref} + u^2_{syref})^{1/2} e^{j(\rho_r + \phi_u)}, \tag{4.1-53}$$

where $\bar{u}'_{sref}$ has been substituted by the expression given in eqn (4.1-52). Thus by considering eqns (4.1-49), (4.1-50), and (4.1-51), the three-phase stator voltage reference values can be put into the following form

$$u_{sAref} = \mathrm{Re}(\bar{u}_{sref}) = (u^2_{sxref} + u^2_{syref})^{1/2} e^{j(\rho_r + \phi_u)} \tag{4.1-54}$$

$$u_{sBref} = \mathrm{Re}(a^2\bar{u}_{sref}) = (u^2_{sxref} + u^2_{syref})^{1/2} e^{j(\rho_r + \phi_u - 2\pi/3)} \tag{4.1-55}$$

$$u_{sCref} = \mathrm{Re}(a\bar{u}_{sref}) = (u^2_{sxref} + u^2_{syref})^{1/2} e^{j(\rho_r + \phi_u + 2\pi/3)} \tag{4.1-56}$$

which could be utilized directly to obtain the three-phase reference voltages.

It is also possible to have an implementation where the decoupling and the transformation of the stator reference voltage components from the rotor-flux-oriented reference frame into the stationary reference frame is performed in one step. For this purpose eqns (4.1-44) and (4.1-45) are substituted into eqns (4.1-47) and (4.1-48) respectively, and the reference values of the stator current components (i_{sxref}, i_{syref}) are used in the decoupling circuit instead of their actual values (i_{sx}, i_{sy}). The following two new voltage equations are then obtained:

$$u_{sDref} = (R_s + L_s p)\cos\rho_r\, i_{sxref} - (R_s + L'_s p)\sin\rho_r i_{syref} \tag{4.1-57}$$

$$u_{sQref} = (R_s + L_s p)\sin\rho_r\, i_{sxref} - (R_s + L'_s p)\cos\rho_r i_{syref}, \tag{4.1-58}$$

an implementation of which is shown in Fig. 4.13. The inputs to the circuit shown in Fig. 4.13 are the reference values of the flux- and torque-producing currents (i_{sxref}, i_{syref}) and the angle ρ_r. The angle ρ_r is obtained from i_{sxref}, i_{syref} and the monitored rotor speed, as described below. The outputs of the circuit shown in Fig. 4.13 are the direct- and quadrature-axis stator voltage references in the stationary reference frame (u_{sDref} and u_{sQref}).

The angular frequency of the stator voltages is ω_{mr} and is obtained from i_{syref} by utilizing eqns (4.1-25) and (4.1-26). It follows from eqn (4.1-25) that if $|\bar{i}_{mr}|$ is constant, $|\bar{i}_{mr}| = i_{sx}$ and $\omega_{mr} = \omega_r + i_{sy}/(T_r i_{sx})$. Thus in accordance with the

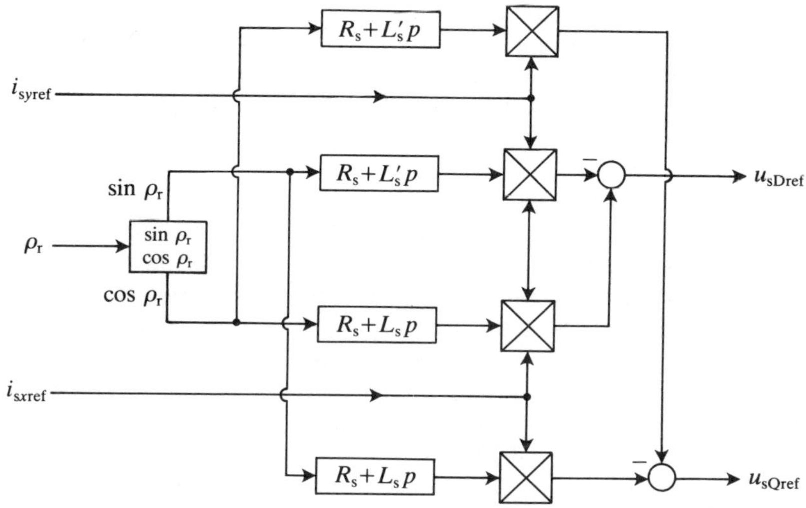

Fig. 4.13. Circuit to obtain reference voltages.

discussion presented above, by using the reference values of the stator currents instead of the actual values, finally

$$\omega_{mr} = \omega_r + \frac{i_{syref}}{T_r i_{sxref}} \tag{4.1-59}$$

is obtained. In Fig. 4.12 eqn (4.1-59) is utilized to obtain ω_{mr}, where ω_r is the monitored value of the rotor speed. The integration of ω_{mr} gives the angle ρ_r which is required in the transformations described above. The integration is carried out by the block labelled '$1/p$'. It follows that accurate values for the rotor flux position are only obtained if the rotor time constant used is accurate. Under linear magnetic conditions, this changes, mainly due to the variation of the rotor resistance. However, it is possible to implement relatively simple compensation of the rotor resistance variation because according to experiments performed for the drive described in Fig. 4.12, the actual value of i_{sy} is approximately proportional to the inverse of the actual (correct) value of the rotor resistance.

The implementation shown in Fig. 4.12 uses eqns (4.1-44) and (4.1-45) and eqn (4.1-45) has been obtained by neglecting the term $L_s' \mathrm{d}i_{sy}/\mathrm{d}t$, where L_s' is the transient stator inductance. This term can be neglected in a standard cage induction machine, where the transient inductance of the stator is much smaller than the magnetizing inductance, which is large. In this case the transient inductance of the stator is approximately equal to the sum of the stator and rotor leakage inductances. However, in induction machines for servo-applications, the magnetizing inductance is smaller than for normal applications, and in this case the effect of L_s' should be incorporated in the controller. If this is not done and there is a step change in the torque reference, there can be an unwanted overshoot in the electromagnetic torque. A solution to this problem can be obtained in several ways, and two of these are now briefly discussed.

The first solution can be obtained by adding the missing term $L_s' \, \mathrm{d}i_{sy}/\mathrm{d}t$ to eqn (4.1-45). In this way a quick response will be obtained, but it is necessary to differentiate the torque-producing stator current component. A second solution can be obtained by incorporating a small delay term which corresponds to the stator transient inductance in the decoupling compensator of Fig. 4.12. In this case the actual torque producing stator current (i_{sy}) will be delayed by the stator transient time constant ($T_s' = L_s'/R_s$) and thus, for accurate decoupling, the torque-producing stator current reference (i_{syref}) should also be delayed by the same time constant. This can be achieved by inserting a new block in Fig. 4.12(a) between points A and B, which contains a first-order lag described by $1/(1+T_s'p)$, by connecting point C to point B and disconnecting points C and A as shown in Fig. 4.12(b). With this implementation, improved dynamic response is obtained and the unwanted overshoot in the electromagnetic torque is reduced. Compared to the first method, the second method results in a slightly slower response.

It is important to note that the implementation shown in Fig. 4.12 utilizes the fact that the rotor flux is constant, and thus is not suitable under field-weakening conditions. However, it can be extended to operate under field weakening. For this purpose eqns (4.1-7), (4.1-8), and (4.1-25) must be considered, but assuming that $|\bar{\imath}_{mr}|$ is not constant. In this case when the expression for $|\bar{\imath}_{mr}|$ obtained from eqn (4.1-25) is substituted into the direct-axis voltage equation defined by eqn (4.1-7), the resulting direct-axis voltage equation will contain three terms. These are the direct-axis voltage across the stator impedance, which is equal to $(R_s + L_s p)i_{sx}$, the direct-axis rotational voltage component $-\omega_{mr} L_s' i_{sy}$ and finally an extra term $-L_m^2 \, p^2 i_{sx}/(R_r + L_r p)$, where $p = \mathrm{d}/\mathrm{d}t$. This third component contains the second time derivative of the direct-axis stator current component i_{sx}. Thus to get a direct relationship between u_{sx} and i_{sx}, the rotational component must be eliminated (this was achieved previously when $|\bar{\imath}_{mr}|$ was assumed to be constant) and since the third component disturbs the simplified structure of the x axis, it should also be cancelled. It should be noted that when $|\bar{\imath}_{mr}| = \text{constant}$, it follows by considering eqn (4.1-25) that this term is only present in the equations in the form of $i_{sx} = |\bar{\imath}_{mr}|$. The required cancellation of the unwanted third component can be performed by the application of such a direct-axis stator current (i_{sx}) controller, which ensures that $pi_{sx} = 0$, i.e. the rate of change of the direct-axis stator current in the rotor-flux-oriented reference frame is zero. For this purpose it is also possible to use a model reference-frame proportional-output error-feedback controller and the actual control current at the output of this controller will replace the current i_{sxref} in

Fig. 4.12, which is related to i_{sx} through the transfer function $1/(1+T_s p)$, where T_s is the stator time constant. It follows from eqn (4.1-25) that $|\bar{i}_{mr}| = i_{sx}/(1+T_r p)$ where T_r is the rotor time constant and this current has to replace i_{sxref} in the terms $L_s i_{sxref}$ and $i_{syref}/(T_r i_{sxref})$ shown in Fig. 4.12.

Under field-weakening operation, it is possible to obtain i_{sxref} similarly to the method shown in Fig. 4.11, or it is possible to determine it from an optimal value of the rotor flux reference $|\bar{\psi}_{rref}| = L_m |\bar{i}_{mref}|$, which has to be dependent on the rotor speed and which gives maximum efficiency of the induction machine subjected to rotor-flux-oriented control. The speed dependency of the rotor flux reference follows from the fact that two loss components, the hysteresis and eddy-current losses, are functions of the stator frequency, which, however, is the sum of the slip frequency and the rotor speed [see eqn (4.1-26)]. However, when deciding the way to obtain i_{sxref}, it should be considered that in the transient state, the main goal is to obtain quick torque response, while efficiency optimization is more important in the steady state.

Implementation of the PWM voltage-source inverter-fed induction machine with no speed sensor and utilizing the actual values of i_{sx} *and* i_{sy}

In the present section an implementation is described for the rotor-flux-oriented control of a voltage-source PWM inverter-fed induction machine, which contains a closed-loop control of the rotor speed but which requires no speed sensor and which is effective even at very low speeds. The on-line rotor speed estimation is based on utilizing Model Reference Adaptive Control (MRAC). The main assumptions used to obtain the models of the induction machine are the same as those used in previous sections and the effects of magnetic saturation are neglected.

The principle of using MRAC for rotor speed identification: Model Reference Adaptive Control systems form one type of adaptive system. It is a great advantage that they lead to relatively simple implementations with high-speed adaptation. In the present section a so-called Parallel Model Reference Adaptive Control system will be utilized for the rotor speed identification in the rotor-flux-oriented control system of a voltage-source inverter-fed induction machine. The basic scheme of the Parallel Model Reference Adaptive Control system is shown in Fig. 4.14.

In Fig. 4.14 **u** is the input column vector (control input) and a reference model of the induction machine and an adjustable system are connected in parallel. The adjustable system corresponds to the mathematical model (parallel rotor speed estimation model) of the vector-controlled induction machine, whose speed is driven by the rotor speed adaptation algorithm which implements the rotor speed identification. The reference model specifies the desired rotor speed (ω_r) and in the adjustable model, the speed to be estimated ($\hat{\omega}_r$) is an unknown parameter. Furthermore, in Fig. 4.14, **y** is the output column vector of the reference model and $\hat{\mathbf{y}}$ is the output column vector of the

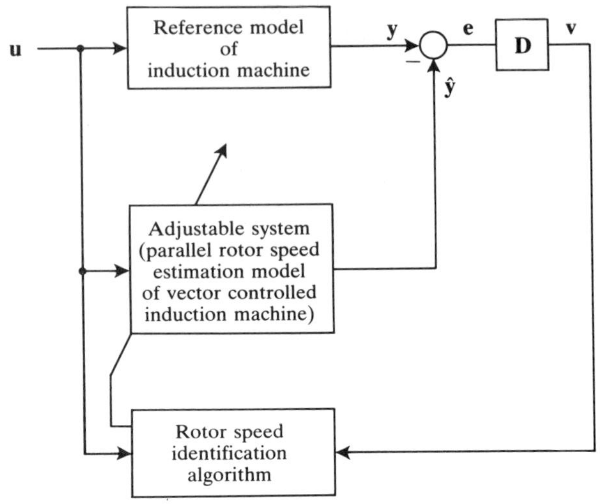

Fig. 4.14. Basic scheme of the Parallel Model Reference Adaptive Control system.

adjustable system. Their difference gives the column vector of state errors **e**. This error is used by the rotor speed identification mechanism to modify the rotor speed of the adjustable system so that the error between the output of the reference model and that of the adjustable model should be zero (output error method). When the rotor speed to be estimated ($\hat{\omega}_r$) is changed in the adjustable model in such a way that the difference between the output of the reference model and the output of the adjustable model is zero, the estimated rotor speed will be equal to the actual rotor speed (ω_r). The error signal actuates the rotor speed identification algorithm which makes this error converge asymptotically to zero. The rotor speed identifier is designed in such a way that quick and stable response should be obtained. In Fig. 4.14 **D** is a so-called linear compensator and $\mathbf{v} = \mathbf{De}$ is the generalized error. **D** is used to obtain stable response.

The model reference rotor speed identifier system can be represented by an equivalent non-linear feedback system which comprises a feedforward time-invariant linear subsystem and a feedback non-linear time-varying subsystem. This is shown in Fig. 4.15.

The input to the linear time-invariant system is **u**, its output is **v**; the output variable of the nonlinear time invariant feedback subsystem is **w**, and $\mathbf{u} = -\mathbf{w}$. The identification algorithm is then chosen according to Popov's hyperstability theorem whereby the transfer function matrix of the linear time-invariant subsystem must be strictly positive real and the non-linear time-varying feedback subsystem has to satisfy Popov's integral inequality.

There are many ways of selecting the output signals of the reference model and the adjustable model. In the next section these will correspond to the rotor

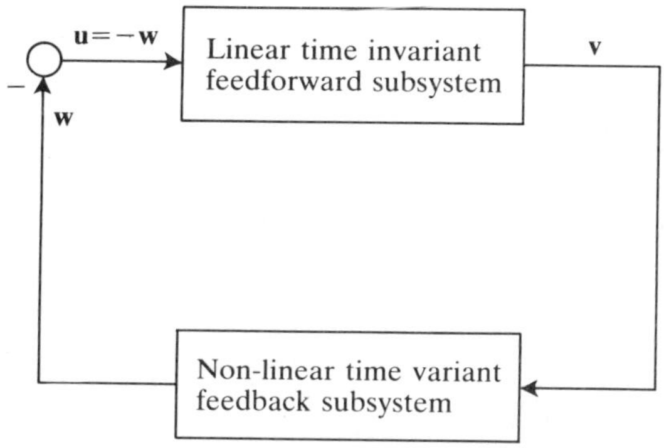

Fig. 4.15. Equivalent non-linear time-varying feedback system.

magnetizing currents and to the so-called modified rotor magnetizing currents, which are introduced instead of the rotor magnetizing currents in order to obtain greater accuracy.

General description of the implementation of the drive: Figure 4.16 shows the schematic of the rotor-flux-oriented control of the induction machine supplied by a voltage-source PWM inverter. There is a similarity between the drives shown in Fig. 4.11 and in Fig. 4.16, but in Fig. 4.16 there is no speed sensor and the speed estimation is obtained by utilizing MRAC as described above.

In the scheme described by Fig. 4.16, there is a closed-loop control of the flux-producing stator current (i_{sx}), the torque-producing stator current (i_{sy}) and the rotor speed. The reference value of the flux-producing stator current (i_{sxref}) is compared with its actual value (i_{sx}) and the error signal is fed into a current controller, a PI controller, as in the control system of Fig. 4.11. The output of the current controller is the signal $\hat{u}_{sx}$. The reference value of the torque-producing stator current (i_{syref}) and its actual value are compared and their difference serves as input to the corresponding current controller, the output of which is the signal $\hat{u}_{sy}$. The signals $\hat{u}_{sx}$ and $\hat{u}_{sy}$ are added to the corresponding outputs of the decoupling circuit, whose role is similar to the decoupling circuit in Fig. 4.11 and the outputs of which are the two-axis rotational voltage components (expressed in the reference frame fixed to the rotor flux linkage space phasor) $-u_{dx}$ and u_{dy}. Thus the reference values of the two-axis components of the stator voltages in the rotor-flux-oriented reference frame (u_{sxref}, u_{syref}) are obtained. As in Fig. 4.11, these component voltages are then transformed into the two-axis stator voltage references of the stationary reference frame (u_{sDref}, u_{sQref}) by the application of the transformation $e^{j\rho_r}$ [see eqn (4.1-46)], where ρ_r is the space angle of the rotor flux-linkage space phasor with respect to the direct axis of the stationary reference frame. This angle is obtained by the integration of ω_{mr}, (this is shown in the block '$1/p$'), where ω_{mr} is the speed of the rotor-flux space phasor and is obtained from the sum of the

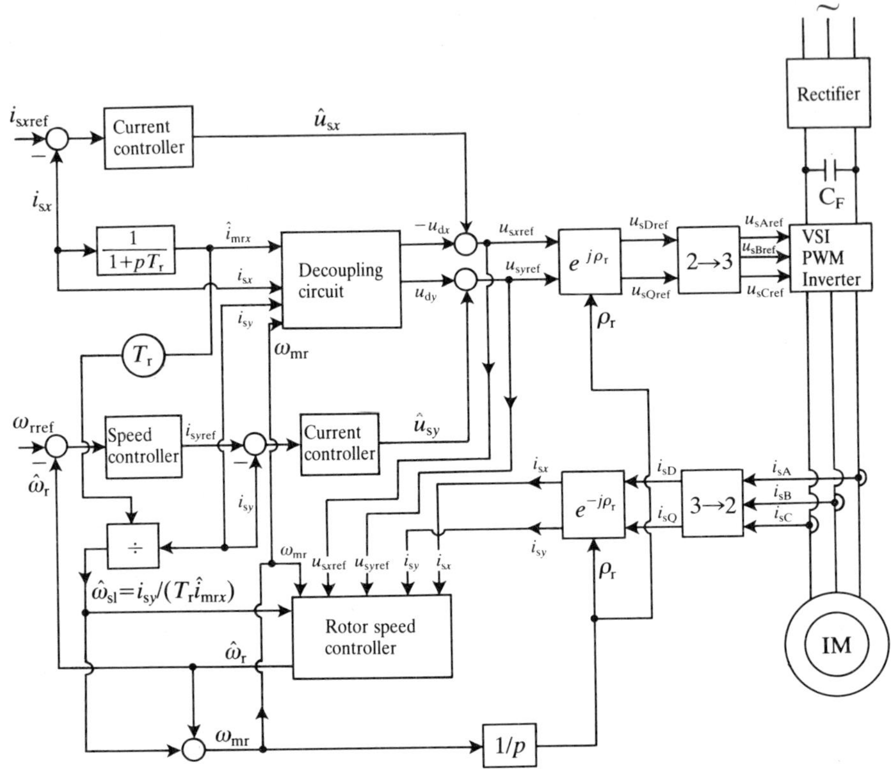

Fig. 4.16. Schematic of the rotor-flux-oriented control of a voltage-source PWM inverter-fed induction machine without a speed sensor.

estimated values of the rotor speed ($\hat{\omega}_r$) and slip speed ($\hat{\omega}_{sl}$) as described later in this section. The corresponding two-axis stator voltage equations are identical to eqns (4.1-47) and (4.1-48). These component voltages are transformed into their three-phase values (u_{sAref}, u_{sBref}, u_{sCref}) by the application of the two-phase to three-phase transformation; the required equations are identical to eqns (4.1-49)–(4.1-51). Of course, it is again possible to perform the two transformations in one step and in this case the compound transformation is described by eqns (4.1-54)–(4.1-56). The application of such a combined transformation can have advantages when the transformations are performed by software.

The rotor speed is also controlled in a closed loop. Thus the reference value of the rotor speed (ω_{rref}) is compared with the estimated value of the rotor speed ($\hat{\omega}_r$) and the error signal is fed into the speed controller, a PI controller. The output of the speed controller is the quadrature-axis stator current reference in the rotor-flux-oriented reference frame (i_{syref}). It should be noted that the estimated value of the rotor speed is obtained by the application of MRAC as described below. The output of the estimator block is the estimated value of the rotor speed, and this is obtained by utilizing the input signals u_{sxref}, u_{syref}, i_{sx}, i_{sy}, ω_{mr} and the estimated value of the slip speed $\hat{\omega}_{sl}$.

There are four inputs to the decoupling circuit shown in Fig. 4.16; these are the stator current components i_{sx} and i_{sy}, the speed of the rotor-flux space phasor (ω_{mr})—which is obtained from the sum of the estimated value of the rotor speed ($\hat{\omega}_r$) and the estimated value of the slip speed ($\hat{\omega}_{sl}$)—and the estimated value of the direct-axis rotor magnetizing-current component ($\hat{i}_{mrx}$). As in Fig. 4.11, the stator current components i_{sx} and i_{sy} are obtained from the monitored three-phase stator currents i_{sA}, i_{sB}, and i_{sC} by the application of two transformations. The first transformation is the three-phase to two-phase transformation and yields the two-axis stator currents i_{sD} and i_{sQ} of the stationary reference frame. The second transformation utilizes $e^{-j\rho_r}$, and finally gives the two-axis components of the stator currents in the rotor-flux-oriented reference frame.

Decoupling circuit: The equations for the output voltages of the decoupling circuit are similar to those which describe the equations of the decoupling circuit shown in Fig. 4.11, but there are some differences. The new equations can be obtained by considering eqns (4.1-9) and (4.1-10), and it follows that the rotational voltages in the direct and quadrature axes of the rotor-flux-oriented reference frame are

$$u_{dx} = -\omega_{mr} L_s' i_{sy} \qquad (4.1\text{-}60)$$

$$u_{dy} = \omega_{mr} L_s' i_{sx} + \omega_{mr}(L_s - L_s')\hat{i}_{mrx}, \qquad (4.1\text{-}61)$$

where L_s' is the stator transient inductance of the induction machine, L_s is the stator inductance, and $\hat{i}_{mrx}$ is the estimated value of the direct-axis rotor magnetizing current component expressed in the reference frame fixed to the rotor flux, which rotates at speed ω_{mr}. It follows from eqn (4.1-25) that $\hat{i}_{mrx}$ can be obtained as

$$\hat{i}_{mrx} = \frac{i_{sx}}{1 + T_r p}, \qquad (4.1\text{-}62)$$

where T_r is the rotor time constant and $p = \mathrm{d}/\mathrm{d}t$. In Fig. 4.16, $\hat{i}_{mrx}$ is obtained from eqn (4.1-62) and therefore the block $1/(1 + T_r p)$ is present between i_{sx} and $\hat{i}_{mrx}$.

It follows from eqns (4.1-60) and (4.1-61) that they contain the speed of the rotor-flux space phasor ω_{mr}, but since there is no speed sensor, it has to be obtained from the estimated value of the rotor speed ($\hat{\omega}_r$) and the estimated value of the slip speed ($\hat{\omega}_{sl}$) and in accordance with eqn (4.1-26), it is equal to $\hat{\omega}_r + \hat{\omega}_{sl}$. The estimation of these two quantities will be discussed in the next section.

Estimation of the rotor speed and the slip speed: (a) Deriving a circuit to obtain the components of the modified rotor magnetizing-current space phasor: In this section a circuit is obtained which will yield as its outputs the actual values of the modified rotor magnetizing-current components expressed in the rotor-flux-oriented reference frame, since these signals will appear as the inputs of a circuit which will be used to obtain the estimated value of the rotor speed. The space phasor of the modified rotor magnetizing currents will all be defined.

The rotor magnetizing-current space phasor has been defined in eqn (4.1-1) and the space-phasor voltage equation of the stator in the rotor-flux-oriented reference frame has been given as eqn (4.1-2) which contains the space phasors of the stator and rotor currents. By utilizing eqn (4.1-1), the space phasor of the rotor currents can be expressed in terms of the space phasors of the stator currents and the rotor magnetizing currents, and when this is substituted into eqn (4.1-2), the following space-phasor equation is obtained

$$(p+j\omega_r)\bar{i}_{mr} = \frac{L_r}{L_m^2}[\bar{u}_{s\psi r} - (R_s + L_s'p)\bar{i}_{s\psi r} - j\omega_{mr}L_s'\bar{i}_{s\psi r}], \qquad (4.1\text{-}63)$$

where $\bar{u}_{s\psi r}$ and $\bar{i}_{s\psi r}$ have been defined in eqns (4.1-4) and (4.1-3) respectively. Thus it is possible to obtain the real- and imaginary-axis components of eqn (4.1-63) and, if the machine parameters R_s, L_s', L_r, and L_m are known and ω_{mr}, i_{sx}, i_{sy} are monitored together with the stator voltage components, which are the reference values u_{sxref}, u_{syref}, then it is possible to determine the components of $\bar{i}_{mr}$ (which in general are i_{mrx} and i_{mry}). However, since the effects of offset errors on pure integral elements cannot be neglected and therefore satisfactory accuracy cannot be reached by utilizing this method, a modified rotor magnetizing-current space phasor ($\bar{i}'_{mr}$) will be defined, for which it is possible to obtain more accurate values of its two-axis components (i'_{mrx}, i'_{mry}).

The relationship of the modified magnetizing-current space phasor ($\bar{i}'_{mr}$) and the unmodified rotor magnetizing current space phasor ($\bar{i}_{mr}$) is

$$(p+j\omega_{mr})\bar{i}_{mr} = \left(p + \frac{1}{T} + j\omega_{mr}\right)\bar{i}'_{mr}, \qquad (4.1\text{-}64)$$

where T is a suitably chosen positive time constant which is changed in accordance with ω_{mr}. Thus from eqns (4.1-63) and (4.1-64), the following differential equation is obtained for the modified rotor magnetizing-current space phasor, if $\bar{u}_{s\psi r}$ is replaced by its reference value $\bar{u}_{s\psi rref} = u_{sxref} + ju_{syref}$,

$$\left(p + \frac{1}{T} + j\omega_r\right)\bar{i}'_{mr} = \frac{L_r}{L_m^2}[\bar{u}_{s\psi rref} - (R_s + L_s'p)\bar{i}_{s\psi r} - j\omega_{mr}L_s'\bar{i}_{s\psi r}]. \quad (4.1\text{-}65)$$

Resolution of eqn (4.1-65) into its real and imaginary components yields

$$\begin{bmatrix} i'_{mrx} \\ i'_{mry} \end{bmatrix} = \begin{bmatrix} p+1/T & -\omega_{mr} \\ -\omega_{mr} & p+1/T \end{bmatrix}^{-1} \frac{L_r}{L_m^2}\left\{\begin{bmatrix} u_{sxref} \\ u_{syref} \end{bmatrix} - \begin{bmatrix} R_s + L_s'p & -\omega_{mr}L_s' \\ \omega_{mr}L_s' & R_s + L_s'p \end{bmatrix}\begin{bmatrix} i_{sx} \\ i_{sy} \end{bmatrix}\right\}. \qquad (4.1\text{-}66)$$

Fig. 4.17 shows an implementation of eqn (4.1-66), where the inputs are u_{sxref}, u_{syref}, i_{sx}, i_{sy}, and ω_{mr}.

The circuit shown in Fig. 4.17 will be utilized when the estimated value of the rotor speed is obtained in a later section, when it is shown that the actual values of the modified rotor magnetizing current components i'_{mrx} and i'_{mry} are the inputs of the circuit which is used to obtain the estimated value of the rotor speed.

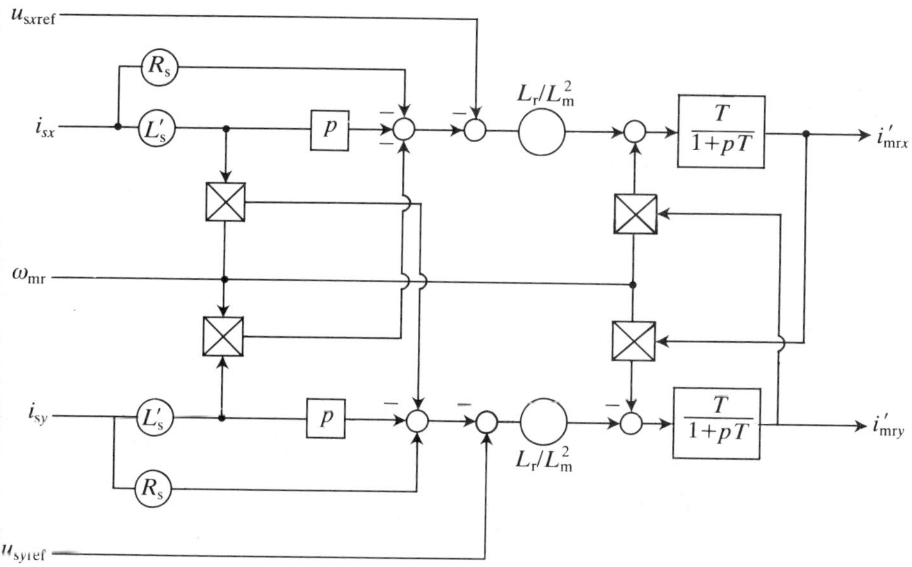

Fig. 4.17. Circuit to obtain i'_{mrx} and i'_{mry}.

(*b*) *State-variable equation of the reference model in the MRAC system:* When the MRAC system is used to obtain a circuit for the estimation of the rotor speed, it is very convenient to obtain the effective adaptation laws by considering the state-variable equations of the reference model and the adjustable model. For this purpose the state-variable voltage equations of these two systems will be derived.

It is possible to obtain the state-variable form of the voltage equations of the reference model used in the MRAC system by using the newly defined modified rotor magnetizing-current space phasor. For this purpose the rotor voltage equation given by eqn (4.1-21) is used, where similarly to eqn (4.1-22), in general, the rotor flux-linkage space phasor is defined as $\bar{\psi}_{r\psi r} = L_m \bar{i}_{mr}$. When this expression is substituted into eqn (4.1-21), the rotor voltage equation will contain the rotor-current space phasor $\bar{i}_{r\psi r}$ and the magnetizing-current space phasor $\bar{i}_{mr}$. However, to obtain an equation which contains the stator-current space phasor $\bar{i}_{s\psi r}$ and i_{mr}, the rotor-current space phasor in the rotor-flux-oriented reference frame ($\bar{\imath}_{r\psi r}$) is expressed in terms of these two currents by utilizing eqn (4.1-1) and finally the following differential equation is obtained for the rotor magnetizing-current space phasor,

$$p\bar{\imath}_{mr}=\frac{\bar{\imath}_{s\psi r}}{T_r}-\bar{\imath}_{mr}\left[\frac{1}{T_r}+j(\omega_{mr}-\omega_r)\right], \tag{4.1-67}$$

which is very similar to eqn (4.1-24). By considering eqn (4.1-64), eqn (4.1-67) can be put into the form,

$$p\bar{\imath}'_{mr}=\frac{\bar{\imath}_{s\psi r}}{T_r}-\bar{\imath}'_{mr}\left[\frac{1}{T_r}+\frac{1}{T}+j(\omega_{mr}-\omega_r)\right]. \tag{4.1-68}$$

Thus by resolving into real and imaginary components, eqns (4.1-67) and (4.1-68) yield the following system of differential equations for the reference model in the state-variable form

$$p\begin{bmatrix} i'_{mrx}\\ i'_{mry}\\ i_{mrx}\\ i_{mry}\end{bmatrix}=\begin{bmatrix} -1/T & \omega_{mr} & -1/T_r & -\omega_r\\ -\omega_{mr} & -1/T & \omega_r & -1/T_r\\ 0 & 0 & -1/T_r & \omega_{mr}-\omega_r\\ 0 & 0 & -(\omega_{mr}-\omega_r) & -1/T_r\end{bmatrix}\begin{bmatrix} i'_{mrx}\\ i'_{mry}\\ i_{mrx}\\ i_{mry}\end{bmatrix}+\frac{1}{T_r}\begin{bmatrix}1&0\\0&1\\1&0\\0&1\end{bmatrix}\begin{bmatrix} i_{sx}\\ i_{sy}\end{bmatrix}. \tag{4.1-69}$$

State-variable equation of the adjustable model in the MRAC system: It is possible to use eqn (4.1-69) to obtain the state-variable form of the voltage equations which describe the adjustable model. For this purpose instead of the actual variables i'_{mrx}, i'_{mry}, i_{mrx}, and i_{mry}, their estimated values $\hat{\imath}'_{mrx}$, $\hat{\imath}'_{mry}$, $\hat{\imath}_{mrx}$, and $\hat{\imath}_{mry}$ are present and the actual rotor speed ω_r is replaced by its estimated value $\hat{\omega}_r$. Thus the following state-variable form of the voltage equations is obtained

$$p\begin{bmatrix} \hat{\imath}'_{mrx}\\ \hat{\imath}'_{mry}\\ \hat{\imath}_{mrx}\\ \hat{\imath}_{mry}\end{bmatrix}=\begin{bmatrix} -1/T & \omega_{mr} & -1/T_r & -\hat{\omega}_r\\ -\omega_{mr} & -1/T & \hat{\omega}_r & -1/T_r\\ 0 & 0 & -1/T_r & \omega_{mr}-\hat{\omega}_r\\ 0 & 0 & -(\omega_{mr}-\hat{\omega}_r) & -1/T_r\end{bmatrix}\begin{bmatrix} \hat{\imath}'_{mrx}\\ \hat{\imath}'_{mry}\\ \hat{\imath}_{mrx}\\ \hat{\imath}_{mry}\end{bmatrix}+\frac{1}{T_r}\begin{bmatrix}1&0\\0&1\\1&0\\0&1\end{bmatrix}\begin{bmatrix} i_{sx}\\ i_{sy}\end{bmatrix}. \tag{4.1-70}$$

Equation (4.1-70) will be used to obtain the circuit for the estimated rotor speed.

(d) Slip frequency estimation: Since the adjustable model is subjected to rotor-flux-oriented control, ω_{mr} is controlled according to the following equation, which is similar to eqn (4.1-26), to yield $\hat{\imath}_{mry}=0$

$$\omega_{mr}=\hat{\omega}_{sl}+\hat{\omega}_r=\hat{\omega}_r+\frac{i_{sy}}{T_r\hat{\imath}_{mrx}}, \tag{4.1-71}$$

where in the slip frequency term the estimated value of the direct-axis rotor magnetizing current expressed in the rotor-flux-oriented reference frame is present; this has been given by eqn (4.1-62). Thus, the substitution of eqn (4.1-62) into eqn (4.1-71) yields

$$\omega_{mr} = \hat{\omega}_r + i_{sy}\frac{1+T_r p}{T_r i_{sx}}. \tag{4.1-72}$$

It follows that in Fig. 4.16, ω_{mr} is obtained by adding $\hat{\omega}_{sl}$ and $\hat{\omega}_r$ and the slip frequency is obtained by dividing i_{sy} by $(T_r\hat{\imath}_{mrx})$, where T_r is the rotor time constant.

(e) Derivation of a circuit to obtain the estimated values of the modified rotor magnetizing current components: In the present section a circuit is obtained which yields the direct- and quadrature-axis components of the estimated modified rotor magnetizing current in the rotor-flux-oriented reference frame ($\hat{\imath}'_{mrx}$ and $\hat{\imath}'_{mry}$). It is necessary to obtain these current components, because they are inputs to the circuit, the output of which is the estimated value of the rotor speed.

If the estimated value of the quadrature-axis rotor magnetizing current in the rotor-flux-oriented reference frame is zero, $\hat{\imath}_{mry}=0$, it follows from eqn (4.1-70) that the estimated values of the modified rotor magnetizing currents in the rotor-flux-oriented reference frame can be obtained from the following two differential equations, if in the second equation $\hat{\omega}_r$ is expressed in terms of ω_{mr} by using eqn (4.1-71):

$$p\hat{\imath}'_{mrx} = \frac{i_{sx}-\hat{\imath}_{mrx}}{T_r} + \omega_{mr}\hat{\imath}'_{mry} - \frac{\hat{\imath}'_{mrx}}{T} \tag{4.1-73}$$

$$p\hat{\imath}'_{mry} = -\omega_{mr}\hat{\imath}'_{mrx} + \omega_{mr}\hat{\imath}_{mrx} - \frac{\hat{\imath}'_{mry}}{T}. \tag{4.1-74}$$

Equations (4.1-73) and (4.1-74) can be realized by the circuit shown in Fig. 4.18. The inputs to the circuit in Fig. 4.18 are the direct-axis stator current

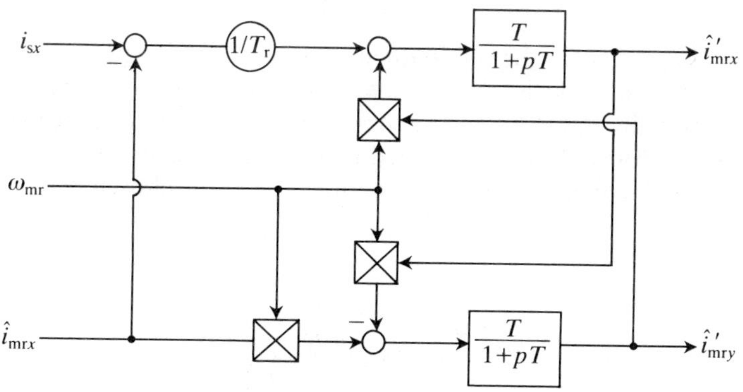

Fig. 4.18. Circuit to obtain $\hat{\imath}'_{mrx}$ and $\hat{\imath}'_{mry}$.

expressed in the rotor-flux-oriented reference frame (i_{sx}), the speed of this reference frame (ω_{mrx}) and the estimated value of the direct-axis component of the rotor magnetizing current, also expressed in the rotor-flux-oriented reference frame ($\hat{i}_{mrx}$). It should be noted that this current is present as the input of the decoupling circuit shown in Fig. 4.16, where it is obtained from i_{sx} by considering the time delay due to the rotor time constant.

(*f*) *Rotor speed estimator:* It will be shown that a rotor speed estimator circuit can be designed the inputs of which are the stator voltage references u_{sxref}, u_{syref}, the actual values of the two-axis stator currents (i_{sx} and i_{sy}) and the speed of the rotor-flux space phasor (ω_{mr}). However, since it has been shown in Figs 4.17 and 4.18 respectively that the actual and estimated values of the two-axis components of the modified rotor magnetizing currents (i'_{mrx}, i'_{mry} and $\hat{i}'_{mrx}$ and $\hat{i}'_{mry}$) can be obtained from u_{sxref}, u_{syref}, i_{sx}, i_{sy}, ω_{mr} and $\hat{i}_{mrx}$ (the latter is closely related to i_{sx}), and it has been shown in Fig. 4.16 how the estimated slip speed ($\hat{\omega}_{sl}$) is obtained from i_{sy} and $\hat{i}_{mrx}$, in the present section that part of the rotor speed estimation circuit will be shown which utilizes as its inputs the actual and estimated values of the modified magnetizing currents and the estimated quantities $\hat{i}_{mrx}$ and $\hat{\omega}_{sl}$.

There are many ways to obtain the rotor speed identification algorithm by utilizing eqns (4.1-69) and (4.1-70), but now a method will be followed which uses the small-signal forms of these state-variable equations. For this purpose the state-variable equations are linearized around an operating point where the rotor speed is ω_{r0} and the speed of the rotor flux-linkage phasor is ω_{mr0}. Thus by subtracting the linearized state-variable equations of the adjustable model from the linearized state-variable equations of the reference model, the following state-variable form of the error equations is obtained:

$$p\begin{bmatrix}\Delta i'_{mrx}-\Delta\hat{i}'_{mrx}\\ \Delta i'_{mry}-\Delta\hat{i}'_{mry}\\ \Delta i_{mrx}-\Delta\hat{i}_{mrx}\\ \Delta i_{mry}\end{bmatrix}=\begin{bmatrix}-1/T & \omega_{mr0} & -1/T_r & -\omega_{r0}\\ -\omega_{mr0} & -1/T & \omega_{r0} & -1/T_r\\ 0 & 0 & -1/T_r & \omega_{sl0}\\ 0 & 0 & -\omega_{sl0} & -1/T_r\end{bmatrix}\begin{bmatrix}\Delta i'_{mrx}-\Delta\hat{i}'_{mrx}\\ \Delta i'_{mry}-\Delta\hat{i}'_{mry}\\ \Delta i_{mrx}-\Delta\hat{i}_{mrx}\\ \Delta i_{mry}\end{bmatrix}+\begin{bmatrix}0\\1\\0\\1\end{bmatrix}\hat{i}_{mrx0}(\Delta\omega_r-\Delta\hat{\omega}_r), \tag{4.1-75}$$

where $\omega_{sl0}=\omega_{mr0}-\omega_{r0}$ is the slip frequency at the operating point and $\hat{i}_{mrx0}$ is the estimated value of the direct-axis rotor magnetizing current at the operating point. In eqn (4.1-75), the input variable is the difference between the small-signal form of the rotor speed ($\Delta\omega_r$) and the estimated rotor speed ($\Delta\hat{\omega}_r$). The small-signal form of the generalized error is chosen as

$$\Delta\mathbf{e}=\mathbf{D}\begin{bmatrix}\Delta i'_{mrx}-\Delta\hat{i}'_{mrx}\\ \Delta i'_{mry}-\Delta\hat{i}'_{mry}\\ \Delta i_{mrx}-\Delta\hat{i}_{mrx}\\ \Delta i_{mry}\end{bmatrix}. \tag{4.1-76}$$

In eqn (4.1-76) matrix **D** has to be determined so as to satisfy Popov's hyperstability theorem, owing to the following considerations.

In the discussion of MRAC general principles, it has been noted that when the rotor speed to be estimated ($\hat{\omega}_r$) is changed in the adjustable model in such a way that the difference between the output of the reference model and the output of the adjustable model is zero, the estimated rotor speed will be equal to the actual rotor speed (ω_r). The error signal actuates the rotor speed identification algorithm which makes this error converge to zero. The rotor speed algorithm is chosen in such a way that quick and stable response should be obtained. For this purpose the rotor speed identifier system is represented by an equivalent non-linear feedback system which consists of a feedforward time-invariant linear subsystem with a feedback non-linear time-varying subsystem and the identification algorithm is chosen according to Popov's hyperstability theorem. According to this, the transfer function matrix of the linear time-invariant subsystem must be strictly positive real and the non-linear time-varying feedback subsystem has to satisfy Popov's integral inequality, according to which

$$\int_0^t \mathbf{v}^T \mathbf{w} \, dt \geq 0 \tag{4.1-77}$$

where $\mathbf{v}$ is the output column vector of the linear time-invariant feedforward subsystem, T denotes the transpose and $-\mathbf{w}$ is the input column vector of the linear time-invariant feedback subsystem.

By defining the small-signal transfer function $\Delta h = \Delta\mathbf{v}/(-\Delta\mathbf{w})$, where it follows from eqn (4.1-75) that $-\Delta\mathbf{w} = i_{mrx0}(\Delta\omega_r - \Delta\hat{\omega}_r)$, and where the generalized error $\Delta\mathbf{v}$, is defined as

$$\Delta\mathbf{v} = [\Delta\omega_{mr} - \Delta\hat{\omega}_r \quad p + 1/T_r \quad 0 \quad 0] \begin{bmatrix} \Delta i'_{mrx} - \Delta\hat{i}'_{mrx} \\ \Delta i'_{mry} - \Delta\hat{i}'_{mry} \\ \Delta i_{mrx} - \Delta\hat{i}_{mrx} \\ \Delta i_{mry} \end{bmatrix}, \tag{4.1-78}$$

where $\Delta\omega_{mr}$ is the small-signal value of the speed of the rotor-flux space phasor. It follows from eqns (4.1-75) and (4.1-78) that the transfer function Δh takes the following form

$$\Delta h = \frac{p^2 + \dfrac{p}{T} + \omega_{mr0}}{\left(p + \dfrac{1}{T}\right)^2 + \omega_{mr0}^2}. \tag{4.1-79}$$

This is strictly positive real when $\omega_{mr0} \neq 0$. If in eqn (4.1-78) the generalized error is chosen to be equal to $\Delta\mathbf{e}$, the transfer functions of the linear time-invariant subsystem $(\Delta i'_{mrx} - \Delta\hat{i}'_{mrx})/(-\Delta\mathbf{w})$ and $(\Delta i'_{mry} - \Delta\hat{i}'_{mry})/(-\Delta\mathbf{w})$ will not be strictly positive real.

The adaptation algorithm can be obtained by considering Popov's integral inequality given by eqn (4.1-77). For the linearized system the application of eqn (4.1-77) yields the following inequality, if $\Delta\mathbf{w}$ is replaced by $\hat{\imath}_{mrx0}(\Delta\hat{\omega}_r - \Delta\omega_r)$,

$$\int_0^t \Delta\mathbf{v}^T\Delta\mathbf{w}\,dt = \int_0^t \Delta\mathbf{v}^T\hat{\imath}_{mrx0}(\Delta\hat{\omega}_r - \Delta\omega_r)\,dt \geq 0, \tag{4.1-80}$$

where $\Delta\mathbf{v}$ is given by eqn (4.1-78). However, it follows from the theory of Parallel Model Reference Adaptive Control systems, that since in general, in the system of Fig. 4.14, the input column vector of the reference model is a non-zero vector and the state vector of the adjustable system is also a non-zero vector, for all values of t (time), the required condition that the error should converge asymptotically to zero implies that in the adaptation mechanism there must be at least one integrator. Thus the adaptation law for the rotor speed is as follows

$$\Delta\hat{\omega}_r = \int_0^t \psi(\Delta\mathbf{v})\,dt + \Delta\hat{\omega}_r(0), \tag{4.1-81}$$

where in general ψ is a non-linear function which depends on the generalized error and $\hat{\omega}_r(0)$ is the initial value of the estimated rotor speed. Substitution of eqn (4.1-81) into eqn (4.1-80) yields

$$\int_0^t \Delta\mathbf{v}^T\hat{\imath}_{mrx0}\left[\int_0^t \psi(\Delta\mathbf{v})\,dt + \Delta\hat{\omega}_r(0) - \Delta\omega_r\right]dt \geq 0. \tag{4.1-82}$$

Equation (4.1-82) can be used to obtain $\psi(\Delta\mathbf{v})$ and when this is known, its substitution into eqn (4.1-81) will yield the required adaptation algorithm for $\Delta\hat{\omega}_r$.

The unknown function ψ can be determined by utilizing the following inequality

$$\int_0^t [pf(t)]kf(t)\,dt = \frac{k[f^2(t) - f^2(0)]}{2} \geq \frac{-kf^2(0)}{2}, \tag{4.1-83}$$

where $f(t)$ is a general function of time, $[pf(t)]$ is its first time derivative and k is a positive constant. By utilizing eqn (4.1-83), in eqn (4.1-82)

$$[pf(t)] = \Delta\mathbf{v}^T\hat{\imath}_{mrx0} \tag{4.1-84}$$

and

$$kf(t) = \int_0^t \psi(\Delta\mathbf{v})\,dt + \Delta\hat{\omega}_r(0) - \Delta\omega_r. \tag{4.1-85}$$

By considering eqn (4.1-84), differentiation of eqn (4.1-85) with respect to time gives

$$\psi(\Delta\mathbf{v}) = k\Delta\mathbf{v}^T\hat{\imath}_{mrx0}, \tag{4.1-86}$$

where $\Delta\mathbf{v}$ has been defined in eqn (4.1-78). Thus the substitution of eqn (4.1-78) into eqn (4.1-86) yields

$$\psi(\Delta\mathbf{v})=k\hat{i}_{\mathrm{mr}x0}\left[(\Delta\omega_{\mathrm{mr}}-\Delta\hat{\omega}_{\mathrm{r}})(\Delta i'_{\mathrm{mr}x}-\Delta\hat{i}'_{\mathrm{mr}x})+\left(p+\frac{1}{T_{\mathrm{r}}}\right)(\Delta i'_{\mathrm{mr}y}-\Delta\hat{i}'_{\mathrm{mr}y})\right], \tag{4.1-87}$$

where $k>0$. Equation (4.1-87) is a so-called integral adaptation.

By rewriting the small-signal values into their corresponding large-signal values, substitution of eqn (4.1-87) into eqn (4.1-81) finally yields

$$\hat{\omega}_{\mathrm{r}}=\hat{\omega}_{\mathrm{r}}(0)+k\int_0^t\left\{\left[(\omega_{\mathrm{mr}}-\hat{\omega}_{\mathrm{r}})(i'_{\mathrm{mr}x}-\hat{i}'_{\mathrm{mr}x})+\left(p+\frac{1}{T_{\mathrm{r}}}\right)(i'_{\mathrm{mr}y}-\hat{i}'_{\mathrm{mr}y})\right]\mathrm{d}t\right\} \tag{4.1-88}$$

where $\hat{i}_{\mathrm{mr}x}$ is the estimated value of the direct-axis component of the rotor magnetizing current expressed in the rotor-flux-oriented reference frame. This current component is an input of the decoupling circuit shown in Fig. 4.16. Furthermore, in eqn (4.1-88) $\hat{\omega}_{\mathrm{r}}(0)$ is the initial value of the estimated rotor speed and the term $\omega_{\mathrm{mr}}-\hat{\omega}_{\mathrm{r}}$ is the estimated value of the slip frequency ($\hat{\omega}_{\mathrm{sl}}$). Equation (4.1-88) can be realized by the circuit shown in Fig. 4.19.

In the circuit shown in Fig. 4.19 the inputs are the estimated value of the slip speed $\hat{\omega}_{\mathrm{sl}}$, and the estimated and actual values of the modified rotor magnetizing-current components ($\hat{i}'_{\mathrm{mr}x}$, $\hat{i}'_{\mathrm{mr}y}$, $i'_{\mathrm{mr}x}$, and $i'_{\mathrm{mr}y}$). Integration is performed in the block labelled '$1/p$'. The output of the circuit is the estimated value of the rotor speed.

Since according to eqn (4.1-79) the transfer function of the generalized error variable depends on the operating point value of the speed of the rotor flux ($\omega_{\mathrm{mr}0}$) and thus to reduce the effect of $\omega_{\mathrm{mr}0}$ on the damping factor of the output

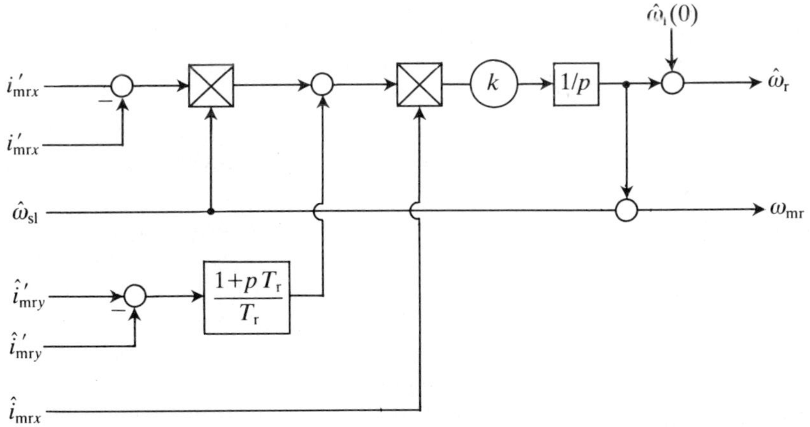

Fig. 4.19. Circuit to obtain the estimated rotor speed.

error, the time constant T (which was introduced in the definition of the modified rotor magnetizing-current space phasor) is controlled as the inverse of the absolute value of ω_{mr0}. Furthermore, it should be noted that the rotor speed identification algorithm has been obtained by utilizing a linearized model, and the real system is non-linear. To ensure that the results obtained from the linearized model are valid, the gain of the rotor speed identifier must be sufficiently high.

Implementation utilizing a speed sensor and the actual values of i_{sx}, i_{sy} *and rotor resistance identification*

As has been discussed above, in practice the rotor resistance can change due to skin effects and to temperature variation. Thus if in the vector control scheme an inaccurate value of this parameter is used, inaccurate commands for the torque- and flux-producing stator currents will result and the determined slip frequency will be incorrect. This will result in degraded performance and unwanted oscillations in the rotor flux and torque. Therefore it is very important to have an accurate estimation of the rotor resistance.

In this section an implementation of the rotor-flux-oriented control of a voltage-source PWM inverter-fed induction machine is described, which utilizes a speed sensor and which contains an on-line estimator of the rotor resistance. The rotor resistance identification to be described is based on the theory of MRAC systems and is valid under all speed and load conditions. The main assumptions used to obtain the model of the induction machine are the same as those used in previous sections, but the rotor resistance can vary. Again, the effects of magnetic saturation are neglected.

The principle of using MRAC for rotor resistance identification: The approach to obtain an identification algorithm for the rotor resistance will be similar to that used above to obtain an estimation of the rotor speed with the aid of MRAC. Thus when a Parallel MRAC system is used for the identification of the rotor resistance in the rotor-flux-oriented control system of an induction machine, a reference model corresponds to the induction machine and an adjustable model to the vector-controlled machine, in which the rotor resistance ($\hat{R}_r$) is an unknown parameter. A similar scheme has been described in Fig. 4.14. If the rotor inductance L_r is a known constant, then the rotor time constant $\hat{T}_r = L_r/\hat{R}_r$ can be considered as the parameter which has to be estimated. As above, the output error method will be used for the identification of the rotor time constant $\hat{T}_r$. Thus when the rotor time constant to be estimated is changed in the adjustable model in such a way that the difference between the output of the reference model and the output of the adjustable model is zero, the estimated rotor time constant will be equal to its actual value ($T_r = L_r/R_r$). The error signal actuates the rotor time constant identification algorithm which makes this error converge asymptotically to zero. The rotor time constant identifier is designed to yield a quick and stable represented by an equivalent non-linear feedback system, which contains a feedforward time-invariant linear subsystem and a feedback non-linear, time-varying subsystem. The rotor time constant identification algorithm is then chosen by satisfying Popov's hyperstability theorem, according to which the transfer-function matrix of the linear time-invariant system must be strictly positive real and the non-linear time-varying system has to satisfy Popov's integral inequality.

(*a*) *Implementation of the drive:* Figure 4.20 shows the schematic of the rotor-

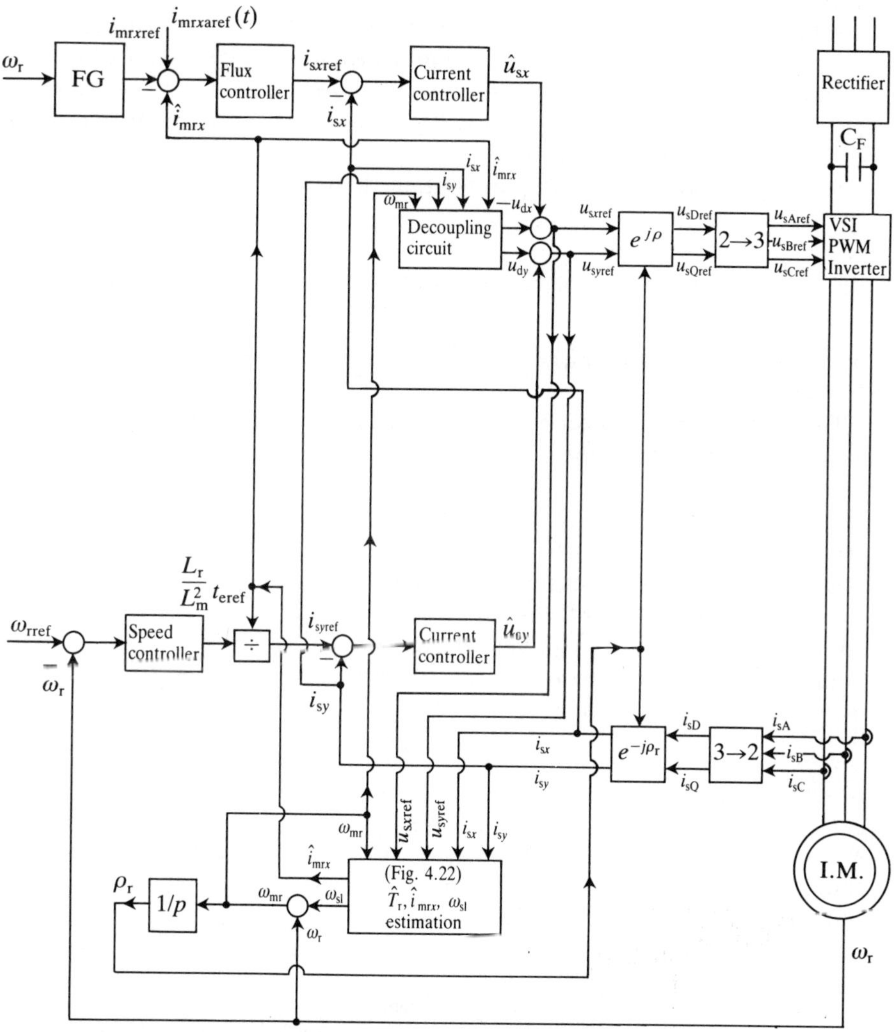

Fig. 4.20. Schematic of the rotor-flux-oriented control of a voltage-source inverter-fed induction machine with speed sensor and rotor time constant estimator.

flux-oriented control of the induction machine supplied by a voltage-source PWM inverter. The rotor speed is monitored with the aid of a speed sensor. This scheme is similar to the one presented in Fig. 4.11, but now the rotor time constant is estimated by utilizing MRAC as described above.

In Fig. 4.20 the input to function generator FG is the monitored rotor speed (ω_r), and the output of FG is the reference value of the direct-axis rotor magnetizing-current component expressed in the reference frame rotating with the rotor flux-linkage space phasor (i_{mrxref}). The task of FG is to achieve field weakening, so that below base speed it gives a constant on its output and above base speed it gives a value which is inversely proportional to the absolute value of the rotor speed. The reference value i_{mrxref} is then added to the direct-axis component of the alternating rotor magnetizing-current component $i_{mrxrefa}(t)=I_{mrxrefa}\sin(\omega_a t)$ and their sum is compared with the estimated value of the direct-axis component of the rotor magnetizing-current space phasor ($\hat{i}_{mrx}$). Thus $i_{mrxref}+i_{mrxrefa}(t)-\hat{i}_{mrx}$ is the input to the flux controller, which is a PI controller.

The estimated current $\hat{i}_{mrx}$ is obtained similarly to that given by eqn (4.1-62), so that

$$\hat{i}_{mrx}=\frac{i_{sx}}{1+\hat{T}_r p}, \tag{4.1-89}$$

but in eqn (4.1-89), the estimated value of the rotor time constant ($\hat{T}_r$) is present.

The reason for the injection of the sinusoidal a.c. current $i_{mrxrefa}(t)$ is to ensure that the rotor time constant can be estimated even if the torque-producing stator current component (i_{sy}) is zero. This is now discussed in detail. As shown previously, when rotor-flux-oriented control is performed, the stator currents are separated into flux- and torque-producing current components (i_{sx}, i_{sy}). To ensure that the quadrature-axis component of the rotor magnetizing-current (or rotor flux-linkage) space phasor is zero, similarly to that shown by eqn (4.1-71), the speed of the special reference frame fixed to the rotor magnetizing-current space phasor (ω_{mr}) must be controlled so as to satisfy the equation,

$$\omega_{mr}=\omega_r+\frac{i_{sy}}{\hat{T}_r\hat{i}_{mrx}}, \tag{4.1-90}$$

where ω_r is the monitored rotor speed, $\hat{T}_r$ is the estimated rotor time constant and $\hat{i}_{mrx}$ is defined by eqn (4.1-89). The second term on the right-hand side of eqn (4.1-90) is the angular slip speed ω_{sl}.

$$\omega_{sl}=\frac{i_{sy}}{\hat{T}_r\hat{i}_{mrx}}. \tag{4.1-91}$$

It follows that if $i_{sy}=0$, the angular slip frequency is zero. Furthermore, if

i_{sx} = constant, it follows from eqn (4.1-89) that $i_{sx} = \hat{i}_{mrx}$ = constant. The following differential equation holds for the estimated value of the rotor magnetizing current ($\hat{\bar{i}}_{mr} = \hat{i}_{mrx} + j\hat{i}_{mry}$), which could be obtained in the same way as eqn (4.1-67), and which is identical to eqn (4.1-67), but with T_r replaced by $\hat{T}_r$,

$$p\hat{\bar{i}}_{mr} = \frac{\bar{i}_{s\psi r}}{\hat{T}_r} - \hat{i}_{mr}\left(\frac{1}{\hat{T}_r} + j\omega_{sl}\right), \tag{4.1-92}$$

where $\bar{i}_{s\psi r} = i_{sx} + ji_{sy}$. It therefore follows from the imaginary-axis component of eqn (4.1-92), that if $i_{sy} = 0$ and $\omega_{sl} = 0$, then $\hat{i}_{mry} = 0$. Thus it can be concluded that when the quadrature-axis stator current (i_{sy}) is zero, and the direct-axis stator current (i_{sx}) is constant, the direct-axis component of the estimated rotor magnetizing-current space phasor ($\hat{i}_{mrx}$) is constant and the quadrature-axis component of the estimated rotor magnetizing current space phasor ($\hat{i}_{mry}$) is zero and they do not depend on the rotor time constant. Therefore in this case the rotor time constant cannot be determined by utilizing the components of the rotor magnetizing current (or rotor flux-linkage space phasor). However, when the sinusoidal current component $i_{mrxrefa}(t) = I_{mrxrefa}\sin(\omega_a t)$ is injected into the direct-axis stator current channel as shown in Fig. 4.20, even when the quadrature-axis stator current (i_{sy}) is zero and thus when the angular slip frequency (ω_{sl}) is zero, according to eqn (4.1-89) $\hat{i}_{mrx}$ will not be constant, but will depend on the rotor time constant; therefore the rotor time constant can be estimated.

Since as a result of the injected a.c. current $i_{mrxrefa}(t)$, i_{sx} will contain an a.c. component, $i_{sxa}(t)$, when i_{sy} is not zero, it follows from eqn (4.1-43) that the electromagnetic torque will contain the term,

$$t_e = \tfrac{3}{2}P\frac{L_m^2}{L_r}\,i_{sxa}(t)i_{sy}, \tag{4.1-93}$$

and thus it can be seen that torque ripple is caused by $i_{sxa}(t)$. However, to avoid the unwanted torque ripple, the reference value of the quadrature-axis stator current (i_{syref}) is controlled as

$$i_{syref} = c\,\frac{t_{eref}}{\hat{i}_{mrx}}, \tag{4.1-94}$$

where $c = L_r/(PL_m^2)$ is a constant, t_{eref} is the reference value of the electromagnetic torque and $\hat{i}_{mrx}$ is the estimated value of the direct-axis component of the rotor magnetizing-current space phasor, as defined in eqn (4.1-89).

In Fig. 4.20 the output of the flux controller is the reference value of the flux producing stator current (i_{sxref}). This is compared with its actual value (i_{sx}) and the error is fed into a current controller, also a PI controller. The output of this controller is $\hat{u}_{sx}$, similarly to that shown in Fig. 4.11. This is added to the appropriate output of the decoupling circuit, which has the same role as in Fig. 4.11. The decoupling circuit can be described by the following equations, which are obtained from eqns (4.1-60) and (4.1-61)

$$u_{dx} = -\omega_{mr} L'_s i_{sy} \tag{4.1-95}$$

$$u_{dy} = \omega_{mr}[L'_s i_{sx} + (1-\sigma)L_s \hat{\imath}_{mrx}], \tag{4.1-96}$$

where σ is the resultant leakage constant and it can be expressed as $(1-L_m^2/(L_s L_r)$ and $\hat{\imath}_{mrx}$ is defined in eqn (4.1-89). The inputs to the decoupling circuit are i_{sx}, i_{sy}, ω_{mr} and $\hat{\imath}_{mrx}$, and its outputs are $-u_{dx}$ and u_{dy}. When $\hat{u}_{sx}$ is added to u_{dx}, u_{sxref} is obtained, which is the direct-axis stator voltage reference expressed in the rotor-magnetizing-current-oriented reference frame. Similarly, u_{dy} is added to $\hat{u}_{sy}$ and thus u_{syref} is obtained, which is the quadrature-axis stator reference voltage in the rotor-magnetizing-current-oriented reference frame. The derivation of $\hat{u}_{sy}$ is now described.

The reference value of the rotor speed (ω_{rref}) is compared with its monitored value (ω_r) and the error serves as input to the speed controller, which is a PI controller. The output of the speed controller is a signal proportional to the torque reference (t_{eref}) and when, according to eqn (4.1-94) this is divided by $\hat{\imath}_{mrx}$, the quadrature-axis stator reference current i_{syref} is obtained. This is compared with the actual value of the quadrature-axis stator current (i_{sy}) and the error is fed into the quadrature-axis stator current controller, again a PI controller. The output of this is $\hat{u}_{sy}$.

As previously, the voltage references u_{sxref} and u_{syref} are first transformed into the voltage components of the stationary reference frame (u_{sDref}, u_{sQref}) by the application of the transformation $e^{j\rho_r}$, where ρ_r is the space angle of the rotor magnetizing-current space phasor with respect to the real-axis of the stationary reference frame. This angle is related to the speed of the rotor-magnetizing-current-oriented reference frame by $\omega_{mr} = p\rho_r$, where p is the differential operator. Thus if ω_{mr} is known, ρ_r can be obtained by integration, and in Fig. 4.20 this is shown in the block labelled '1/p'. The two-axis voltage references are transformed into their three-phase values (u_{sAref}, u_{sBref}, u_{sCref}) by the application of the two-phase to three-phase transformation, as discussed previously.

The stator currents (i_{sA}, i_{sB}, i_{sC}) are monitored, and first their two-axis values (i_{sD}, i_{sQ}) are obtained by the application of the three-phase to two-phase transformation and these components are then transformed into the direct- and quadrature-axis components i_{sx}, i_{sy} by the application of the inverse transformation $e^{-j\rho_r}$.

By utilizing eqns (4.1-90) and (4.1-91), ω_{mr} is obtained by adding the actual rotor speed (ω_r) and the slip speed ω_{sl}.

There are two more quantities which have to be derived, these are ω_{sl} and $\hat{\imath}_{mrx}$ and they depend on the rotor time constant, which has to be estimated. In the next section the identification of $\hat{T}_r$, and the determination of $\hat{\imath}_{mx}$ and ω_{sl}

will be discussed. It will be shown that they can be obtained by using a circuit, the inputs of which are u_{sxref}, u_{syref}, i_{sx}, i_{sy}, and ω_{mr}, as shown in Fig. 4.20.

(*b*) *Estimation of the rotor time constant, slip speed and direct-axis rotor magnetizing current:* A circuit is described which gives as its outputs the estimated value of the rotor time constant ($\hat{T}_r$), the estimated value of the direct-axis component of the rotor magnetizing current space phasor ($\hat{i}_{mrx}$) and the slip speed ω_{sl}. It is assumed that apart from the rotor resistance, all the other parameters of the machine are constant.

The circuit utilizes the direct- and quadrature-axis components of the so-called modified rotor magnetizing-current space phasor (i'_{mrx}, i'_{mry}), which have been introduced above and which are closely related to the direct- and quadrature-axis rotor magnetizing-current components i_{mrx}, i_{mry}. As noted previously, the reason for the utilization of the modified rotor magnetizing-current components is that they can be obtained more accurately then the 'normal' rotor magnetizing current components. Equation (4.1-66) gives the necessary equations to obtain these components and in Fig. 4.17 a circuit has been shown for the derivation of i'_{mrx} and i'_{mry}, where the inputs to the circuit are u_{sxref}, u_{syref}, i_{sx}, i_{sy} and ω_{mr}; ω_{mr} will be obtained by the technique discussed in the present section. However, since this circuit does not contain the rotor resistance, it can be used, without any modifications in the circuit to be described in the present section.

As noted earlier, the theory of MRAC will be used to obtain the estimated value of the rotor time constant, and the rotor time constant estimator system will be represented by an equivalent non-linear feedback system, which comprises a feedforward time-invariant subsystem and a feedback non-linear time-varying subsystem. The analysis of these subsystems yields the required identification algorithm.

(i) State-variable small-signal equations of the equivalent non-linear feedback system: It follows from eqn (4.1-67) that the rotor magnetizing-current space phasor ($\bar{i}_{mr}$) can be expressed in terms of the stator-current space phasor ($\bar{i}_{s\psi r}$), slip frequency (ω_{sl}) and the actual value of the rotor time constant (T_r) as

$$p\bar{i}_{mr} = \frac{\bar{i}_{s\psi r}}{T_r} - \bar{i}_{mr}\left(\frac{1}{T_r} + j\omega_{sl}\right), \qquad (4.1\text{-}97)$$

where $\omega_{sl} = \omega_{mr} - \omega_r$. When eqn (4.1-97) is linearized around an operating point, where the rotor speed is ω_{r0}, the speed of the reference frame is ω_{mr0} and the corresponding value of the rotor magnetizing-current space phasor is $\bar{i}_{mr0}$, the following small signal equation is obtained:

$$p\Delta\bar{i}_{mr} = \frac{\Delta\bar{i}_{s\psi r}}{T_r} - \Delta i_{mr}\left(\frac{1}{T_r} + j\omega_{sl0}\right) - j\bar{i}_{mr0}\Delta\omega_{sl}, \qquad (4.1\text{-}98)$$

where $\Delta\omega_{sl}=\Delta\omega_{mr}-\Delta\omega_r$ and $\Delta\omega_{mr}$, and $\Delta\omega_r$ are the small-signal values of the speed of the rotor flux-linkage space phasor and rotor speed respectively and $\omega_{sl0}=\omega_{mr0}-\omega_{r0}$. In eqn (4.1-98) T_r can be assumed to be constant, since it is the ratio of the rotor inductance to the rotor resistance and the rotor inductance has been assumed to be constant and the rate of change of the rotor resistance is very slow.

Furthermore, by linearizing eqn (4.1-64) around the operating point described above, the following small-signal equation is obtained for the modified rotor magnetizing-current space phasor, in terms of the small-signal value of the rotor magnetizing-current space phasor ($\Delta\bar{\imath}_{mr}$):

$$\Delta\bar{\imath}'_{mr}=(p+j\omega_{mr0})\Delta\bar{\imath}_{mr}-j\Delta\omega_r\frac{\bar{\imath}'_{mr0}-\bar{\imath}_{mr0}}{p+1/T+j\omega_{mr0}}, \tag{4.1-99}$$

where $\bar{\imath}'_{mr0}$ is the value of the modified rotor magnetizing-current space phasor at the operating point and T is a time constant first introduced in eqn (4.1-64).

The small-signal equations of the corresponding estimated values are obtained from eqns (4.1-98) and (4.1-99) as follows, if the small-signal values of the space phasors $\Delta\bar{\imath}_{mr}$ and $\Delta\bar{\imath}'_{mr}$ are replaced by their estimated values $\Delta\hat{\bar{\imath}}_{mr}$ and $\Delta\hat{\bar{\imath}}'_{mr}$ and if the rotor time constant T_r is substituted by its estimated value $\hat{T}_r$:

$$p\Delta\hat{\bar{\imath}}_{mr}=\frac{\Delta\hat{\bar{\imath}}_{s\psi r}}{\hat{T}_r}-\Delta\hat{\bar{\imath}}_{mr}\left(\frac{1}{\hat{T}_r}+j\omega_{sl0}\right)-j\bar{\imath}_{mr0}\Delta\omega_{sl} \tag{4.1-100}$$

$$\Delta\hat{\bar{\imath}}'_{mr}=(p+j\omega_{mr0})\Delta\hat{\bar{\imath}}_{mr}-j\Delta\omega_r\frac{\bar{\imath}'_{mr0}-\bar{\imath}_{mr0}}{p+1/T+j\omega_{mr0}}. \tag{4.1-101}$$

The space-phasor form of the state error equation describing the equivalent non-linear feedback system is obtained by subtracting eqn (4.1-100) from eqn (4.1-98) and thus

$$p(\Delta\bar{\imath}_{mr}-\Delta\hat{\bar{\imath}}_{mr})=\left(\frac{1}{T_r}-\frac{1}{\hat{T}_r}\right)(\Delta\bar{\imath}_{s\psi r}-\Delta\hat{\bar{\imath}}_{mr})+\left(\frac{1}{T_r}+j\omega_{sl0}\right)(\Delta\bar{\imath}_{mr}-\Delta\hat{\bar{\imath}}_{mr}). \tag{4.1-102}$$

Furthermore, the space-phasor form of the state error equation describing the linear time-variant subsystem of the equivalent non-linear feedback system is obtained by subtracting eqn (4.1-101) from eqn (4.1-99) to yield

$$\Delta\bar{\imath}'_{mr}-\Delta\hat{\bar{\imath}}'_{mr}=\frac{(p+j\omega_{mr0})(\Delta\bar{\imath}_{mr}-\Delta\hat{\bar{\imath}}_{mr})}{p+1/T+j\omega_{mr0}}. \tag{4.1-103}$$

From eqns (4.1-102) and (4.1-103), it follows that the small-signal space-phasor forms of the input and the output of the non-linear time-variant subsystem of the equivalent non-linear feedback system are

$$\Delta\mathbf{v}=\Delta\bar{\imath}'_{mr}-\Delta\hat{\bar{\imath}}'_{mr} \tag{4.1-104}$$

and

$$\Delta\mathbf{w} = -\left(\frac{1}{T_r} - \frac{1}{\hat{T}_r}\right)(\Delta\bar{i}_{s\psi r} - \Delta\hat{\bar{i}}_{mr}) \quad (4.1\text{-}105)$$

respectively.

Resolution of eqns (4.1-102) and (4.1-103) into their real- and imaginary-axis components give the following small-signal form of the two-axis equations if it is considered that $\Delta\bar{i}_{mr} = \Delta i_{mrx} + j\Delta i_{mry}$, $\Delta\bar{i}'_{mr} = \Delta i'_{mrx} + j\Delta i'_{mry}$, and $\Delta\bar{i}_{s\psi r} = \Delta i_{sx} + j\Delta i_{sy}$:

$$p\begin{bmatrix}\Delta i_{mrx} - \Delta\hat{i}_{mrx}\\ \Delta i_{mry} - \Delta\hat{i}_{mry}\end{bmatrix} = \begin{bmatrix}1/T_r & \omega_{sl0}\\ -\omega_{sl0} & -1/T_r\end{bmatrix}\begin{bmatrix}\Delta i_{mrx} - \Delta\hat{i}_{mrx}\\ \Delta i_{mry} - \Delta\hat{i}_{mry}\end{bmatrix} + \left(\frac{1}{T_r} - \frac{1}{\hat{T}_r}\right)\begin{bmatrix}\Delta i_{sx} - \Delta\hat{i}_{mrx}\\ \Delta i_{sy} - \Delta\hat{i}_{mry}\end{bmatrix} \quad (4.1\text{-}106)$$

and

$$\begin{bmatrix}\Delta i'_{mrx} - \Delta\hat{i}'_{mrx}\\ \Delta i'_{mry} - \Delta\hat{i}'_{mry}\end{bmatrix} = \frac{1}{(p+1/T)^2 + \omega_{mr0}^2}\begin{bmatrix}p+1/T & \omega_{mr0}\\ \omega_{mr0} & p+1/T\end{bmatrix} \begin{bmatrix}p & -\omega_{mr0}\\ \omega_{mr0} & p\end{bmatrix}\begin{bmatrix}\Delta i_{mrx} - \Delta\hat{i}_{mrx}\\ \Delta i_{mry} - \Delta\hat{i}_{mry}\end{bmatrix}, \quad (4.1\text{-}107)$$

where i'_{mrx}, i'_{mry} are the two-axis components of the modified rotor magnetizing current expressed in the reference frame fixed to the rotor flux-linkage space phasor, i_{mrx} and i_{mry} are the two-axis components of the rotor magnetizing current in the same reference frame and the symbol ^ denotes estimated values. Furthermore, it follows from eqns (4.1-104) and (4.1-105) that the two-axis forms of the input and output of the non-linear time-variant subsystem of the equivalent non-linear feedback system are

$$\Delta\mathbf{v} = \begin{bmatrix}\Delta i'_{mrx} - \Delta\hat{i}'_{mrx}\\ \Delta i'_{mry} - \Delta\hat{i}'_{mry}\end{bmatrix} \quad (4.1\text{-}108)$$

and

$$\Delta\mathbf{w} = -\left(\frac{1}{T_r} - \frac{1}{\hat{T}_r}\right)\begin{bmatrix}(\Delta i_{sx} - \Delta\hat{i}_{mrx})\\ (\Delta i_{sy} - \Delta\hat{i}_{mry})\end{bmatrix} \quad (4.1\text{-}109)$$

respectively. It has been noted in connection with eqn (4.1-81) that when the output error method is used in a Parallel Model Reference Adaptive Control system, the adaptation mechanism must contain at least one integrator to ensure that the error converges asymptotically to zero. Thus

$$\frac{1}{\hat{T}_r} = \int_0^t \psi \, dt + \frac{1}{\hat{T}_r(0)}, \quad (4.1\text{-}110)$$

where $\hat{T}_r(0)$ is the initial value of the estimated rotor time constant and ψ is in general a non-linear function of the errors $\Delta e_x = \Delta i'_{mrx} - \Delta\hat{i}'_{mrx}$ and $\Delta e_y =$

$\Delta i'_{mry}-\Delta\hat{i}'_{mry}$. The non-linear function ψ will be determined below and will eventually be used to obtain the algorithm for the rotor time constant identification.

(ii) Rotor time constant identification algorithm: As discussed earlier, the identification algorithm of the rotor time constant is selected in such a way that the equivalent non-linear feedback system should satisfy Popov's hyperstability theorem. Thus two conditions must be satisfied. According to the first condition, the transfer-function matrix of the linear time-invariant subsystem must be strictly positive real. It can be shown by considering eqns (4.1-98) and (4.1-99) that this condition can be satisfied if the time constant T is chosen to be a positive constant. According to the second condition, the non-linear time-varying block must satisfy Popov's integral inequality, which has been given by eqn (4.1-77), but of course now $\mathbf{v}$ and $\mathbf{w}$ correspond to $\Delta\mathbf{v}$ and $\Delta\mathbf{w}$ defined by eqns (4.1-108) and (4.1-109) respectively. Thus it follows from eqns (4.1-77), (4.1-108), (4.1-109), and (4.1-110) that the following equation, which contains the non-linear function ϕ, must hold:

$$\int_0^t \Bigg\{[(\Delta i'_{mrx}-\Delta\hat{i}'_{mrx})(\Delta i_{sx}-\Delta\hat{i}_{mrx})+(\Delta i'_{mry}-\Delta\hat{i}'_{mry})(\Delta i_{sy}-\Delta\hat{i}_{mry})]$$

$$\left[\int_0^t \psi\,\mathrm{d}t+\left(\frac{1}{\hat{T}_r(0)}-\frac{1}{T_r}\right)\right]\Bigg\}\,\mathrm{d}t\geq 0. \qquad (4.1\text{-}111)$$

By utilizing the inequality given by eqn (4.1-83) for a general function $f(t)$ in eqn (4.1-111),

$$[pf(t)]=(\Delta i'_{mrx}-\Delta\hat{i}'_{mrx})(\Delta i_{sx}-\Delta\hat{i}_{mrx})+(\Delta i'_{mry}-\Delta\hat{i}'_{mry})(\Delta i_{sy}-\Delta\hat{i}_{mry}) \qquad (4.1\text{-}112)$$

and

$$kf(t)=\int_0^t \psi\,\mathrm{d}t+\frac{1}{\hat{T}_r(0)}-\frac{1}{T_r}, \qquad (4.1\text{-}113)$$

where $k>0$. With eqn (4.1-112), differentiation of eqn (4.1-113) with respect to time gives

$$\psi=k[(\Delta i'_{mrx}-\Delta\hat{i}'_{mrx})(\Delta i_{sx}-\Delta\hat{i}_{mrx})+(\Delta i'_{mry}-\Delta\hat{i}'_{mry})(\Delta i_{sy}-\Delta\hat{i}_{mry})]; \qquad (4.1\text{-}114)$$

where k is a positive constant.

Substitution of eqn (4.1-114) into eqn (4.1-113) yields the following equation if all the small signal quantities are replaced by the corresponding large signal quantities:

$$\frac{1}{\hat{T}_r}-\frac{1}{\hat{T}_r(0)}=k\int_0^t [(i'_{mrx}-\hat{i}'_{mrx})(i_{sx}-\hat{i}_{mrx})$$

$$+(i'_{mry}-\hat{i}'_{mry})(i_{sy}-\hat{i}_{mry})]\,\mathrm{d}t. \qquad (4.1\text{-}115)$$

$\hat{T}_r$ is the estimated value of the rotor time constant, $\hat{T}_r(0)$ is the initial value of the rotor time constant, i_{sx}, i_{sy}, i_{mrx}, and i_{mry} are the direct- (x-axis) and quadrature-axis (y-axis) components of the stator currents and the rotor magnetizing currents respectively, but expressed in the reference frame fixed to the rotor flux-linkage space phasor and the symbol ^ denotes an estimated value, i'_{mrx} and i'_{mry} are the direct- and imaginary-axis components of the modified rotor magnetizing-current space phasor expressed in the same special reference frame. When the identification of the rotor time constant is performed according to eqn (4.1-115), the length of time of the identification will vary according to the value chosen for k and, since the modified rotor magnetizing currents are obtained from the rotor magnetizing currents [see eqn (4.1-64)], it will also depend on the time constant T. When the gain of the integrator (k) is increased, the identification time decreases, but for large values of k the identification system becomes more sensitive to noise. To ensure that the results obtained from the small-signal model are valid for the actual model, the gain of the integrator must be sufficiently high. It follows that initially it is preferable to use a large value of k and to replace this after the initial identification. Since the slip frequency is controlled according to eqn (4.1-91), the estimated value of the quadrature-axis rotor magnetizing current ($\hat{i}_{mry}$) becomes zero. Equation (4.1-115) can be implemented by the circuit shown in Fig. 4.21.

In the circuit shown in Fig. 4.21 the inputs are the estimated values of the

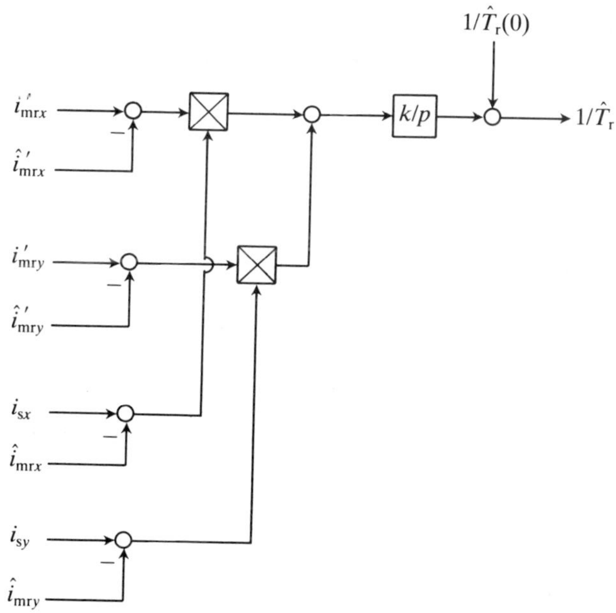

Fig. 4.21. Circuit to obtain the estimated rotor time constant.

modified rotor magnetizing-current components ($\hat{\imath}'_{mrx}$, $\hat{\imath}'_{mry}$), the actual values of the rotor magnetizing-current components (i'_{mrx}, i'_{mry}), the estimated values of the rotor magnetizing currents ($\hat{\imath}_{mrx}$, $\hat{\imath}_{mry}$) and the actual values of the stator currents i_{sx}, i_{sy}. Integration is performed in the block labelled 'k/p'. The output of the circuit is $1/\hat{T}_r$ where $\hat{T}_r$ is the estimated value of the rotor time constant. Since it has been assumed that the rotor inductance (L_r) is constant, the inverse of the estimated rotor time constant is proportional to the estimated rotor resistance ($\hat{R}_r$) where the proportionality factor (c) is equal to the reciprocal value of the rotor inductance.

The circuit shown in Fig. 4.21 is utilized in the block shown in Fig. 4.22, which yields as its outputs $\hat{T}_r$, $\hat{\imath}_{mrx}$, and ω_{sl}. Further details of this block are described below, where it is shown how the input signals to this block can be obtained.

(iii) Circuit to obtain the estimated values of the rotor time constant, the direct-axis rotor magnetizing current and the slip frequency: In the present section the implementation of the block shown in Fig. 4.20 is described, which yields as its outputs the estimated values of the rotor time constant ($\hat{T}_r$) and the direct-axis rotor magnetizing current ($\hat{\imath}_{mrx}$), together with the slip speed (ω_{sl}). This is shown in Fig. 4.22. It can be seen that the circuit shown in Fig. 4.22 requires input signals which can be obtained from four circuits.

In Fig. 4.22 the inputs to the first circuit are the reference stator voltages u_{sxref}, u_{syref}, the direct- and quadrature-axis stator currents i_{sx}, i_{sy} and the speed of the rotor flux-linkage space phasor (ω_{mr}). These are obtained as described earlier. The outputs of this circuit are the modified rotor magnetizing current components i'_{mrx} and i'_{mry}. Such a circuit has been described in Fig. 4.17.

The inputs to the second circuit shown in Fig. 4.22 are the stator current

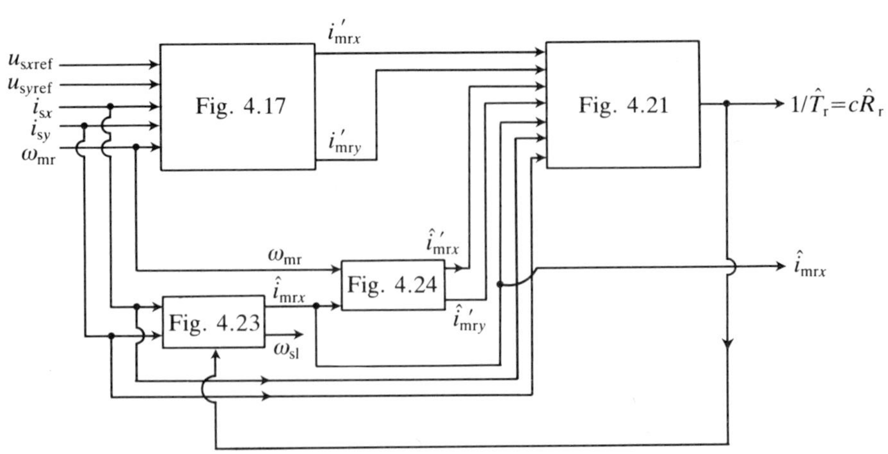

Fig. 4.22. Circuit to obtain $\hat{T}_r$, $\hat{\imath}_{mrx}$, and ω_{sl}.

components i_{sx} and i_{sy}, together with the estimated reciprocal value of the rotor time constant $(1/\hat{T}_r)$ and the outputs of this circuit are the estimated value of the direct-axis rotor magnetizing current $(\hat{i}_{mrx})$ and the slip frequency (ω_{sl}). This block can be implemented by the circuit shown in Fig. 4.23. This circuit is based on eqns (4.1-89) and (4.1-91) and the signal for $1/\hat{T}_r$ is obtained from the output of the circuit shown in Fig. 4.21.

The inputs of the third circuit shown in Fig. 4.22 are the signals ω_{mr} (which is also an input to the circuit shown in Fig. 4.17) and the estimated value of the direct-axis rotor magnetizing current $(\hat{i}_{mrx})$ and its outputs are the estimated values of the direct- and quadrature-axis components of the modified rotor magnetizing currents ($\hat{i}'_{mrx}$ and $\hat{i}'_{mry}$). The circuit to obtain these two currents is shown in Fig. 4.24. This circuit can be obtained by considering eqn (4.1-64), but by replacing the space phasors of the rotor magnetizing current and the modified rotor magnetizing current ($\bar{i}_{mr}$ and $\bar{i}'_{mr}$) by their estimated values $\hat{\bar{i}}_{mr}$ and $\hat{\bar{i}}'_{mr}$ respectively and by resolving the space-phasor quantities into their direct- and quadrature-axis components. Thus it follows that

$$p\hat{i}_{mrx}-\omega_{mr}\hat{i}_{mry}=\left(p+\frac{1}{T}\right)\hat{i}'_{mrx}-\omega_{mr}\hat{i}'_{mry} \tag{4.1-116}$$

and

$$\omega_{mr}\hat{i}_{mrx}+p\hat{i}_{mry}=\omega_{mr}\hat{i}'_{mrx}+\left(p+\frac{1}{T}\right)\hat{i}'_{mry}. \tag{4.1-117}$$

However, when the control is performed according to eqns (4.1-89) and (4.1-91), the quadrature-axis component of the estimated rotor magnetizing current $(\hat{i}_{mry})$ is zero. Thus eqns (4.1-116) and (4.1-117) can be realized by the circuit shown in Fig. 4.24.

Finally it should be noted that the circuit shown in Fig. 4.22 contains a fourth circuit, which has been shown in Fig. 4.21.

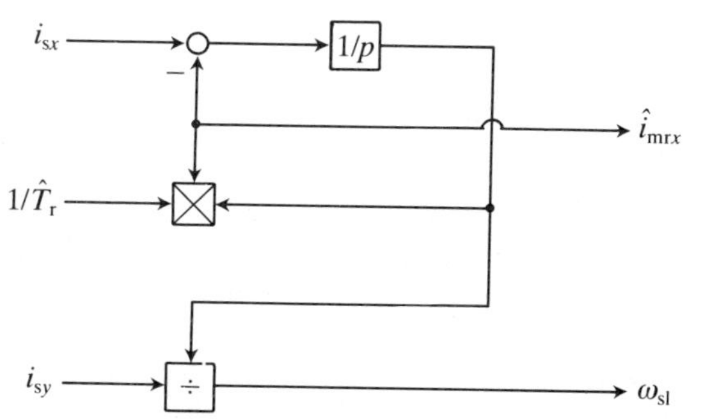

Fig. 4.23. Circuit to obtain $\hat{i}_{mrx}$ and ω_{sl}.

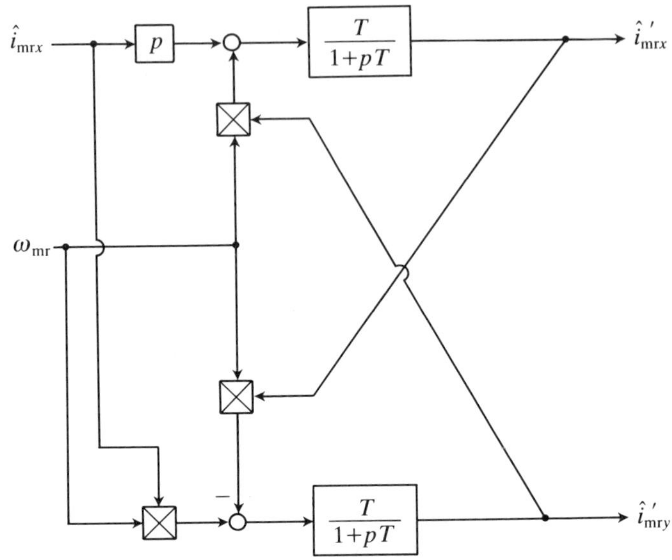

Fig. 4.24. Circuit to obtain $\hat{i}'_{mrx}$ and $\hat{i}'_{mry}$.

4.1.2 Control of an induction machine with impressed currents

In contrast to inverters which function as voltage sources, inverters which function as current sources are becoming the main power sources for high-performance a.c. machine drives. In the lower-power region PWM inverters with fast current control are employed. In this case, because of the chopping mode of the inverter and the high gain of the current loops, the a.c. machine follows the current reference signals very quickly. At higher power levels the Current Source Inverter (CSI) is used where a supply-side controlled rectifier is connected through a large d.c. link reactor to the motor-side inverter. Its major advantages are:

- its simplicity;
- its inherent ability for regeneration (by reversing the d.c. link voltage) and reversal (by electronic reversal of the phase sequence of the machine currents, which is carried out by changing the sequence of operation of the inverter switches);
- it does not require silicon-controlled rectifiers with high switching speeds—it can use converter-grade thyristors. (As shown later, this follows because commutation causes voltage spikes superimposed on the nearly sinusoidal stator voltages and these can be reduced by the application of large commutating capacitors or by reducing the transient inductance of the stator. Thus commutation is relatively slow and therefore it is not necessary to use thyristors with high switching speed);

- the inverter will recover from a short-circuit across any two of its output terminals and is undamaged by the misfire of the output thyristors (because there is a large d.c. link reactor which prevents quick changes of the link current).

However, the current response of the conventional CSI with auto-sequential commutation is much slower than that of the current-controlled PWM voltage inverter. Furthermore, at low speed there can exist unwanted torque pulsations (this will be discussed later).

By the application of the conventional CSI, rapid control of both the phase and amplitude of an a.c. current cannot be achieved. In this case, the amplitude of the a.c. currents is determined by the magnitude of the d.c. link current, and as a result of the large filter inductance, the current response to an input command is greatly influenced by the d.c. link parameters (rectifier phase-control delay time, rated value of the rectifier output voltage, gain of the d.c. link current loop, inductance of the d.c. link). Although the response can be minimized by establishing high gain in the current-regulator loop of the controlled rectifier, the output of which is the d.c. link current, the large gain can result in control problems because of the saturation of the output voltage of the controlled rectifier and ripple current instabilities at light load. Thus the current response of the conventional CSI is slower than for the current-controlled PWM inverter. It should be noted that when a so-called notched autosequentially commutated inverter is employed, where the d.c. link current is kept constant and thus the influence of the d.c. link parameters on the dynamics of the system response can be neglected, rapid control of both the phase and amplitude of the a.c. current can be achieved.

In this section the rotor-flux-oriented control of induction machines will be discussed for four cases. The induction machine can be supplied by: (i) a voltage-source PWM inverter with fast current control; (ii) a conventional current-source inverter with auto-sequential commutation (CSI); (iii) a cycloconverter with fast current control; or (iv) impressed rotor currents. The main assumptions are those used in Section 4.1.1, but unless stated otherwise, the time lag of the inverter is neglected although it could be considered by using the technique described in Section 3.1.1.

When a high dynamic performance induction machine drive with impressed currents is designed, ideally the following requirements should be satisfied:

- Smooth speed response without cogging or torque pulsations at low speed (it will be shown that sometimes it is difficult to satisfy this requirement);

- Smooth speed reversals under any torque condition;

- Capability of four-quadrant operation (this can be achieved by changing the phase sequence of the stator currents and the polarity of the d.c. link voltage);

- Operation of the drive with constant full torque below base speed and above base speed with reduced flux (field-weakening operation).

There are many control schemes known for induction machine drives with impressed currents, but rotor-flux-oriented control has emerged as one of the most frequently used techniques. The application of this technique yields fast dynamic response and most of the requirements mentioned above can also be satisfied. Similarly to that discussed in Section 4.1.1, in order to obtain high dynamic performance, the stator currents of the machine are transformed into flux- and torque-producing current components (i_{sx}, i_{sy}). These are defined in eqn (4.1-3), and Fig. 2.16 shows the relationship between various quantities in the reference frames fixed to the stator and rotor flux-linkage space phasors. Below base speed the modulus of the rotor magnetizing current ($|\bar{i}_{mr}|$), defined by eqn (4.1-1), is maintained at its maximum possible value but is limited by magnetic saturation. Above base speed, $|\bar{i}_{mr}|$ is reduced (field-weakening operation). The electromagnetic torque is controlled by the quadrature-axis stator current i_{sy}, (see, for example, eqn (2.1-197) and Section 2.1.8].

Control of the induction machine supplied by a current-controlled PWM inverter

In this section the rotor-flux-oriented control of an induction machine supplied by a voltage-source inverter with fast current control will be described. The inverter can be a transistorized inverter with high switching frequency, such as the one shown in Fig. 3.5b. Two types of implementation will be given, which use the direct method and the indirect method. As discussed at the beginning of this chapter, when the direct method is used (flux-feedback control), the space angle of the rotor flux-linkage space phasor is obtained by direct measurements (e.g. by using Hall sensors) or by using a so-called flux model. However, when the indirect method is used (feedforward control), the space angle of the rotor flux-linkage space phasor is obtained as the sum of the monitored rotor angle (θ_r) and the computed reference value of the slip angle (θ_{sl}), where the slip angle gives the position of the rotor flux-linkage space phasor relative to the rotor (or more precisely relative to the direct axis of the reference frame fixed to the rotor).

Implementation which utilizes the direct method: Figure 4.25 shows the schematic of the direct implementation of the rotor-flux-oriented control of an induction machine supplied by a current-controlled PWM inverter. This is simpler than the implementation shown in Fig. 4.12 for a voltage-controlled PWM inverter, since now the stator currents are impressed by fast current control loops and thus the scheme does not utilize the stator voltage equations and there is no decoupling circuit in the implementation of Fig. 4.25.

In Fig. 4.25 the same cascade control structure is utilized as in Fig. 4.12. The monitored value of the rotor speed (ω_r) is integrated (the integration is denoted by $1/p$) to give the actual value of the rotor angle (θ_r). This is

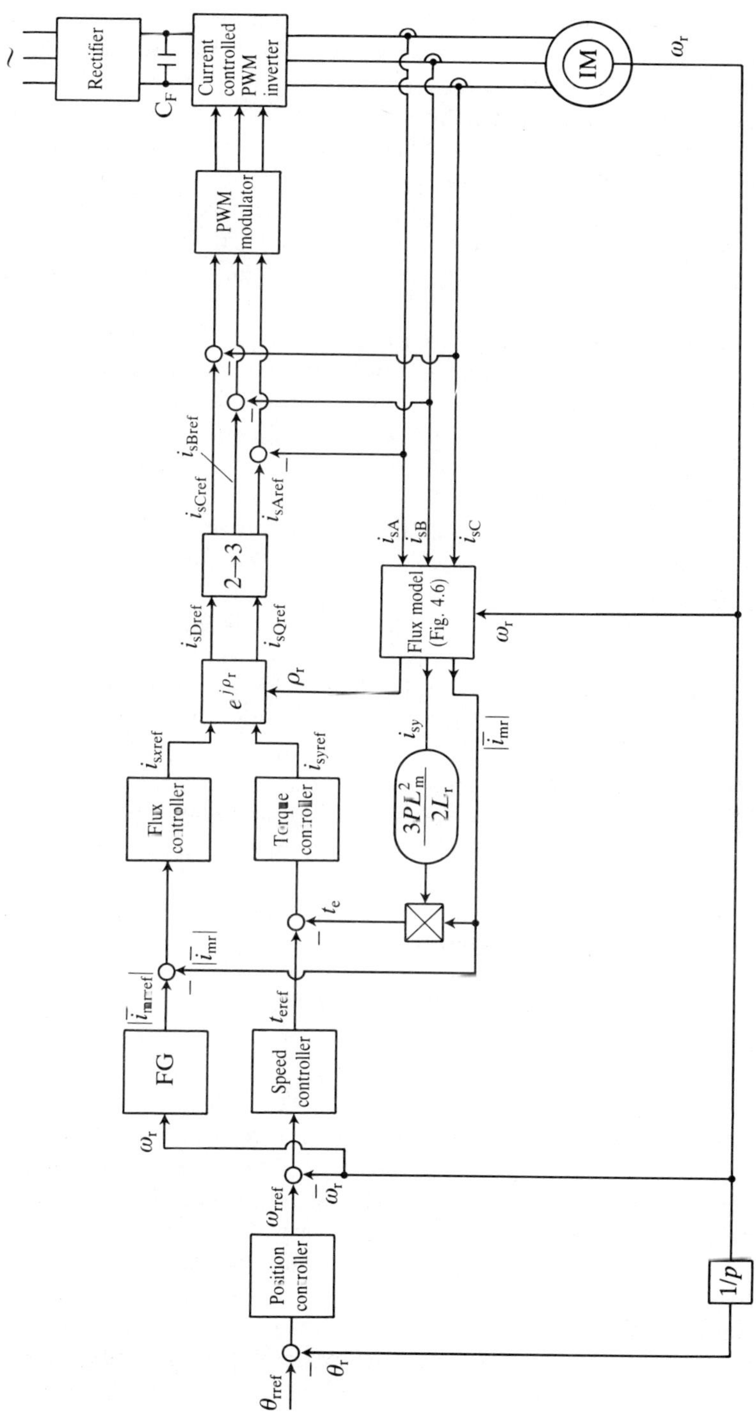

Fig. 4.25. Schematic of the direct implementation of the rotor-flux-oriented control of an induction machine supplied by a current-controlled PWM inverter.

compared with its reference value (θ_{rref}) and the resulting error serves as the input to the position controller, which is a PI controller. The output of this is the reference value of the rotor speed (ω_{rref}). When this is compared with the monitored value of the rotor speed, the error signal is supplied to the input of the speed controller, also a PI controller, and the output of which is the reference value of the electromagnetic torque (t_{eref}). Comparison of t_{eref} and the actual value of the torque t_e gives an error which serves as input to the torque controller, again a PI controller, and its output is the reference value of the quadrature-axis stator current expressed in the rotor-flux-oriented reference frame (i_{syref}).

The direct-axis stator current reference (i_{sxref}), which is expressed in the rotor-flux-oriented reference frame, is obtained as the output of the flux controller (PI controller) the input of which is the difference between the reference value of the rotor magnetizing current $|\bar{\imath}_{\text{mrref}}|$ and the actual value of the rotor magnetizing current $|\bar{\imath}_{\text{mr}}|$. The reference current $|\bar{\imath}_{\text{mrref}}|$ is obtained as the output of the function generator FG, which allows field weakening to be implemented. Thus the input of FG is the monitored rotor speed and below base speed FG gives a constant value of $|\bar{\imath}_{\text{mrref}}|$, while above base speed $|\bar{\imath}_{\text{mrref}}|$ is inversely proportional to the rotor speed.

In accordance with eqn (4.1-3), the stator current references i_{sxref} and i_{syref} are first transformed into the two-axis stator current references of the stationary reference frame (i_{sDref}, i_{sQref}) by the application of the transformation $e^{j\rho_r}$ where ρ_r is the angle of the rotor magnetizing-current space phasor with respect to the direct-axis (sD) of the stationary reference frame. For convenience Fig. 4.26 shows the relationship of the space phasor of the stator currents in the stationary and in the special, rotor-flux-oriented reference frame. The space phasor of the rotor magnetizing current is also shown. Figure 4.26 is similar to Fig. 2.16.

Thus by resolving eqn (4.1-3) into its real and imaginary axes components, but using reference values, it follows that

$$i_{\text{sDref}} = \cos\rho_r i_{\text{sxref}} - \sin\rho_r i_{\text{syref}} \tag{4.1-118}$$

$$i_{\text{sQref}} = \sin\rho_r i_{\text{sxref}} + \cos\rho_r i_{\text{syref}}. \tag{4.1-119}$$

These two-axis current references are then transformed into their three-phase reference values (i_{sAref}, i_{sBref}, i_{sCref}) by the application of the two-phase to three-phase transformation indicated by the block labelled '2→3'. These are used, together with the monitored three-phase currents (i_{sA}, i_{sB}, i_{sC}), to obtain the gate signals necessary for the inverter which supplies the induction machine.

As in eqns (4.1-49), (4.1-50), and (4.1-51) in the case of voltages, it is possible to obtain the three-phase reference currents by simply considering that in the absence of zero-sequence currents, the projections of the current space phasor

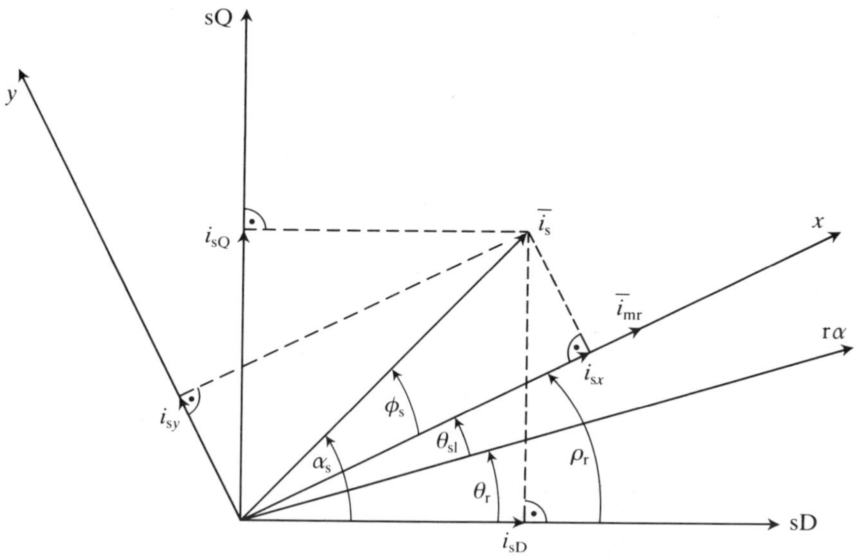

Fig. 4.26. Stator-current and rotor-current space phasors in the stationary and other reference frames.

on the corresponding axes yield the instantaneous values of the phase currents; thus

$$i_{sAref} = \text{Re}(\bar{i}_{sref}) = \text{Re}(i_{sDref} + ji_{sQref}) = i_{sDref}, \tag{4.1-120}$$

where $\bar{i}_{sref}$ is the reference value of the stator-current space phasor in the stationary reference frame,

$$i_{sBref} = \text{Re}(a^2\bar{i}_{sref}) = \frac{-i_{sDref}}{2} + \sqrt{3}\,\frac{i_{sQref}}{2} \tag{4.1-121}$$

and

$$i_{sCref} = \text{Re}(a\bar{i}_{sref}) = \frac{-i_{sDref}}{2} - \sqrt{3}\,\frac{i_{sQref}}{2} \tag{4.1-122}$$

Of course it is also possible to utilize the Euler forms of the space-phasor equations. Thus the space phasor of the stator current references in the rotor-flux-oriented reference frame can be expressed as

$$\bar{i}'_{sref} = i_{sxref} + ji_{syref} = (i^2_{sxref} + i^2_{syref})^{1/2} e^{j\phi_s}, \tag{4.1-123}$$

where

$$\phi_s = \tan^{-1} \frac{i_{syref}}{i_{sxref}} \tag{4.1-124}$$

and ϕ_s is the space angle of the stator-current space phasor with respect to the real axis of the rotor-flux-oriented reference frame, as shown in Fig. 4.26.

From eqn (4.1-3) or Fig. 4.26 in the stationary reference frame the space phasor of the stator reference currents is obtained as

$$\bar{\imath}_{sref} = \bar{\imath}'_{sref}\, e^{j\rho_r} = (i^2_{sxref} + i^2_{syref})^{1/2} e^{j(\rho_r + \phi_s)}, \tag{4.1-125}$$

where $(\rho_r + \phi_s) = \alpha_s$ and α_s is the space angle of the stator-current space phasor shown in Fig. 4.26. In eqn (4.1-125) $\bar{\imath}'_{sref}$ has been substituted by the expression given in eqn (4.1-123). Thus from eqns (4.1-120), (4.1-121), (4.1-122), and (4.1-125), the three-phase current references can be put into the form,

$$i_{sAref} = \mathrm{Re}(\bar{\imath}_{sref}) = (i^2_{sxref} + i^2_{syref})^{1/2} e^{j(\rho_r + \phi_s)} \tag{4.1-126}$$

$$i_{sBref} = \mathrm{Re}(a^2 \bar{\imath}_{sref}) = (i^2_{sxref} + i^2_{syref})^{1/2} e^{j(\rho_r + \phi_s - 2\pi/3)} \tag{4.1-127}$$

$$i_{sCref} = \mathrm{Re}(a \bar{\imath}_{sref}) = (i^2_{sxref} + i^2_{syref})^{1/2} e^{j(\rho_r + \phi_s + 2\pi/3)}. \tag{4.1-128}$$

Equations (4.1-126), (4.1-127) and (4.1-128) could be used directly to obtain the three-phase reference currents.

In Fig. 4.25 the monitored stator currents, together with the monitored rotor speed, are inputs to the flux model which has been described in Fig. 4.6. This contains the rotor time constant T_r. The outputs of the flux model are the rotor magnetizing current ($|\bar{\imath}_{mr}|$), the torque-producing stator current (i_{sy}) and the spatial position of the rotor flux-linkage space phasor (ρ_r). In accordance with eqn (4.1-43), the electromagnetic torque (t_e) is obtained by multiplying i_{sy} by the constant $3PL^2_m/(2L_r)$, where P is the number of pole pairs, L_m is the magnetizing inductance of the machine and L_r is the self-inductance of the rotor.

The behaviour of the induction machine subjected to rotor-flux-oriented control shown in Fig. 4.25 is similar to that of the separately excited d.c. machine. The stator frequency is equal to the first time derivative of ρ_r. It should be noted that since the flux model uses the rotor-time constant, the accuracy of the output signals of the flux model depends on the rotor time constant. Accurate values of T_r can be obtained, for example, by using Model Reference Adaptive Control to establish an on-line rotor time constant estimator.

Implementation which utilizes the indirect method: The implementation of indirect rotor-flux-oriented control of an induction machine supplied by a current-controlled PWM inverter is similar to that of the direct method, except that the space angle of the rotor magnetizing-current space phasor (ρ_r), is obtained as the sum of the rotor angle (θ_r) and the reference value of the slip angle (θ_{sl}). These angles are shown in Fig. 4.26 and the required equations follow from eqn (4.1-59), according to which the speed of the rotor magnetizing current space phasor is

$$\omega_{mr} = \omega_r + \omega_{slref} \tag{4.1-129}$$

where ω_r is the rotor speed,

$$\omega_r = \frac{d\theta_r}{dt}, \tag{4.1-130}$$

and ω_{slref} is the reference value of the slip frequency,

$$\omega_{slref} = \frac{i_{syref}}{T_r i_{sxref}} \tag{4.1-131}$$

where i_{sxref} and i_{syref} are the reference values of the direct- and quadrature-axis stator currents in the rotor-flux-oriented reference frame. Furthermore

$$\omega_{mr} = \frac{d\rho_r}{dt}. \tag{4.1-132}$$

Thus it follows from eqns (4.1-129)–(4.1-132) that

$$\rho_r = \int \omega_{mr}\, dt = \int (\omega_r + \omega_{slref})\, dt = \theta_r + \theta_{slref}. \tag{4.1-133}$$

As shown in Fig. 4.26, the slip angle θ_{slref} gives the position of the rotor magnetizing-current space phasor with respect to the direct axis ($r\alpha$-axis) of the reference frame fixed to the rotor. Thus by considering eqns (4.1-130) and (4.1-131)

$$\rho_r = \theta_r + \int \omega_{slref}\, dt = \theta_r + \int \frac{i_{syref}}{T_r i_{sxref}}\, dt, \tag{4.1-134}$$

which serves as the basis for obtaining the angle ρ_r when the indirect method is used in the implementation of the rotor-flux-oriented control of the induction machine. According to eqn (4.1-134) the division of the direct- and quadrature-axis stator currents (i_{sxref}, i_{syref}) is controlled by the slip frequency ω_{sl} and the two reference currents are used to determine the required slip frequency. When the rotor angle and the reference value of the slip frequency angle are added, the position of the rotor magnetizing-current space phasor is obtained. To obtain accurate values of ρ_r, the addition of the two angles must be performed very accurately; this can be done digitally, which also has the advantage of avoiding drift problems associated with analog implementation.

The schematic of the drive is shown in Fig. 4.27. In Fig. 4.27(a) the implementation uses Cartesian coordinates and in Fig. 4.27(b) polar coordinates are used. However, for simplicity, in Fig.4.27(b) only that part of the

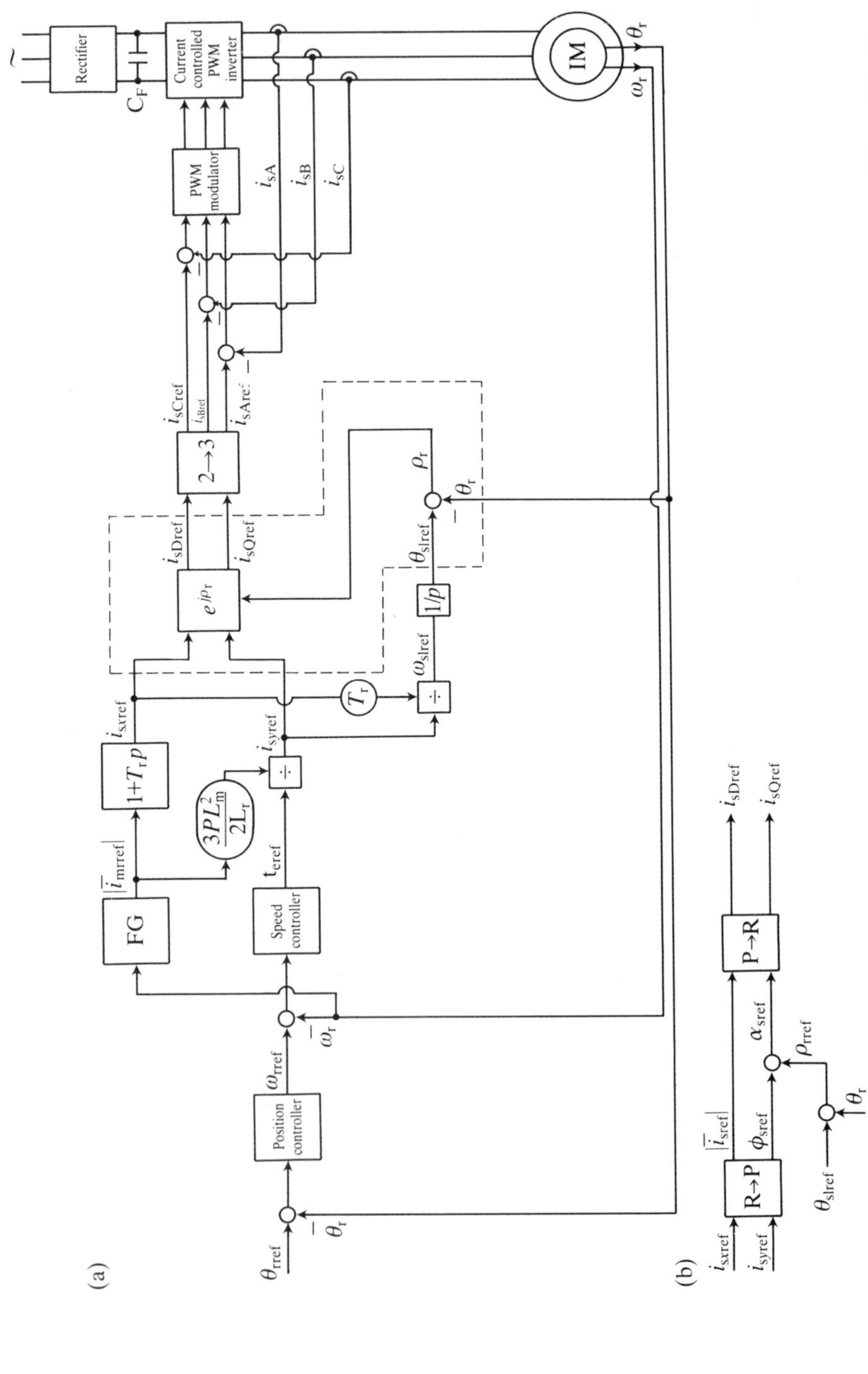

Fig. 4.27. Schematic of the indirect rotor-flux-oriented control of a current-controlled PWM inverter-fed induction machine. (a) In Cartesian coordinates. (b) In polar coordinates.

scheme is shown which corresponds to the part shown in Fig. 4.27(a) within the dashed line.

In Fig. 4.27 an incremental rotor position sensor is used to obtain the rotor angle (θ_r) and the rotor speed (ω_r). The actual value of the rotor angle (θ_r) is compared with its reference value (θ_{rref}) and the resulting error serves as the input to the position controller, which is a PI controller. The output of the position controller is the reference value of the rotor speed (ω_{rref}). This is compared with the monitored value of the rotor speed and the error is supplied as the input to the speed controller, which is also a PI controller. Its output is the reference value of the electromagnetic torque (t_{eref}) and in accordance with eqn (4.1-43) this is divided by the constant $3PL_m^2/(2L_r)$ to yield the reference value of the stator current expressed in the rotor-flux-oriented reference frame (i_{syref}).

To enable field weakening to be performed, the monitored rotor speed serves as the input to the function generator FG shown in Fig. 4.27, the output of which is the reference value of the rotor magnetizing current ($|\bar{i}_{mrref}|$. Below base speed FG gives a constant value of $|\bar{i}_{mrref}|$, while above base speed $|\bar{i}_{mrref}|$, is inversely proportional to the rotor speed.

It follows from eqn (4.1-26), that, when the actual currents are replaced by their reference values,

$$i_{sxref} = (1 + T_r p)|\bar{i}_{mrref}| \tag{4.1-135}$$

and eqn (4.1-135) is used in the implementation shown in Fig. 4.27 to derive the reference value of the direct-axis stator current component expressed in the rotor-flux-oriented reference frame (i_{sxref}).

As described in the previous section, the stator current references i_{sxref}, i_{syref} are first transformed into the two-axis stator current references of the stationary reference frame (i_{sDref}, i_{sQref}) by the application of the transformation $e^{j\rho_r}$, where ρ_r is the angle of the rotor magnetizing current space phasor with respect to the direct-axis (sD) of the stationary reference frame. These two-axis current references are then transformed into their three-phase reference values (i_{sAref}, i_{sBref}, i_{sCref}) by the application of the two-phase to three-phase transformation indicated by the block labelled '2→3'. The corresponding equations have been given in the previous section. The reference stator currents together with the monitored three-phase stator currents (i_{sA}, i_{sB}, i_{sC}) are used to obtain the gate signals necessary for the inverter which supplies the induction machine.

Finally eqn (4.1-134) is utilized to obtain the space angle ρ_r from i_{sxref}, i_{syref} and the monitored value of the rotor angle. The block labelled '$1/p$' performs the required integration of the slip frequency.

In Fig. 4.27(b) the reference currents i_{sxref} and i_{syref} are converted by a rectangular-to-polar converter into the modulus of the stator-current space phasor ($|\bar{i}_{sref}|$) and also into the phase-angle of the stator-current space phasor with respect to the real axis of the rotor magnetizing-current reference frame

(ϕ_{sref}), which according to Fig. 4.26 is equal to $\alpha_{\text{sref}} - \rho_{\text{rref}}$. Thus when ρ_{rref} is added to this angle, α_{sref} is obtained, which according to Fig. 4.26 is the phase-angle of the stator-current space phasor with respect to the real axis of the stationary reference frame. Thus $|\bar{i}_{\text{sref}}|$ and α_{sref} are inputs to a polar-to-rectangular converter, the outputs of which are i_{sDref} and i_{sQref}. In accordance with Fig. 4.26, the angle ρ_{rref} is obtained by adding the rotor angle to the slip frequency angle and the latter is obtained by the integration of the angular slip frequency. The slip angle is obtained in the same way as in Fig. 4.27(a).

The indirect implementation given in Fig. 4.27 is similar to the implementation used in a CSI-fed induction machine drive with the conventional slip frequency control. In that case the inverter frequency command is obtained as the sum of the slip frequency command and the rotational frequency of the motor, and the slip frequency command is obtained from the measured d.c. link current by utilizing a function generator. This function generator ensures that below base speed the magnetizing flux of the machine should be constant. In contrast to maintaining constant magnetizing flux, in rotor-flux-oriented control the rotor flux is kept constant. Furthermore, since the traditional slip frequency control method of the CSI-fed induction machine is based on steady-state equations, during transient operation (i.e. when the machine is accelerated or decelerated) the angle between the stator-current space phasor and the rotor magnetizing-current space phasor is not controlled correctly. In contrast to this, when rotor-flux-oriented control is performed, the correct relationship is achieved between these two space phasors even under transient conditions.

Control of the induction machine supplied by a conventional Current-Source Inverter (CSI)

In this section the rotor-flux-oriented control of an induction machine supplied by a conventional CSI with auto-sequential commutation is described. Again two implementations are discussed which correspond to the direct and indirect methods of obtaining the space angle of the rotor flux-linkage space phasor.

Figure 4.28 shows the schematic of a conventional six-step CSI with auto-sequential commutation. In Fig. 4.28, a.c. power is first converted to d.c., which is then inverted by a square-wave six-step inverter to produce variable-frequency a.c. currents. Thus the supply-side mains-commutated thyristor converter provides, through a high-inductance filter, the smooth d.c. link current (i_{D}). This is supplied to the machine-side inverter. To maintain a constant direct link current specified by the reference value i_{Dref}, there is a current control loop, where i_{Dref} is compared with i_{D} and the difference serves as input to the current controller (which is a PI controller), whose output is used to control the firing angle of the supply-side converter. The phase-controlled rectifier can be replaced by an uncontrolled (diode) rectifier followed by a d.c. chopper to produce the required variable d.c. voltage

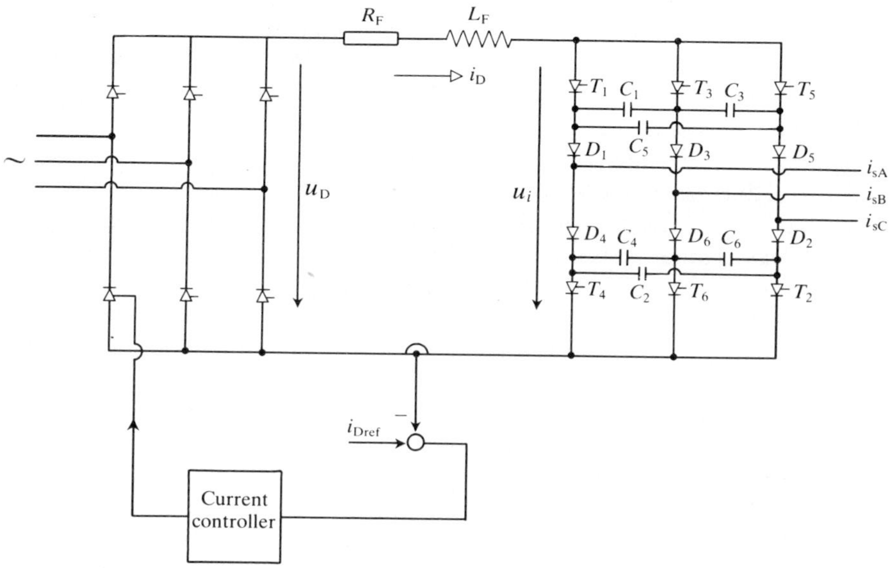

Fig. 4.28. Schematic of conventional six-step CSI with auto-sequential commutation.

source. Furthermore, it should also be noted that when the CSI supplies the stator windings of an induction machine, the d.c. link current is influenced by the back e.m.f. of the induction machine and under transient conditions the application of the PI current controller in the d.c. link current loop cannot give satisfactory current response. However, this problem can be overcome by the application of a function generator which compensates for the motor back e.m.f. This will be discussed in more detail below.

In Fig. 4.28, the machine-side converter contains six force-commutated thyristors. The function of the capacitors is to effect the successful commutation of current from each thyristor to the next thyristor in the same upper or lower row, i.e. thyristor T_1 is extinguished when T_3 is fired, T_2 is turned-off when T_4 is fired etc. The six diodes decouple the commutating capacitors (C_1–C_6) from the input and output terminals of the machine-side converter, except during commutation. Thus they allow the capacitors to remain charged to relatively high voltages. During commutation, the capacitors become charged to peak voltages, which are mainly determined by the magnitude of the motor current to be commutated. Thus the inverter can commutate satisfactorily over a wide range of output voltage and frequency.

As a result of the commutation, there are voltage spikes superimposed on the almost sinusoidal stator voltages. The amplitudes of these are proportional to the link current and to the square root of the stator transient inductance (L_s') and are inversely proportional to the square root of the commutating capacitance. They can be reduced by increasing the capacitance of the commutation capacitors. Since the capacitor voltages appear directly on the semiconductor components, this increases their required voltage rating even when the fundamental component of the voltage waveform is relatively small. These voltage spikes are one of the limiting factors of the horsepower rating of CSI-fed induction machine drives. Another limiting factor in extending the horsepower rating of the machine is the rather limited speed range, which is again due to commutation problems. Although generally it is desirable to increase the output frequency of the inverter to the highest possible value, since this can result in a reduction of the size and cost of the machine, and there exist solid-state devices with high switching frequency, the commutation period for the CSI (which is affected by the commutating capacitor and the commutating inductance L_s') is significantly lower than for a voltage-source inverter of equal rating.

The d.c. link current is switched through the inverter thyristors to produce three-phase, six-stepped line currents. Each thyristor conducts for $2\pi/3$ and at any instant one upper and one lower thyristor is conducting. If it is assumed that only two phases conduct at any instant, six distinct modes of operation result. A more detailed description of the operation of the machine-side converter will now be given. For this purpose it is assumed that the diodes and the thyristors are ideal switches, that the d.c. link current is constant and does not contain ripples, and that commutation of the currents is completed in less than one-sixth of a period (T) of a stator current. The operation can be divided into three stages. If initially two thyristors, T_1 and T_2 conduct, as shown in Fig. 4.29(a), the stator currents of the machine are $i_{sA}=i_D$, $i_{sB}=0$, $i_{sC}=-i_D$ and thus from the definition of the space phasor of the stator currents $\bar{i}_s$ [eqn (2.1-4)],

$$\bar{i}_s=\frac{2i_D(1-a^2)}{3}=\frac{2i_D e^{j\pi/6}}{\sqrt{3}}. \quad (4.1\text{-}136)$$

Thus the space phasor of the stator currents is in position 1 in Fig. 4.29(e). The first stage begins at the instant when thyristor T_3 is triggered and simultaneously the gating signal is removed from T_1. Thus capacitor C_1 (which was charged at a previous commutation) begins to discharge and turns off T_1. As soon as T_1 turns off, the d.c. link current flows through capacitor C_1 and the series connected capacitors C_3 and C_5 [Fig. 4.29(b)]. The first stage continues until diode D_3 starts to conduct and at this instant the second stage begins [Fig. 4.29(c)]. Thus there is three-phase conduction, where none of the stator currents are zero, so the space phasor of the stator currents must move from the position 1 in the direction of position 2 shown in Fig. 4.29(e). During the second stage, the current gets diverted from the capacitor bank, diode D_1 and stator phase sA to diode D_3 and stator phase sB, and the second stage ends when the current in the capacitor bank is zero [D_1 is blocked, T_3, D_3 and T_2, D_2 conduct—see Fig. 4.29(d)]. Since $i_{sA}=0$, $i_{sB}=i_D$ and $i_{sC}=-i_D$, it

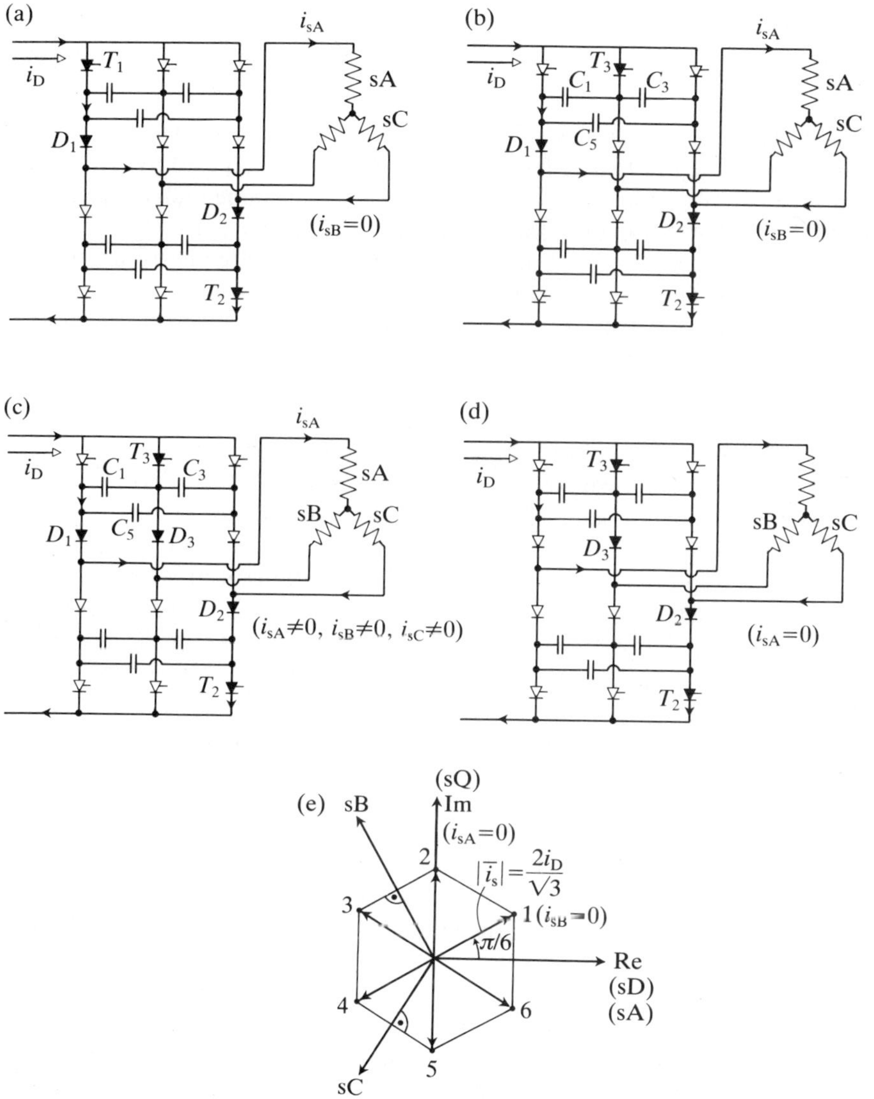

Fig. 4.29. Commutation process. (a) Initial stage of commutation. (b) First stage of commutation. (c) Second stage of commutation. (d) Third stage of commutation. (e) Locus of stator-current space phasor.

follows by considering eqn (2.1-4) that the space phasor of the stator currents is

$$\bar{i}_s = \frac{2a - a^2}{3} = \frac{j2i_D}{\sqrt{3}} \tag{4.1-137}$$

and this corresponds to position 2 of the stator-current space phasor in Fig. 4.29(e). The third stage lasts until thyristor T_4 is triggered to turn off T_2 and the d.c. link current flows across T_3, D_3, the machine and T_2, D_2.

It follows that the locus of the stator currents over one period of the stator currents is a hexagon. During one-sixth of a cycle, the space phasor remains in a fixed position and during three-phase conduction it rotates by $2\pi/6$ along the hexagon. The locus is a closed curve (hexagon) and is symmetrical with respect to the origin of the stator reference frame as a result of the periodicity. It also follows that the magnitude of the three-phase currents is controlled by the regulation of the d.c. link current. The frequency of the three-phase currents is determined by the rate of switching of the inverter thyristors.

When the CSI shown in Fig. 4.28 supplies the stator windings of an induction machine, it should be noted that at low speed there can be undesirable torque pulsations, which are mainly due to the non-sinusoidal stator currents produced by the inverter. The pulsations occur at six times the operating frequency and at very low frequency the frequency of the pulsations is reduced so much that the rotor moves in steps instead of rotating smoothly. This unsmooth rotation makes this drive unsuitable for servo-drive applications which require continuous position control.

In the case of the CSI-fed induction machine, one possibility for torque-ripple control is the pulse-width modulation of the stator currents by multiple commutations of the inverter in each half cycle. When this technique is used, the stator currents are notched at specific points to eliminate certain time harmonics. Each notch added to the current waveforms allows the elimination of a specific time harmonic, i.e. an appropriate double-notched waveform at each quarter cycle can result in the elimination of the fifth and seventh time harmonics in the stator currents and thus in the elimination of the sixth-harmonic torque. Pulse-width modulation can be achieved by the application of an angle loop, where the loop acts as a switching regulator which advances or reverses the sense of rotation of the space phasor of stator currents. A stator-current-angle control loop is utilized in the implementation of a rotor-flux-oriented control of a CSI-fed induction machine described in the next section.

Implementation which utilizes the direct method: In the present section the rotor-flux-oriented control of a CSI-fed induction machine is described, where the angle of the rotor magnetizing-current space phasor is obtained by using the so-called direct method.

As discussed above, in the CSI-fed induction machine the d.c. link current is determined by the supply-side converter. Since the modulus of the stator-current space phasor ($|\bar{\imath}_s|$) must be proportional to the d.c. link current (i_D), this is also determined by the supply-side converter. However, the space angle of the stator-current space phasor, α_s, which is also shown in Fig. 4.26, is determined by the machine-side inverter. Both $|\bar{\imath}_s(t)|$ and $\alpha_s(t)$ are continuous functions. By considering that one of the three stator currents is always zero outside the commutation intervals, the locus of the stator-current space phasor is a six-pointed star, whose radius is proportional to the d.c. link current, and according to eqns (4.1-136) and (4.1-137) is equal to $|\bar{\imath}_s| = (2i_D)/\sqrt{3}$. It has been shown that when currents flow in all the three stator phases, the locus is characterized by a hexagon which connects the end points of the six-pointed star.

Outside the commutation interval the switching state of the inverter is characterized by $e^{j\beta}$, where the angle β changes discontinuously in increments of $\pm\pi/3$. The firing signals for the inverter can be derived by the application of a six-step bidirectional ring counter, which is stepped clockwise and anticlockwise. This method is utilized in Fig. 4.30.

The outer control loops of the drive shown in Fig. 4.30 are similar to those presented in Fig. 4.25. The instantaneous value of the angular rotor speed (ω_r) is obtained together with the instantaneous value of the rotor angle (θ_r). This is compared with the reference value of the rotor angle (θ_{rref}) and the resulting error is the input to the position controller, which is a PI controller. The output of this is the reference value of the rotor speed (ω_{rref}). When this is compared with the monitored value of the rotor speed (ω_r), the resulting error is supplied to the input of the speed controller, also a PI controller, the output of which is the reference value of the electromagnetic torque (t_{eref}). Comparison of the torque reference with its actual value (t_e) gives an error, which serves as input to the torque controller (PI controller) of which the output is the reference value of the quadrature-axis stator current expressed in the rotor-flux-oriented reference frame (i_{syref}).

The direct-axis stator current reference (i_{sxref}), which is also expressed in the rotor-flux-oriented reference frame, is obtained as the output of the flux controller (PI controller), the input of which is the difference between the reference value of the rotor magnetizing current ($|\bar{\imath}_{mrref}|$) and its actual value ($|\bar{\imath}_{mr}|$). The reference current $|\bar{\imath}_{mrref}|$ is obtained as the output of the function generator FG, which allows field weakening to be implemented. Therefore the input of FG is the monitored rotor speed and below base speed FG outputs a constant value of $|\bar{\imath}_{mrref}|$, while above base speed, $|\bar{\imath}_{mrref}|$ is inversely proportional to the rotor speed. The two-axis stator current references are then inputs to the R–P block which performs the rectangular-to-polar conversion and its outputs are the modulus ($|\bar{\imath}_{sref}|$) and the space angle (ϕ_{sref}) of the reference value of the stator-current space phasor in the rotor-flux-oriented reference frame. The angle (ϕ_s) has also been shown in Fig. 4.26.

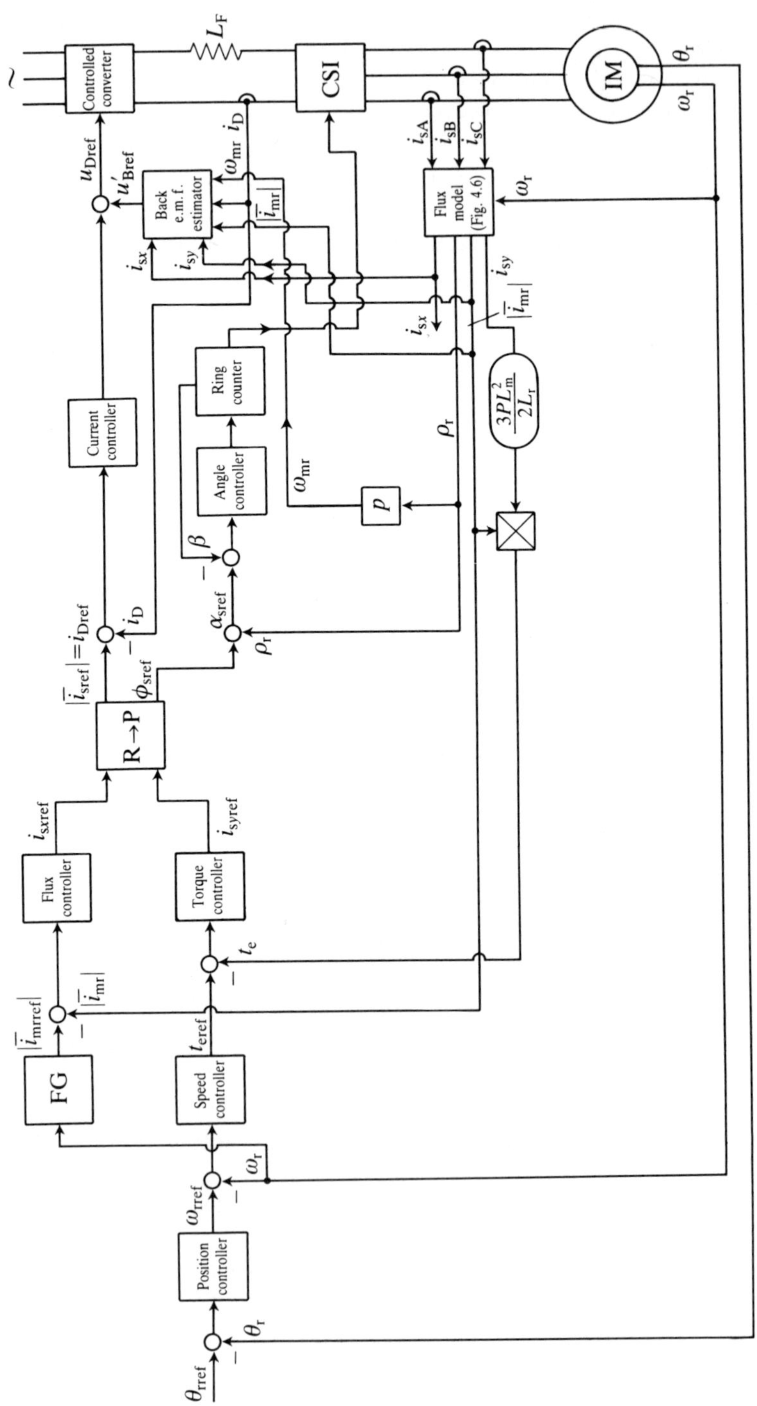

Fig. 4.30. Schematic of the direct rotor-flux-oriented control of a CSI-fed induction machine.

In Fig. 4.30 the monitored stator currents, together with the monitored rotor speed are inputs to the flux model. This has been described in Fig. 4.6. The output of the flux model are the rotor magnetizing current ($|\bar{i}_{mr}|$), the torque-producing stator current (i_{sy}) and the spatial position of the rotor flux linkage space phasor (ρ_r). In accordance with eqn (4.1-43), the electromagnetic torque is obtained by multiplying the torque-producing current by the constant $3PL_m^2/(2L_r)$, where P is the number of pole pairs, L_m is the magnetizing inductance of the machine and L_r is the self-inductance of the rotor.

In Fig. 4.30 the angle ϕ_{sref} is first added to the angle ρ_r to yield the space angle of the reference value of the stator-current space phasor with respect to the real-axis of the stator reference frame (α_{sref}). This angle has also been shown in Fig. 4.26. The angle α_{sref} is then compared with the angle β described above, which corresponds to the state of the inverter, and the resulting error serves as the input to the angle controller, which is a PI controller. With the aid of the angle-control loop, a six-step ring counter is switched forward and backward between two adjacent states. Another advantage of using the angle-control loop is that, at low speed, the torque pulsations, which have been discussed in the previous section and are caused by the interaction of the approximately sinusoidal rotor magnetizing currents and the non-sinusoidal stator currents, will be reduced because of the reduced distortions in the stator currents. However, as discussed in the previous section, the CSI-fed induction machine is not suitable for servo-drive applications, where at very low speed controlled torque is required. In contrast to this, in the voltage-source PWM inverter-fed induction machine there is no low-speed cogging effect.

As discussed above, the amplitude of the stator currents is controlled through the d.c. link current (i_D) and conventionally a PI controller is used for this purpose. However, with this type of control the d.c. link current is influenced by the back e.m.f. of the motor. During transient conditions the conventional control cannot give satisfactory current response and, as a consequence, the performance of the drive deteriorates. The influence of the back e.m.f. can be reduced by increasing the gain of the PI current controller, which results in the narrowing of the bandwidth of the d.c. link current loop. However, this will not yield the desired current response since, as will be shown below, the back e.m.f. is a function of several variables whose dynamics are faster or as fast as that of the d.c. link current. Furthermore, the bandwidth of the d.c. link current loop is determined by a large time constant [$T_D = L/R$ where R and L will be defined in eqns (4.1-145) and (4.1-146) respectively] and also by the stability requirements on the current response. A solution to this problem and thus improved current response can be obtained if, in addition to the PI current controller, a function generator is incorporated in the control to compensate for the back e.m.f.s of the induction machine. The details of the design of the required back e.m.f. estimator will now be discussed. For this purpose the voltage equation of the d.c. link, which is valid under transient conditions, has to be obtained, and it has to contain the back e.m.f. of the machine but referred to the inverter input.

First the space phasor of the back e.m.f. of the machine is obtained. It is assumed that there are no time harmonics in the d.c. link current and voltage, and none on the output of the inverter, apart from the fundamental. The effects of magnetic saturation are also neglected. Equation (4.1-6), the space-phasor form of the stator voltage equation in the rotor-flux-oriented reference frame is

$$\bar{u}_{s\psi r}=(R_s+L_s'p)\bar{i}_{s\psi r}+j\omega_{mr}L_s'\bar{i}_{s\psi r}+(L_s-L_s')(p+j\omega_{mr})\bar{i}_{mr}, \qquad (4.1\text{-}138)$$

where $p=d/dt$, R_s and L_s are the resistance and self inductance of a stator winding respectively, L_s' is the transient inductance of the stator, $\bar{u}_{s\psi r}$ and $\bar{i}_{s\psi r}$ are the stator voltage and current space phasors in the rotor-flux-oriented reference frame and $\bar{i}_{mr}$ is the rotor magnetizing-current space phasor in the same reference frame ($\bar{i}_{mr}=|\bar{i}_{mr}|$). ω_{mr} is the speed of the reference frame and is equal to the sum of the rotor speed and the slip speed. The third term on the right-hand side of eqn (4.1-138) gives the space phasor of the back e.m.f.s of the induction machine,

$$\bar{u}_B=(L_s-L_s')(p+j\omega_{mr})|\bar{i}_{mr}|. \qquad (4.1\text{-}139)$$

This will be utilized in the back e.m.f. estimator, but it has to be referred to the inverter input.

The dynamic voltage equation of the d.c. link will now be obtained. It follows from Fig. 4.28 that the output voltage of the controlled rectifier (u_D) and the input voltage of the inverter (u_i) are related by

$$u_D=(R_F+L_Fp)i_D+u_i, \qquad (4.1\text{-}140)$$

where R_F and L_F are the resistance and inductance of the filter (smoothing reactor) and i_D is the d.c. link current. Furthermore, by assuming that there are no losses in the inverter (and thus the power into the inverter is equal to the output power of the inverter), it follows from eqn (2.1-87) that

$$u_i i_D=\tfrac{2}{3}\,\mathrm{Re}(\bar{u}_{s\psi r}\bar{i}^{*}_{s\psi r}) \qquad (4.1\text{-}141)$$

where the asterisk denotes the complex conjugate and Re denotes the real part of a complex quantity.

If the expression for the voltage space phasor defined by eqn (4.1-138) is substituted into eqn (4.1-141), but its third term is replaced by u_B, the following equation is obtained after some rearrangement of the various terms,

$$\frac{\tfrac{2}{3}\,\mathrm{Re}(\bar{u}_B\bar{i}^{*}_{s\psi r})}{i_D}=u_i-\frac{2}{3}(R_s+L_s'p)\frac{|\bar{i}_s|^2}{i_D}=u_B'. \qquad (4.1\text{-}142)$$

The second term after the first equal sign of eqn (4.1-142) contains the ohmic and leakage voltage drops referred to the inverter input and the term on the left-hand-side is the motor back e.m.f. referred to the inverter input (u_B'). When the expression for $\bar{u}_B$ given by eqn (4.1-139) is substituted into the left-hand side of eqn (4.1-142), and substituting $\bar{i}_{s\psi r}=i_{sx}+ji_{sy}$,

$$u'_B=\frac{\frac{2}{3}\,\mathrm{Re}(\bar{u}_B\bar{i}^*_{s\psi r})}{i_D}=\tfrac{2}{3}\,L_s(1-\sigma)\,(p|\bar{i}_{mr}|i_{sx}+\omega_{mr}|\bar{i}_{mr}|\,i_{sy})/i_D, \quad (4.1\text{-}143)$$

where σ is the resultant leakage constant, $\sigma=1-L_m^2/(L_sL_r)$. It can be seen that u'_B depends on the level of the rotor magnetizing current (rotor flux level), flux- and torque-producing stator currents, the d.c. link current and also on the rotor speed (this is contained in ω_{mr}). Thus by taking u_i from eqn (4.1-140) and substituting into eqn (4.1-142), the following equation is obtained, which describes the dynamics of the d.c. link in terms of u'_B:

$$u_D=(R+Lp)i_D+u'_B, \quad (4.1\text{-}144)$$

where u'_B is defined by eqn (4.1-143) and

$$R=R_F+\frac{2}{3}\left(R_s\frac{|\bar{i}_s|^2}{i_D^2}\right) \quad (4.1\text{-}145)$$

$$L=L_F+\frac{2}{3}\left(L'_s\frac{|\bar{i}_s|^2}{i_D^2}\right). \quad (4.1\text{-}146)$$

It follows from eqn (4.1-146) that u'_B acts as a feedback in the drive. The back e.m.f. can be compensated by the application of a back e.m.f. estimator, which is also shown in Fig. 4.30, and which is based on eqn (4.1-143).

In general, it follows from eqn (4.1-143), that the inputs to the back e.m.f. estimator shown in Fig. 4.30 are the d.c. link current (i_D), the flux-producing stator current component (i_{sx}), the torque-producing stator current component (i_{sy}) and the speed of the rotor flux (ω_{mr}). When the back e.m.f. estimator is used together with the PI current controller in the d.c. link loop, the d.c. link current is no longer disturbed by the back e.m.f. of the machine and its transient behaviour is only determined by the PI current controller and a first-order lag, whose time constant is equal to L/R, where L and R are defined in eqns (4.1-145) and (4.1-146) respectively. When the rotor flux is constant, $|\bar{i}_{mr}|$=constant, eqn (4.1-143) takes a very simple form,

$$u'_B=\frac{2}{3}L_s(1-\sigma)\frac{\omega_{mr}|\bar{i}_{mr}|i_{sy}}{i_D}. \quad (4.1\text{-}147)$$

This can be arranged into various forms for various implementations and, for example, by considering eqn (4.1-43) it follows that the product $|\bar{i}_{mr}|i_{sy}$ is proportional to the electromagnetic torque (t_e) and eqn (4.1-147) takes the form,

$$u'_B=\frac{t_e\omega_{mr}}{Pi_D}, \quad (4.1\text{-}148)$$

where P is the number of pole pairs. In eqn (4.1-148) ω_{mr} can be replaced by using eqn (4.1-26) and thus

$$u'_B = \left[t_e \left(\omega_r + \frac{i_{sy}}{T_r |\bar{i}_{mr}|} \right) \right] \Big/ P i_D, \tag{4.1-149}$$

where ω_r is the rotor speed and T_r is the rotor time constant. It is also possible to use eqn (4.1-43) and to eliminate i_{sy} in eqn (4.1-149) to yield

$$u'_B = \left[t_e \left(\omega_r + \frac{t_e R_r}{c |\bar{i}_{mr}|^2} \right) \right] \Big/ P i_D, \tag{4.1-150}$$

where $c = \frac{3}{2} P L_m^2$ and R_r is the rotor resistance (referred to the stator winding). A very simple realization of the u'_B compensator is obtained if t_e is replaced by its reference value.

It should be noted that when the motor back e.m.f. estimator is incorporated into the drive control, the improvement in the d.c. link current response also results in improved torque and flux responses. Furthermore, back e.m.f.s are also present in the output current control loops of the current-controlled PWM Voltage-Source Inverter-fed induction machine. In this case these e.m.f.s can be obtained by the transformation of u'_B into the actual back e.m.f.s, and current control with e.m.f. compensation can be established, where the compensator can be designed on the basis of the space-phasor equation of the back e.m.f.s.

Implementation which utilizes the indirect method: In this section the rotor-flux-oriented control of a CSI-fed induction machine is described, where the angle of the rotor magnetizing-current space phasor is obtained by using the so-called indirect method. Fig. 4.31 shows the schematic of the drive.

The implementation shown in Fig. 4.31 resembles the implementation shown in Fig. 4.27, which contains the control scheme of the indirect rotor-flux-oriented control of a current controlled PWM inverter-fed induction machine. The outer control loops shown in Fig. 4.31 are the same as those shown in Fig. 4.27, and therefore they are not described here. The reference values of the direct- and quadrature-axis stator currents expressed in the rotor-flux-oriented reference frame (i_{sxref}, i_{syref}) are obtained in the same way as shown in Fig. 4.27. Furthermore the reference value of the slip frequency (ω_{slref}) is obtained identically to that shown in Fig. 4.27, by utilizing i_{sxref} and i_{syref} and the known value of the rotor time constant (T_r), which can also be obtained by on-line identification.

In Fig. 4.31, as in Fig. 4.30, the reference stator currents i_{sxref}, i_{syref} are converted into polar coordinates by a rectangular-to-polar converter, and thus the outputs of the R→P converter are the modulus of the reference value of the stator-current space phasor ($|\bar{i}_{sref}|$) together with its space angle (ϕ_{sref}), which angle is shown in Fig. 4.26 and is the angle of the stator-current space

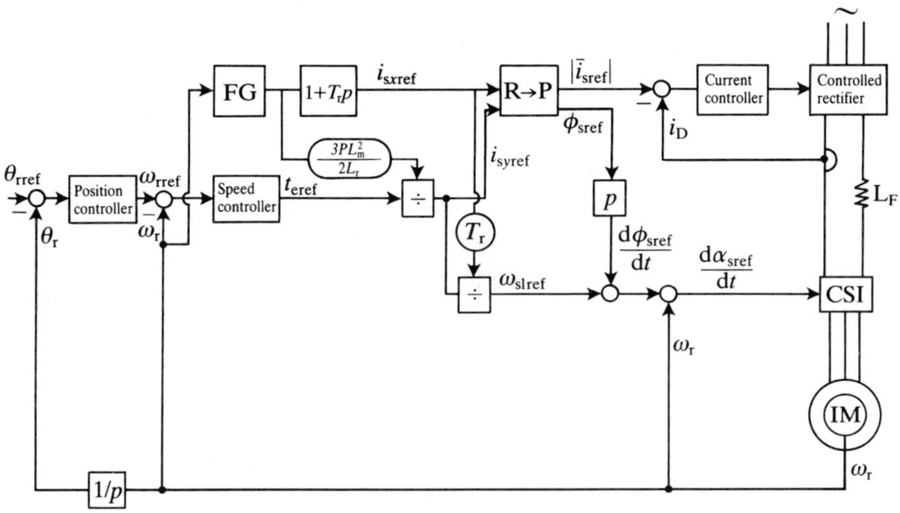

Fig. 4.31. Schematic of the indirect rotor-flux-oriented control of a CSI-fed induction machine.

phasor with respect to the real axis of the rotor-flux-oriented reference frame. It follows from Fig. 4.26 that

$$\alpha_{sref} = \theta_r + \theta_{slref} + \phi_{sref}, \tag{4.1-151}$$

where α_s is the space angle of the stator-current space phasor with respect to the real axis of the stationary reference frame, θ_r is the rotor-angle, which gives the position of the real axis of the reference frame fixed to the rotor with respect to the real axis of the stationary reference frame, θ_{sl} is the slip angle, which gives the position of the rotor flux-linkage space phasor with respect to the real axis of the reference frame fixed to the rotor and ϕ_s is the angle of the stator-current space phasor with respect to the real axis of the rotor-flux-oriented reference frame. The subscript 'ref' denotes the reference value. Thus differentiation of eqn (4.1-151) with respect to time yields

$$p\alpha_{sref} = p\theta_r + p\theta_{slref} + p\phi_{sref} = \omega_1 = \omega_r + \omega_{slref} + p\phi_{sref} \tag{4.1-152}$$

where $p = d/dt$, ω_1 is the required stator frequency, ω_r is the rotor speed and ω_{slref} is the reference value of the slip frequency, which is defined in eqn (4.1-131), repeated here for convenience,

$$\omega_{slref} = \frac{i_{syref}}{T_r i_{sxref}}. \tag{4.1-153}$$

Thus in Fig. 4.31 the stator frequency ω_1 is obtained by considering eqns (4.1-152) and (4.1-153). The supply-side converter is controlled in the same way as in Fig. 4.28.

It should be noted that to obtain lower torque pulsations at low speeds, the technique described in Fig. 4.30 could be utilized. Furthermore, a more accurate value of the slip frequency can be obtained by utilizing the actual values of the flux- and torque-producing stator currents (i_{sx}, i_{sy}) instead of their reference values (i_{sxref}, i_{syref}), since the presence of the d.c. link inductor makes the stator current slow to respond to a change in the reference currents. During transient operation the slip frequency determined from i_{sxref} and i_{syref} differs from the slip frequency which is determined from the actual values of these currents and the difference is large at the beginning of the transient. As a consequence, an inappropriate value of the slip frequency results, and this affects adversely the torque response of the drive and leads to deteriorated performance. The actual values of the stator current: components i_{sx} and i_{sy} can be obtained in several ways, including the utilization of the appropriate transformations of the monitored stator currents i_{sA}, i_{sB}, i_{sC}. In this case, by considering eqns (2.1-112) and (2.1-113), first the direct- and quadrature-axis stator currents in the stationary reference frame are obtained as follows, if it is assumed that there are no zero-sequence currents:

$$i_{sD}=i_{sA} \tag{4.1-154}$$

$$i_{sQ}=\frac{1}{\sqrt{3}}(i_{sB}-i_{sC}). \tag{4.1-155}$$

These are then transformed into the two-axis currents established in the rotor-flux-oriented reference frame by utilizing eqn (2.1-192):

$$i_{sx}=\cos\rho_r i_{sD}+\sin\rho_r i_{sQ} \tag{4.1-156}$$

$$i_{sy}=-\sin\rho_r i_{sD}+\cos\rho_r i_{sQ}. \tag{4.1-157}$$

However, this requires the angle ρ_r, which is the angle of the rotor magnetizing-current space phasor with respect to the real axis of the stationary reference frame. Furthermore, the two transformations required lead to large computation time in the case of a digital implementation or to a complicated hardware in the case of analogue implementation.

However, a much simpler implementation can be obtained if i_{sx} and i_{sy} are derived from the monitored d.c. link current (i_D) and the angle of the stator-current space phasor with respect to the real axis of the rotor-flux-oriented reference frame (ϕ_s). This angle is also shown in Fig. 4.26. Since the commutation period is short in the CSI inverter, it can be neglected and according to Fig. 4.29(e) the modulus of the stator-current space phasor $|\bar{i}_s|$ is proportional to the d.c. link current (i_D),

$$|\bar{i}_s|=\frac{2i_D}{\sqrt{3}}, \tag{4.1-158}$$

and the amplitude of the fundamental current is equal to $(2\sqrt{3}/\pi)i_D$. Thus, from Fig. 4.26, the space phasor of the stator currents in the rotor-flux-oriented reference frame can be expressed as

$$\bar{i}_{s\psi r}=i_{sx}+ji_{sy}=|\bar{i}_s|e^{j\phi_s} \tag{4.1-159}$$

and the following expressions are obtained for the flux- and torque-producing stator current components

$$i_{sx}=ci_D\cos\phi_s \tag{4.1-160}$$

$$i_{sy}=ci_D\sin\phi_s, \tag{4.1-161}$$

where c is a constant. Thus i_{sx} and i_{sy} can be obtained in a simple manner from i_D and ϕ_s and the required transformation is not dependent on any machine parameter. A possible implementation is shown in Fig. 4.32, where ϕ_s is replaced by ϕ_{sref}, which is the space angle of the reference value of the stator-current space phasor.

In Fig. 4.32 i_{sxref} and i_{syref} are converted by a rectangular-to-polar converter into $|\bar{i}_{sref}|$ and ϕ_{sref}. As shown in Fig. 4.31, the slip frequency ω_{sl} is added to $d\phi_{sref}/dt$ and to this is added the monitored rotor speed to yield $\omega_1=d\alpha_{sref}/dt$. The monitored d.c. link current is first multiplied by the constant c, and ci_D and ϕ_{sref} are inputs to a polar-to-rectangular converter, the outputs of which are i_{sy} and i_{sx}. By utilizing eqn (4.1-25), $|\bar{i}_{mr}|$ is obtained from i_{sx} in the block which contains $1/(1+T_rp)$, where $p=d/dt$ and T_r is the rotor time constant. The slip frequency (ω_{sl}) is obtained by utilizing eqn (4.1-26) and thus $\omega_{sl}=i_{sy}/(T_r|\bar{i}_{mr}|)$. The full rotor-flux-oriented control scheme can be obtained by adding the parts of the circuit shown in Fig. 4.31 to Fig. 4.32.

Control of the induction machine supplied by a current-controlled cycloconverter

A cycloconverter-fed induction machine drive is very suitable for high-power,

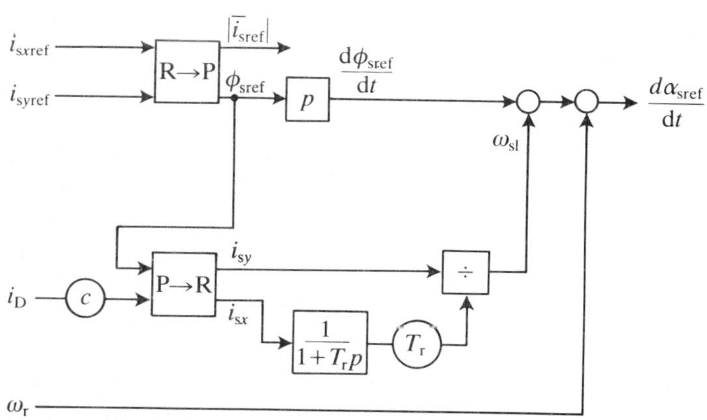

Fig. 4.32. Determination of i_{sx}, i_{sy}, and ω_{sl} by utilizing i_D and ϕ_{sref}.

low-speed applications. Some applications include rolling steel mills, temper mills etc. and the characteristics of rolling mills have been described in Section 3.2.1. These characteristics can be satisfied by the vector-controlled cycloconverter-fed induction machine. Owing to the advantages of the induction machine over the d.c. machine (reliability, lower moment of inertia, ruggedness, absence of commutator and brushes etc.) the cycloconverter-fed induction machine offers additional advantages over d.c. drives, and conventional d.c. drives can be replaced by high-performance vector-controlled cycloconverter-fed induction machines.

As a result of application of line-commutated bridge converters, the output frequency of the cycloconverter varies between 0 and 25 Hz if the converter is supplied at a frequency of 50 Hz and between 0 and 20 Hz if the supply frequency is 60 Hz. This limits the speed range of the cycloconverter, but in low-speed drives this is of no importance. If necessary, the speed range can be increased by increasing the number of phases on the line-side of the cycloconverter, by employing a transformer with the required number of secondary windings between the line and the cycloconverter. However, it is a disadvantage of this technique that as a result of the increased number of phases, the number of thyristors has to be increased in the converter. The high input reactive power of the cycloconverter drive is also a disadvantage, but there are several well-known techniques to reduce this (asymmetrical voltage control, neutral potential bias control, etc.).

In this section the rotor-flux-oriented control of a current-controlled cycloconverter-fed induction machine is described. As described in Section 3.2.1 for the cycloconverter-fed salient-pole synchronous machine, each stator phase of the induction machine is supplied by two antiparallel three-phase six-pulse bridge converters. These bridge converters have been shown in Fig. 3.20. The stator currents of the induction machine are separately controlled by fast current loops. Fig. 4.33 shows an implementation of the drive.

The control structure is similar to that used for d.c. machines. The difference between the monitored rotor speed (ω_r) and the reference rotor speed (ω_{rref}) serves as the input to the speed controller, which is a PI controller, and this produces the required torque reference (t_{eref}). The function generator FG has the same role as, for example, in Fig. 4.25; this allows field weakening to be achieved and its output is the reference value of the rotor magnetizing current ($|\bar{\imath}_{mrref}|$). The output of the flux controller (which has the same role as in Fig. 4.25) is the reference value of the flux-producing stator current component (i_{sxref}) and its input is $|\bar{\imath}_{mrref}|-|\bar{\imath}_{mr}|$), where the actual value of the rotor magnetizing current ($|\bar{\imath}_{mr}|$) is obtained by utilizing the flux model shown in Fig. 4.6.

The inputs to the flux model are the rotor speed (ω_r) and the stator currents (in the absence of zero-sequence currents only two line currents have to be monitored). The outputs of the flux model are the flux- and torque-producing

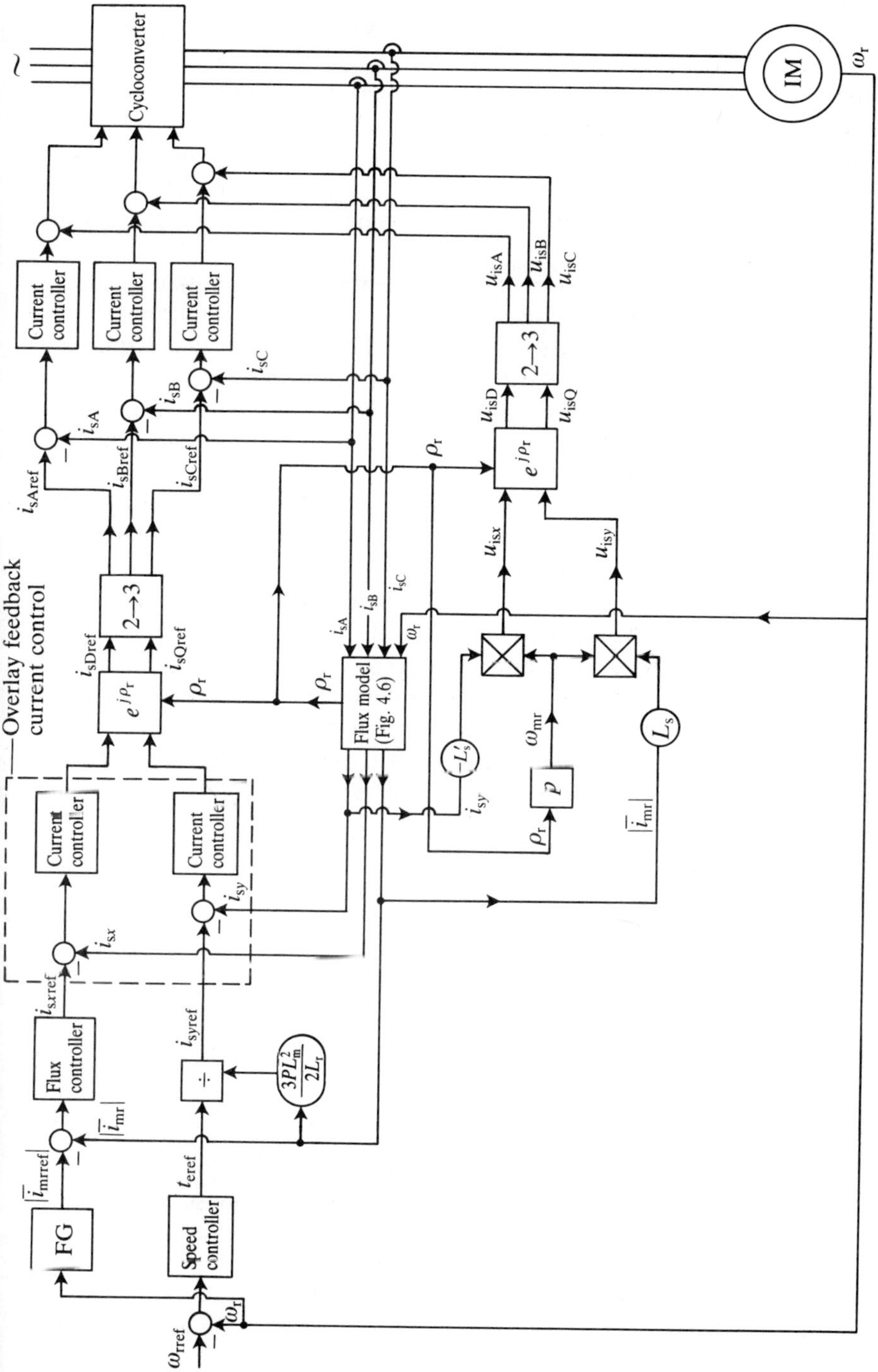

Fig. 4.33. Schematic of the rotor-flux-oriented control of a current-controlled cycloconverter.

stator currents (i_{sx}, i_{sy}), which are established in the rotor-flux-oriented reference frame, ρ_r, which is the angle of the rotor magnetizing-current space phasor with respect to the real axis of the stationary reference frame, and $|\bar{i}_{mr}|$, which is the modulus of the rotor magnetizing-current space phasor. These quantities are also shown in Fig. 4.26. In accordance with eqn (2.1-197), the reference value of the torque-producing stator current (i_{syref}) is obtained by dividing the torque reference by the factor $3P|\bar{i}_{mr}|L_m^2/(2L_r)$, where P is the number of pole pairs and L_m and L_r are the magnetizing inductance and rotor self-inductance respectively.

To obtain high accuracy, overlay current control loops are used to control the flux- and torque-producing stator currents. The outputs of the controllers in these loops are inputs to the transformation block $e^{j\rho_r}$, which is required to obtain the reference values of the direct- and quadrature-axis stator currents (i_{sDref}, i_{sQref}) established in the stationary reference frame, as discussed in connection with Fig. 4.25. These currents are then transformed into their three-phase components by the application of the three-phase to two-phase transformation, which are then compared with their actual values. The resulting errors are inputs to the appropriate current controllers which control the cycloconverter.

It has been noted in Section 3.2.2 that at higher output frequencies of the cycloconverter, induced a.c. voltages will disturb the current control system and there will result phase and amplitude errors, which even in the steady-state cannot be eliminated by the current controllers which control the cycloconverter. However, one possible solution to this is to add feedforward voltages to the outputs of these current controllers. Thus in Fig. 4.33 e.m.f. compensation is realized by feedforward of the induced voltages (u_{isA}, u_{isB}, u_{isC}). These are obtained from their two-axis values (u_{isD}, u_{isQ}) by the application of the two-phase to three-phase transformation. However, these are related to the induced voltages established in the rotor-flux-oriented reference frame (u_{isx}, u_{isy}) by the transformation $e^{j\rho_r}$. The required induced voltages u_{isx} and u_{isy} can be obtained from eqns (4.1-7) and (4.1-8) and by using the fact that in the steady-state $|\bar{i}_{mr}|$=constant and considering eqn (4.1-25), which yields $|\bar{i}_{mr}|=i_{sx}$; so finally

$$u_{isx}=-\omega_{mr}L_s' i_{sy} \tag{4.1-162}$$

$$u_{isy}=\omega_{mr}L_s|\bar{i}_{mr}|. \tag{4.1-163}$$

ω_{mr} is the speed of the rotor-flux-oriented reference frame ($\omega_{mr}=d\rho_r/dt$), and L_s and L_s' are the self-inductance and transient inductance of the stator respectively.

Control of the slip-ring induction machine with impressed rotor currents

Rotor-flux-oriented control can also be used for the slip-ring induction machine with impressed rotor currents. For this purpose the converter in the rotor circuit can be a current-controlled cycloconverter (e.g. a six-pulse cycloconverter as described in the previous section), a pulse-width modulated inverter with current control (again described earlier) or, for example, a static converter cascade. In the last case there is dual converter in the rotor and this is shown in Fig. 4.34.

In the implementation shown in Fig. 4.34 a forced commutated thyristor converter is connected to the slip rings of the induction machine and this is connected via a d.c. reactor to a naturally commutated controlled thyristor bridge converter (special form of static Scherbius cascade). Thus slip power can flow in both directions (to the rotor from the supply and from the rotor to the supply) and the speed of the induction machine can be controlled in the supersynchronous and subsynchronous ranges. Below the synchronous speed in the motoring mode and above the synchronous speed in the regeneration mode the rotor-side converter operates as a rectifier and the line-side converter as an inverter, and slip power is returned to the stator, usually through a transformer. Below the synchronous speed in the regeneration mode and above the synchronous speed in the motoring mode, the rotor-side converter operates as an inverter and the line-side converter operates as a rectifier, and slip power is supplied to the rotor. At the synchronous speed, slip power is taken from the supply to excite the rotor windings, and in this case the machine behaves as a synchronous machine. It should be noted that if the rotor-side converter was a line-commutated bridge converter, near to the synchronous speed, when the a.c. voltage available to commutate the converter is small, the

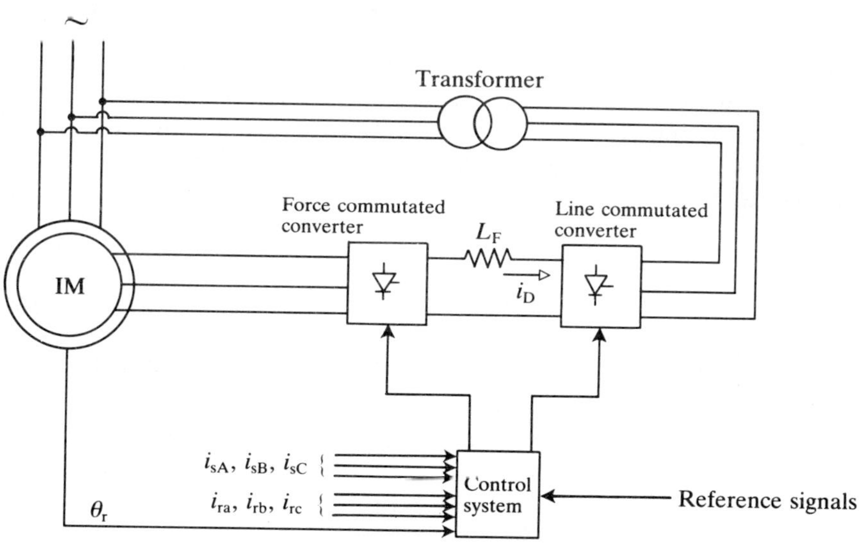

Fig. 4.34. Schematic of the control of a doubly-fed induction machine with converter cascade in the rotor.

commutating ability of the converter would be lost. It is for this reason that a force-commutated rotor-side converter is used in the special form of the static Scherbius cascade shown in Fig. 4.34, and thus full torque capability can be produced near standstill.

Since impressed rotor currents are assumed, the rotor voltage equations can be disregarded from the point of view of the dynamics of the drive under consideration and this results in a simplified dynamic structure of the drive. Thus only the stator voltage equations have to be considered.

The space-phasor form of the stator voltage equation in the rotor-flux-oriented reference frame has been given as eqn (4.1-6) and its two-axis forms have been given as eqn (4.1-7) and eqn (4.1-8). These contain the rotor magnetizing current ($|\bar{\imath}_{mr}|$), which has been defined in eqn (4.1-1). These equations together with the equation of the electromagnetic torque, eqn (4.1-43), could be used to obtain the implementations of the rotor-flux-oriented control of a slip-ring induction machine with impressed rotor currents. However, in contrast to the implementations so far discussed for induction machines, where the rotor currents were not directly measurable quantities, in the slip-ring machine it is possible to monitor these and this results in simpler implementations than, say, in the case of a squirrel-cage machine. If the stator currents and the rotor currents are measured together with the rotor angle (θ_r) then $|\bar{\imath}_{mr}|$ and ρ_r, which is the space angle of the rotor magnetizing-current space phasor with respect to the direct-axis of the stationary reference frame (see Fig. 4.26), are known quantities. Mathematically this follows from eqn (4.1-1), which is now repeated here for convenience,

$$\bar{\imath}_{mr} = \bar{\imath}_{s\psi r} + (1+\sigma_r)\bar{\imath}_{r\psi r}, \tag{4.1-164}$$

where $\bar{\imath}_{s\psi r}$ and $\bar{\imath}_{r\psi r}$ are the space phasors of the stator currents and rotor currents respectively in the rotor-flux-oriented reference frame. By considering eqn (4.1-3) and eqn (2.1-194) they can be expressed in terms of the space phasor of the stator currents and rotor currents expressed in the stationary reference frame ($\bar{\imath}_s$, $\bar{\imath}_r'$) as

$$\bar{\imath}_{s\psi r} = \bar{\imath}_s e^{-j\rho_r} \tag{4.1-165}$$

$$\bar{\imath}_{r\psi r} = \bar{\imath}_r' e^{-j\rho_r}. \tag{4.1-166}$$

Substitution of eqns (4.1-165) and (4.1-166) into eqn (4.1-164) yields

$$\bar{\imath}_{mr} = \left[\bar{\imath}_s + \frac{L_r}{L_m}\bar{\imath}_r'\right] e^{-j\rho_r}. \tag{4.1-167}$$

Thus it follows from eqn (2.1-133) and eqn (4.1-167) that

$$\bar{\imath}_{mr} = |\bar{\imath}_{mr}| e^{j\rho_r} = \bar{\imath}_s + \frac{L_r}{L_m}\bar{\imath}_r' = \bar{\imath}_s + \frac{L_r}{L_m}\bar{\imath}_r e^{j\theta_r}, \tag{4.1-168}$$

where in terms of their two-axis components

$$\bar{i}_s = i_{sD} + ji_{sQ} \tag{4.1-169}$$

$$\bar{i}_r = i_{r\alpha} + ji_{r\beta} \tag{4.1-170}$$

$$\bar{i}_r e^{j\theta_r} = i_{rd} + ji_{rq} \tag{4.1-171}$$

$$\bar{i}_{mr} = i_{mrD} + ji_{mrQ} \tag{4.1-172}$$

and θ_r is the rotor angle. Equations (4.1-168)–(4.1-171) can be used to obtain $|\bar{i}_{mr}|$ and ρ_r from the measured values of the stator and rotor currents and the rotor angle. Physically eqn (4.1-168) gives the rotor magnetizing-current space phasor in the stationary reference frame and, as expected, this contains two components; these are the stator-current space phasor and a component which would be equal to the rotor-current space phasor if there was no rotor leakage. In this case (when the rotor leakage inductance is zero) $L_r = L_m$ and the rotor magnetizing-current space phasor is identical to the conventional magnetizing-current space phasor, which is equal to the sum of the stator- and rotor-current space phasors. Equation (4.1-168) can also be obtained directly by dividing the expression for the rotor flux-linkage space phasor expressed in the rotor reference frame [see eqn (2.1-134)] by the magnetizing inductance and this technique has been used for the derivation of eqn (4.1-28), which gives the same definition for the rotor magnetizing-current space phasor as eqn (4.1-168).

It should be noted that when $|\bar{i}_{mr}|$ and ρ_r are determined from eqn (4.1-168), there is no dependency on the rotor resistance, in contrast to the derivation based on the flux models described in Fig. 4.6 and Fig. 4.7. This is an advantage, since the rotor resistance varies as a result of changes in the temperature and the skin effect. By using eqn (4.1-168), Fig. 4.35 shows a possible implementation for obtaining $|\bar{i}_{mr}|$ and ρ_r.

It follows from Fig. 4.35 that the monitored three-phase stator currents (i_{sA}, i_{sB}, i_{sC}) are first transformed into the real- and imaginary-axis components of the stator-current space phasor expressed in the stationary reference frame ($\bar{i}_s$) by the application of the three-phase to two-phase transformation. The monitored rotor currents (i_{ra}, i_{rb}, i_{rc}) are similarly transformed into the $i_{r\alpha}$, $i_{r\beta}$ components (which are the direct- and quadrature-axis rotor currents in the rotor reference frame) by the application of the three-phase to two-phase transformation. These are then transformed into i_{rd} and i_{rq}, which are the two-axis rotor current components in the rotor reference frame, by the application of the transformation $e^{j\theta_r}$ [see eqn (4.1-171) and eqn (2.1-40)]. Finally by considering eqn (4.1-168) i_{rd} and i_{rq} are multiplied by L_r/L_m and the newly obtained rotor currents are added to i_{sD} and i_{sQ} respectively to yield the two-axis components of the rotor magnetizing-current space phasor in the stationary reference frame (i_{mrD}, i_{mrQ}). These are inputs to a rectangular-to-polar converter the outputs of which are $|\bar{i}_{mr}|$ and ρ_r.

By utilizing the implementation shown in Fig. 4.35, several other quantities can also be obtained which could be required in various implementations of the rotor-flux-oriented control of slip-ring induction machines with impressed rotor currents. Thus when ρ_r is differentiated, ω_{mr}, the speed of the rotor flux-linkage space phasor, is obtained. From the monitored rotor angle (θ_r), and ρ_r, the slip angle θ_{sl} is obtained (this angle is shown in the phasor diagram of Fig. 4.26). It follows from eqn (4.1-165) that the space phasor of the stator currents in the rotor-flux-oriented reference frame is $(i_{sx}+ji_{sy})=(i_{sD}+ji_{sQ})e^{-j\rho_r}$, and resolution into real and imaginary terms gives eqns (4.1-156) and (4.1-157) which could be used to obtain i_{sx} and i_{sy} from i_{sD} and i_{sQ}. The modulus ($|\bar{i}_s|$) and the space angle (α_s) of the stator-current space phasor expressed in the stationary reference frame are obtained in Fig. 4.35 by the application of a rectangular-to-polar converter by using i_{sD} and i_{sQ} as input quantities. The angle α_s is also shown in Fig. 4.26. Finally, in accordance with Fig. 4.26, in Fig. 4.35 the torque angle (ϕ_s), which is the angle of the stator-current space phasor with respect to the real axis of the rotor-flux-oriented reference frame, is obtained as $\phi_s=\alpha_s-\rho_r$.

For the implementation of the rotor-flux-oriented slip-ring machine with impressed rotor currents, when the rotor speed is monitored and when this is

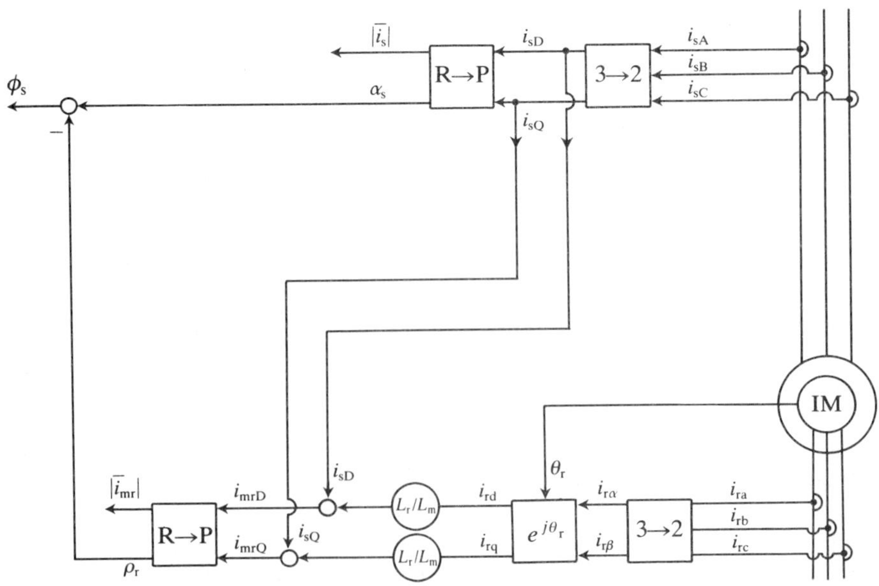

Fig. 4.35. Obtaining $|\bar{i}_{mr}|$, $|\bar{i}_s|$, ρ_r, and ϕ_s by monitoring the stator and rotor currents and the rotor angle.

compared with its reference value, the error signal can serve as input to a speed controller (PI controller) the output of which is the torque reference.

For the rotor-flux-oriented control of the doubly-fed induction machine with the static converter cascade shown in Fig. 4.34, the torque reference can be used to obtain the reference value of the d.c. link current (i_D) which flows across the reactor connected between the two converters. Thus the line-side converter can be controlled by using the closed loop control of the d.c. link current, where the difference between the reference d.c. link current and the actual d.c. link current is fed into a PI controller. The rotor-side converter can be controlled by using the closed loop control of the torque angle. For this purpose a PI controller can be used, the input of which is the difference between the reference actual values of the torque angle (the latter can be obtained with the help of Fig. 4.35). A more detailed description will be given in Section 4.2.6, where stator-flux-oriented control of the same drive will be discussed.

The vector control of the slip-ring motor with impressed rotor currents using a cycloconverter will be discussed in detail in Section 4.2.6, where stator-flux-oriented control is discussed, since it will be shown that in the stator-flux-oriented system, the real- and imaginary-axis components of the rotor currents (i_{rx}, i_{ry}) correspond directly to the two-axis stator currents expressed in the stator-flux-oriented reference frame. Thus control of the i_{rx} current can be used to maintain the reactive stator current at a fixed value (or at a value dependent on the voltage) and the current i_{ry} can be used for torque control. Beside the speed-independent control of active and reactive powers, it is also an advantage of the application of this control scheme that it leads to the avoidance of the instability problems associated with doubly fed machines.

4.2 STATOR-FLUX-ORIENTED CONTROL OF INDUCTION MACHINES

In this part of the chapter stator flux-oriented control will be applied to induction machines. The main assumptions are the same as those used in connection with rotor-flux-oriented control of induction machines and, in particular, the effects of magnetic saturation are neglected. Since there is a large similarity between stator-flux-oriented control and rotor-flux-oriented control, emphasis will only be laid on the differences between these two control techniques.

4.2.1 Expressions for the electromagnetic torque

It follows from Section 2.1.8 that in the stator-flux-oriented reference frame the electromagnetic torque of an induction machine is proportional to the product of the modulus of the stator magnetizing-current space phasor ($|\bar{i}_{ms}|$)

and the torque-producing stator current component (i_{sy}); the torque can be controlled by independent control of these two currents. Thus from eqn (2.1-209)

$$t_e = c|\bar{i}_{ms}|i_{sy}, \qquad (4.2\text{-}1)$$

where $c = 3PL_m/2$ is a constant, P is the number of pole pairs and L_m is the magnetizing inductance of the machine. In eqn (4.2-1) the stator magnetizing-current space phasor can be put in terms of the stator- and rotor-current space phasors expressed in the stator-flux-oriented reference frame ($\bar{i}_{s\psi s}$, $\bar{i}_{r\psi s}$) as, if eqn (2.1-208) is considered,

$$|\bar{i}_{ms}| = \bar{i}_{r\psi s} + \frac{\bar{i}_{s\psi s}L_s}{L_m} = (i_{rx} + ji_{ry}) + (i_{sx} + ji_{sy})\frac{L_s}{L_m}, \qquad (4.2\text{-}2)$$

where L_s is the self inductance of the stator and i_{rx}, i_{ry}, i_{sx}, i_{sy} are the two-axis components of the rotor- and stator-current space phasors respectively, expressed in the stator-flux-oriented reference frame. In eqn (2.1-201) relating the space phasor of the stator currents expressed in the stator (stationary) and stator-flux-oriented reference frames has been given and it follows that

$$\bar{i}_{s\psi s} = i_{sx} + ji_{sy} = (i_{sD} + ji_{sQ})e^{-j\rho_s}, \qquad (4.2\text{-}3)$$

where i_{sD} and i_{sQ} are the direct- and quadrature-axis components of the stator currents in the stationary reference frame and the angle ρ_s gives the position of the stator magnetizing current with respect to the real-axis (sD) of the stationary reference frame (this is shown in Fig. 2.17). The rotor-current space phasor in the reference frame fixed to the rotor can be expressed in terms of its two-axis components as $\bar{i}_r = i_{r\alpha} + ji_{r\beta} = |\bar{i}_r|e^{j\alpha_r}$, and thus it follows from eqn (2.1-206) that the rotor-current space phasor in the stator-flux-oriented reference frame can be expressed as

$$\bar{i}_{r\psi s} = i_{rx} + ji_{ry} = (i_{r\alpha} + ji_{r\beta})e^{j\theta_r}e^{-j\rho_s} = |\bar{i}_r|e^{j(\alpha_r + \theta_r - \rho_s)}. \qquad (4.2\text{-}4)$$

It is evident that the stator and rotor current components i_{sx}, i_{sy}, i_{rx}, i_{ry}, defined above, are not equal to the similar current components defined by eqns (2.1-192) and (2.1-194), which are established in the rotor-flux-oriented reference frame. Only for simplicity is the real axis of the stator-flux-oriented reference frame denoted as x, and its imaginary axis as y, but these should not be confused with the real and imaginary axes of the rotor-flux-oriented reference frame. For better understanding, the stator magnetizing-current space phasor and the rotor-current space phasor are shown in Fig. 4.36, where the new reference frame (x, y) is also shown, together with the angles α_r, θ_r, and ρ_s.

It follows from Fig. 4.36 that the angle $\phi_r = (\alpha_r + \theta_r - \rho_s)$, which is the torque angle, is the angle of the rotor-current space phasor with respect to the stator magnetizing-current space phasor.

It should be noted that in the steady state, the definition of $\bar{i}_{ms}$ (expressed in

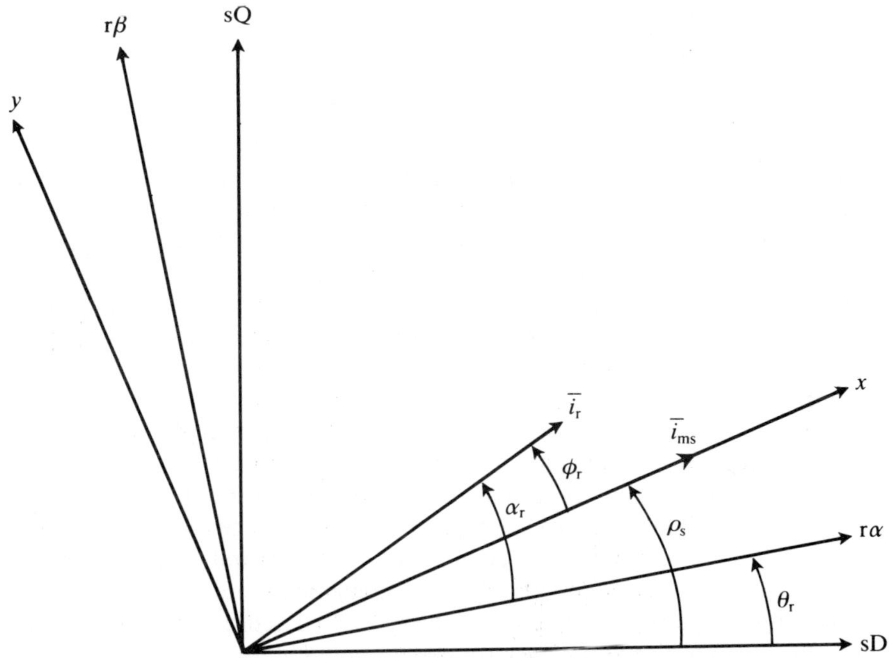

Fig. 4.36. Stator magnetizing-current space phasor and rotor-current space phasor.

the stationary reference frame) can also be obtained by considering Fig. 4.10(b) if the general turns ratio is selected as $a=L_s/L_m$.

A new expression will now be obtained for the electromagnetic torque, which instead of the quadrature-axis stator current (i_{sy}) contains the modulus of the rotor-current space phasor and the torque angle. This will be useful for the implementation of stator-flux-oriented control of doubly fed induction machines. It follows from eqn (4.2-2) that the torque-producing stator current can be expressed as

$$i_{sy}=-i_{ry}\frac{L_m}{L_s} \tag{4.2-5}$$

and i_{ry} can be obtained from eqn (4.2-4) as $i_{ry}=|\bar{i}_r|\sin(\alpha_r+\theta_r-\rho_s)$. By substitution of this expression into eqn (4.2-5) and considering eqn (4.2-1), the following expression is obtained for the electromagnetic torque:

$$t_e=-3|\bar{i}_{ms}|\,|\bar{i}_r|\,\sin(\alpha_r+\theta_r-\rho_s)\frac{L_m^2}{2L_s}. \tag{4.2-6}$$

This expression will be used in Section 4.2.6, where stator-flux-oriented control of a doubly fed three-phase induction machine with impressed rotor currents will be discussed.

4.2.2 Stator voltage equations for the squirrel-cage machine

In this section, the space-phasor form of the stator voltage equation is first obtained in the reference frame fixed to the stator magnetizing flux-linkage space phasor. This will then be resolved into its two-axis components. For this purpose the voltage equations formulated in the general reference frame are directly utilized. These have been given in Section 2.1.7. The stator-current space phasor will be expressed in terms of the stator magnetizing-current space phasor, which has been defined in the previous section. Thus the resulting stator voltage equation will contain the modulus and space angle (or speed) of the stator magnetizing-current space phasor. These quantities are required in the implementation of stator-flux-oriented control. Two alternatives will then be considered, which correspond to a squirrel-cage machine and to a slip-ring induction machine where there is direct access to the rotor currents.

By considering eqns (2.1-148) and (2.1-150), which define the stator voltage and flux linkage space-phasor equations in the general reference frame respectively, and by neglecting the effects of magnetic saturation, the following stator voltage equation is obtained in the special reference frame fixed to the stator linkage space phasor, which rotates at a speed $\omega_{ms}=d\rho_s/dt$ (the angle ρ_s has been defined in the previous section):

$$\bar{u}_{s\psi s}=R_s\bar{\imath}_{s\psi s}+L_s\frac{d\bar{\imath}_{s\psi s}}{dt}+L_m\frac{d\bar{\imath}_{r\psi s}}{dt}+j\omega_{ms}L_s\bar{\imath}_{s\psi s}+j\omega_{ms}L_m\bar{\imath}_{r\psi s}, \tag{4.2-7}$$

where as in eqn (4.2-3), which defines the stator-current space phasor in the stator-flux-oriented reference frame, the stator-voltage space phasor in the same reference frame can be expressed as

$$\bar{u}_{s\psi s}=u_{sx}+ju_{sy}=(u_{sD}+ju_{sQ})e^{-j\rho_s}=\bar{u}_s e^{-j\rho_s}. \tag{4.2-8}$$

In eqn (4.2-7) $\bar{\imath}_{r\psi s}$ is the space phasor of the rotor currents in the stator-flux-oriented reference frame and has been defined in eqn (4.2-4).

It is assumed that the squirrel-cage induction machine is supplied by a voltage-source inverter, as described in Section 4.1.1, and the stator currents cannot be impressed by fast current control loops. Thus to obtain the required reference voltages of the inverter, it is necessary to use the stator voltage equations.

It follows from eqn (4.2-7) that if a stator-flux-oriented control scheme for a squirrel-cage induction machine has to be obtained, the rotor currents have to be eliminated from the equations since these cannot be measured directly. However, it follows from eqn (4.2-2) that

$$\bar{\imath}_{r\psi s}=|\bar{\imath}_{ms}|-\frac{L_s\bar{\imath}_{s\psi s}}{L_m} \tag{4.2-9}$$

and substitution of eqn (4.2-9) into eqn (4.2-7) yields

$$\bar{u}_{s\psi s} = R_s \bar{i}_{s\psi s} + L_m \frac{d|\bar{i}_{ms}|}{dt} + j\omega_{ms} L_m |\bar{i}_{ms}|. \tag{4.2-10}$$

This is much simpler than the corresponding voltage equation expressed in the rotor-flux-oriented reference frame [eqn (4.1-6)]. As a direct consequence of the selection of the stator-flux-oriented reference frame, there is no separate stator-leakage voltage component present in eqn (4.2-10). Furthermore, it does not contain the derivative of the stator-current space phasor, which is present when the rotor-flux-oriented reference frame is used. If the modulus of the stator magnetizing-current space phasor is constant, then eqn (4.2-10) takes an extremely simple form, which does not contain the derivative $d|\bar{i}_{ms}|/dt$. When eqn (4.2-10) is resolved into its two-axis components,

$$i_{sx} = \frac{u_{sx}}{R_s} - \frac{L_m}{R_s} \frac{d|\bar{i}_{ms}|}{dt} \tag{4.2-11}$$

$$i_{sy} = \frac{u_{sy}}{R_s} - \omega_{ms} L_m \frac{|\bar{i}_{ms}|}{R_s}. \tag{4.2-12}$$

These equations are much simpler than eqns-(4.1-7) and (4.1-8). For stator-flux-oriented control, i_{sx} and i_{sy} have to be independently controlled. However, it follows from eqn (4.2-12) that there is an unwanted coupling term and thus u_{sx} and u_{sy} cannot be considered as decoupled variables for the stator flux and electromagnetic torque. The stator current components i_{sx} and i_{sy} can only be independently controlled if eqns (4.2-11) and (4.2-12) are decoupled and the stator currents are indirectly controlled by controlling the terminal voltages of the induction machine. As in Section 4.1.1, it follows from eqns (4.2-11) and (4.2-12) that the required decoupling voltages are

$$u_{dx} = 0$$

$$u_{dy} = \omega_{ms} L_m |\bar{i}_{ms}|. \tag{4.2-13}$$

Thus a very simple decoupling circuit can be established, the inputs of which are ω_{ms} and $|\bar{i}_{ms}|$ and the output of which is u_{dy}. It follows from the simplicity of the stator voltage equations that when an induction machine is supplied by impressed stator voltages, the stator-flux-oriented control scheme will be simpler than the rotor-flux-oriented control scheme. When a squirrel-cage machine is supplied by a converter which functions as a current source, the stator equations can be omitted from the equations of the drive.

4.2.3 Rotor voltage equations for the squirrel-cage induction machine

The rotor voltage equations expressed in the stator-flux-oriented reference frame can be used to obtain the modulus and speed (or phase angle) of the stator flux-linkage space phasor (or stator magnetizing-current space phasor).

The required flux model can be obtained by using the rotor voltage equations formulated in the stator-flux-oriented reference frame.

For this purpose, eqn (2.1-153) is used, which gives the space-phasor form of the rotor voltage equation in the general reference frame, but it is assumed that all the inductance parameters are constant. Thus in the stator-flux-oriented reference frame ($\omega_g=\omega_{ms}$, $\bar{u}_{rg}=\bar{u}_{r\psi s}$, $\bar{\imath}_{sg}=\bar{\imath}_{s\psi s}$, $\bar{\imath}_{rg}=\bar{\imath}_{r\psi s}$) this takes the form,

$$\bar{u}_{r\psi s}=R_r\bar{\imath}_{r\psi s}+L_r\frac{d\bar{\imath}_{r\psi s}}{dt}+L_m\frac{d\bar{\imath}_{s\psi s}}{dt}+j(\omega_{ms}-\omega_r)(L_r\bar{\imath}_{r\psi s}+L_m\bar{\imath}_{s\psi s}), \qquad (4.2\text{-}14)$$

where ω_r is the rotor speed and R_r and L_r are the rotor resistance and rotor self-inductance. Equation (4.2-14) contains the rotor-current space phasor, but since for a squirrel-cage machine this cannot be directly obtained, it is now eliminated. By the substitution of eqn (4.2-9) into eqn (4.2-14) and by considering that, since the rotor windings can be considered as short-circuited windings, $\bar{u}_{r\psi s}=0$,

$$0=R_r\left[|\bar{\imath}_{ms}|-\frac{L_s}{L_m}\bar{\imath}_{s\psi s}\right]+L_r\frac{d|\bar{\imath}_{ms}|}{dt}-\left(\frac{L_s'L_r}{L_m}\right)\frac{d\bar{\imath}_{s\psi s}}{dt}$$
$$+j\omega_{sl}\left[L_r|\bar{\imath}_{ms}|-\left(\frac{L_s'L_r}{L_m}\right)\bar{\imath}_{s\psi s}\right] \qquad (4.2\text{-}15)$$

is obtained, where

$$\omega_{sl}=\omega_{ms}-\omega_r \qquad (4.2\text{-}16)$$

is the angular slip frequency. This is equal to the first time derivative of the slip angle θ_{sl}, which is equal to $\rho_s-\theta_r$, as shown in Fig. 4.36, and is the angle of the stator flux-linkage space phasor (or stator magnetizing-current space phasor) with respect to the real axis (rα) of the rotor reference frame. In eqn (4.2-15) L_s' is the transient stator inductance. It should be noted that eqn (4.2-15) is more complicated than the rotor voltage equation formulated in the rotor-flux-oriented reference frame [eqn (4.1-24)]. This has the consequence that when the induction machine is supplied by impressed stator currents, the implementation of stator-flux-oriented control will be more complicated than the implementation of the rotor-flux-oriented control. Resolution of eqn (4.2-15) into its real- and imaginary-axis components yields

$$\frac{L_m}{L_s'}\frac{d|\bar{\imath}_{ms}|}{dt}+\frac{L_m}{L_sT_r'}|\bar{\imath}_{ms}|=\frac{di_{sx}}{dt}+\frac{i_{sx}}{T_r'}-\omega_{sl}i_{sy} \qquad (4.2\text{-}17)$$

$$\omega_{sl}\left(\frac{L_m|\bar{\imath}_{ms}|}{L_s'}-i_{sx}\right)=\frac{di_{sy}}{dt}+\frac{i_{sy}}{T_r'} \qquad (4.2\text{-}18)$$

where T_r' is the transient rotor time constant.

For an induction machine with impressed stator currents eqn (4.2-17) and (4.2-18) have to be considered when stator-flux-oriented control is used. However, it follows from eqn (4.2-17) that there exists a coupling between the torque-producing stator current component i_{sy} and the stator magnetizing current. Therefore if there is a change in the torque-producing current, and if this is not followed by an appropriate change in i_{sx}, there will exist an unwanted transient in the stator magnetizing current. However, this undesirable coupling can be eliminated by the utilization of a decoupling circuit, as will now be described.

If the stator magnetizing current is regulated by a flux controller (as shown in Fig. 4.37), the input of which is the difference between the reference value ($|\bar{i}_{msref}|$) and the actual value ($|\bar{i}_{ms}|$) of the stator magnetizing current, and it is assumed that the output of this controller (which can be a PI controller) is the current $\hat{i}_{sx}$, then the required decoupling current i_{dx} has to be added to $\hat{i}_{sx}$ to yield the reference value of the stator current along the real axis of the stator-flux-oriented reference frame (i_{sxref}). It follows that

$$i_{sx}=\hat{i}_{sx}+i_{dx} \tag{4.2-19}$$

and the substitution of eqn (4.2-19) into eqn (4.2-17) yields

$$\frac{L_m}{L'_s}\frac{d|\bar{i}_{ms}|}{dt}+\frac{L_m}{L_sT'_r}|\bar{i}_{ms}|=\frac{d\hat{i}_{sx}}{dt}+\frac{\hat{i}_{sx}}{T'_r}+\frac{di_{dx}}{dt}+\frac{i_{dx}}{T'_r}-\omega_{sl}i_{sy}. \tag{4.2-20}$$

It follows from eqn (4.2-20) that $|\bar{i}_{ms}|$ can be decoupled from i_{sy} if the three last terms on the right-hand side of eqn (4.2-20) are zero, i.e.

$$\left(p+\frac{1}{T'_r}\right)i_{dx}=\omega_{sl}i_{sy}, \tag{4.2-21}$$

where $p=d/dt$. Thus it follows from eqn (4.2-21) that the required current i_{dx} can be expressed as

$$i_{dx}=\omega_{sl}i_{sy}\frac{T'_r}{1+T'_rp}. \tag{4.2-22}$$

In eqn (4.2-22) the angular slip frequency ω_{sl} is present. This can be obtained by considering eqn (4.2-18). It follows from eqn (4.2-18) that

$$\omega_{sl}=\frac{(1+T'_rp)L_si_{sy}}{T_r(L_m|\bar{i}_{ms}|-L'_si_{sx})}. \tag{4.2-23}$$

Equation (4.2-22), together with eqn (4.2-23), defines the required decoupling circuit. Thus when the correct machine parameters are used in these equations, the correct value of i_{sx} is obtained for a given value of i_{sy}, and thus

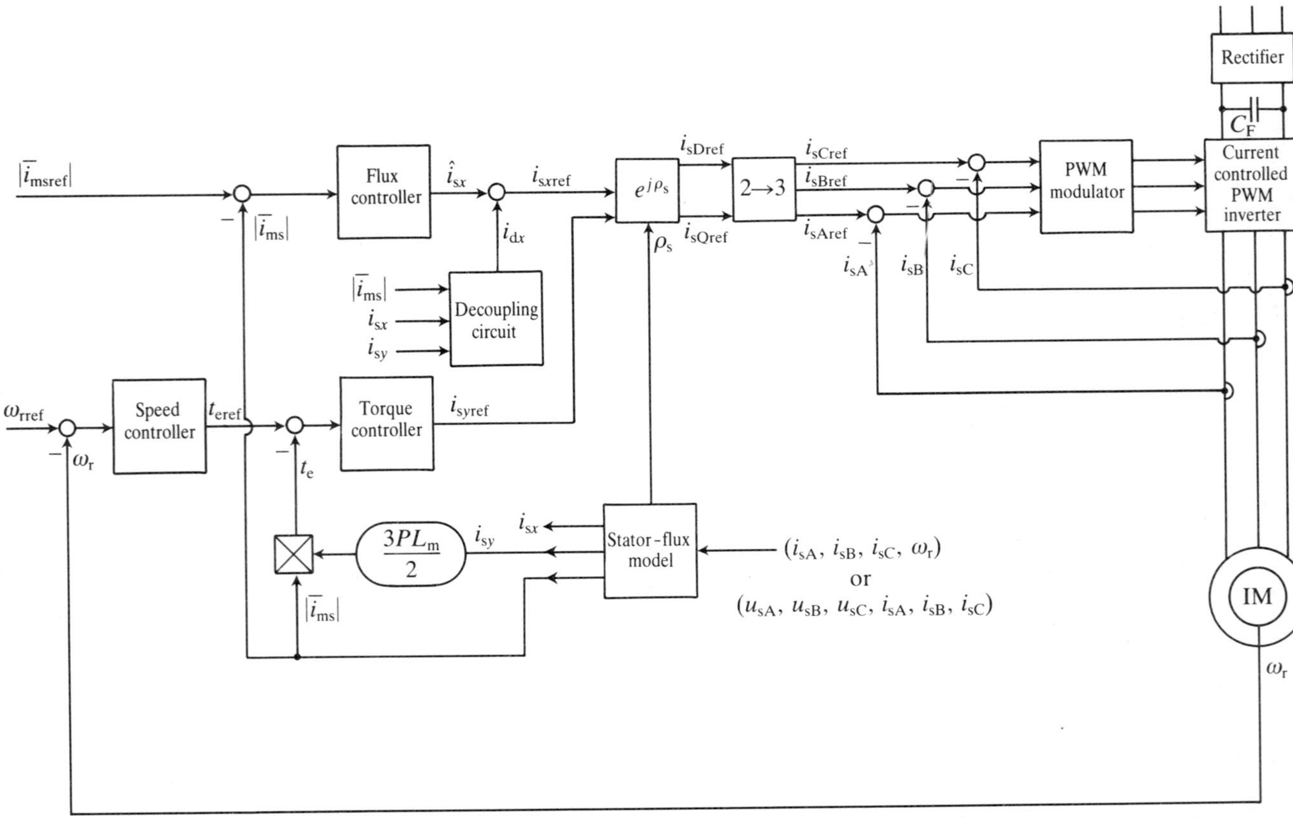

Fig. 4.37. Schematic of the stator-flux-oriented control of an induction machine with a current-controlled PWM inverter.

$|\bar{i}_{ms}|$ is not altered by any changes in the torque reference even during transient operation. By using the decoupling circuit defined by eqns (4.2-22) and (4.2-23), Fig. 4.37 shows as an example the schematic of the stator-flux-oriented control of an induction machine supplied by a current regulated PWM inverter (which has also been used in the implementation shown in Fig. 4.25).

Since this scheme is similar to the one shown in Fig. 4.25 only the differences will be pointed out. The direct-axis stator reference current is obtained as the sum of $\hat{i}_{sx}$ (which is the output of the flux controller) and i_{dx}, which is the output of the decoupling circuit described above. A flux model is used to obtain the modulus ($|\bar{i}_{ms}|$) and the space angle (ρ_s) of the stator magnetizing-current space phasor. The torque-producing stator current component (i_{sy}) is obtained from the two-axis stator currents of the stationary reference frame (i_{sD}, i_{sQ}) by considering eqn (4.2-3). The flux model can use at its inputs the actual stator currents (i_{sA}, i_{sB}, i_{sC}) and the monitored rotor speed (ω_r) and thus by considering eqns (4.2-16), (4.2-17), and (4.2-18) its outputs are i_{sy}, $|\bar{i}_{ms}|$, and ω_{ms} and ρ_s is obtained by the integration of ω_{ms}. There are several other alternatives and if, for example, the stator voltages and currents are monitored, it follows from eqn (2.1-124) that the space phasor of the stator flux linkages ($\bar{\psi}_s$) can be obtained by the integration of $\bar{u}_s - R_s\bar{i}_s$, where $\bar{u}_s$ and $\bar{i}_s$ are the space phasors of the stator voltages and currents respectively in the stationary reference frame (in terms of their two-axis components $\bar{u}_s = \bar{u}_{sD} + ju_{sQ}$, and $\bar{i}_s = i_{sD} + ji_{sQ}$). When $\bar{\psi}_s$ is divided by L_m, the space phasor of the stator magnetizing current in the stationary reference frame is obtained, and its modulus is equal to $|\bar{i}_{ms}|$ and its phase angle is equal to ρ_s. However, when this technique is used, as pointed out previously, at lower stator frequencies the stator resistance will dominate and accurate ohmic voltage-drop compensation must be performed. Therefore this type of solution is not used below 3 Hz with a 50 Hz supply.

4.2.4 Stator voltage equations for the doubly fed induction machine

For a doubly fed machine, it is useful to express $\bar{i}_{s\psi s}$ in terms of $|\bar{i}_{ms}|$ and $\bar{i}_{r\psi s}$. Thus it follows from eqn (4.2-2) that

$$\bar{i}_{s\psi s} = \frac{L_m}{L_s}(|\bar{i}_{ms}| - \bar{i}_{r\psi s}) \tag{4.2-24}$$

and when this is substituted into eqn (4.2-7)

$$\bar{u}_{s\psi s} = \frac{R_s L_m}{L_s}(|\bar{i}_{ms}| - \bar{i}_{r\psi s}) + \frac{L_m \, \mathrm{d}|\bar{i}_{ms}|}{\mathrm{d}t} + j\omega_{ms} L_m |\bar{i}_{ms}|. \tag{4.2-25}$$

This can be put into the form,

$$T_s \frac{\mathrm{d}|\bar{i}_{ms}|}{\mathrm{d}t} + |\bar{i}_{ms}|(1 + j\omega_{ms} T_s) = \bar{u}_{s\psi s} \frac{L_s}{R_s L_m} + \bar{i}_{r\psi s}, \tag{4.2-26}$$

where $T_s = L_s/R_s$ is the stator time constant. By considering the definitions of the stator voltage, stator current and rotor-current space phasors in terms of their two-axis components established in the stator-flux-oriented reference frame [eqns (4.2-3), (4.2-4), and (4.2-8)], resolution of eqn (4.2-26) into its real- and imaginary-axis components yields

$$T_s \frac{d|\bar{i}_{ms}|}{dt} + |\bar{i}_{ms}| = \frac{L_s}{R_s L_m} u_{sx} + i_{rx} \tag{4.2-27}$$

$$\omega_{ms} T_s |\bar{i}_{ms}| = \frac{L_s}{R_s L_m} u_{sy} + i_{ry}. \tag{4.2-28}$$

For the doubly fed induction machine, the stator voltage components can be obtained as follows. By assuming a sinusoidal symmetrical three-phase supply voltage system, with frequency ω_1, the stator-voltage space phasor in the stationary reference frame can be expressed as

$$\bar{u}_s = \sqrt{2} U_s \, e^{j\omega_1 t}, \tag{4.2-29}$$

where U_s is the r.m.s. value of the line-to-neutral voltages and t denotes time. Thus substitution of eqn (4.2-29) into eqn (4.2-8) gives

$$\bar{u}_{s\psi s} = \sqrt{2} U_s \, e^{j(\omega_1 t - \rho_s)} = u_{sx} + j u_{sy} \tag{4.2-30}$$

and thus

$$u_{sx} = \sqrt{2} U_s \cos(\omega_1 t - \rho_s) \tag{4.2-31}$$

$$u_{sy} = \sqrt{2} U_s \sin(\omega_1 t - \rho_s). \tag{4.2-32}$$

Compared with eqns (4.2-11), (4.2-12), which contain the stator currents i_{sx}, i_{sy}, eqns (4.2-27) and (4.2-28) contain the rotor currents i_{rx}, i_{ry} because the rotor currents are subjected to stator-flux-oriented control. As mentioned earlier, two implementations will be discussed, where there is a current-controlled cycloconverter in the rotor of the slip-ring induction machine and, where there is a static converter cascade in the rotor.

4.2.5 Rotor voltage equations for the doubly fed induction machine

By substitution of eqn (4.2-24) into eqn (4.2-14), the space-phasor form of the rotor voltage equation in the stator-flux-oriented reference frame will be

$$\bar{u}_{r\psi s} = (R_r + L_r' p)\bar{i}_{r\psi s} + \frac{L_m^2}{L_s} p|\bar{i}_{ms}| + j\omega_{sl}\left[\frac{L_m^2}{L_s}|\bar{i}_{ms}| + L_r' \bar{i}_{r\psi s}\right], \tag{4.2-33}$$

where it follows from eqn (4.2-16) that $\omega_{sl} = \omega_{ms} - \omega_r$ is the angular slip frequency, ω_{ms} is the speed of the reference frame and ω_r is the rotor speed.

Equation (4.2-33) can be resolved into its direct- (x) and quadrature-axis (y) components and thus

$$T'_r \frac{di_{rx}}{dt} + i_{rx} = \frac{u_{rx}}{R_r} + \omega_{sl} T'_r i_{ry} - (T_r - T'_r) \frac{d|\bar{i}_{ms}|}{dt} \tag{4.2-34}$$

and

$$T'_r \frac{di_{ry}}{dt} + i_{ry} = \frac{u_{ry}}{R_r} - \omega_{sl} T'_r i_{rx} - \omega_{sl}(T_r - T'_r)|\bar{i}_{ms}|. \tag{4.2-35}$$

These equations are very similar to eqns (4.1-7) and (4.1-8) and they can be used to obtain the implementation of stator-flux-oriented control if the rotor currents are not impressed. When these two equations are compared with eqns (4.1-7) and (4.1-8), it can be seen that instead of the stator current components, the rotor current components are now present and the rotor magnetizing-current space phasor is replaced by the stator magnetizing-current space phasor. Furthermore, the stator voltage components are replaced by the rotor voltage components, the stator resistance is replaced by the rotor resistance, the speed of the rotor magnetizing-current space phasor is replaced by the speed of the stator magnetizing-flux-space phasor and finally the stator time constant and stator transient time constant are replaced by the rotor time constant and rotor transient time constant respectively. It also follows from eqns (4.2-34) and (4.2-35) that there are unwanted coupling terms between the rotor circuits along the x and y axes. These can be eliminated by utilizing the following decoupling voltage components, which are obtained by considering the rotational voltages in eqns (4.2-34) and (4.2-35) and assuming constant-stator-flux operation ($|\bar{i}_{ms}|$=constant):

$$u_{drx} = -\omega_{sl} L'_r i_{ry} \tag{4.2-36}$$

$$u_{dry} = \omega_{sl} L'_r i_{rx} + \omega_{sl}(L_r - L'_r)|\bar{i}_{ms}|. \tag{4.2-37}$$

These have to be added to the outputs of the current controllers which control the rotor currents i_{rx} and i_{ry}.

With impressed rotor currents it is not necessary to consider the rotor voltage equations, but to achieve improved dynamic performance, appropriate terms must be added to the outputs of the rotor current controllers, which in the case of $|\bar{i}_{ms}|$=constant compensate for the last term in eqn (4.2-35), which acts as a disturbance for rotor current controllers. Thus it follows from eqn (4.2-35) that in the stator-flux-oriented reference frame the necessary feedforward signal takes the form,

$$\frac{u_{iy}}{R_r} = \omega_{sl}(T_r - T'_r)|\bar{i}_{ms}| \tag{4.2-38}$$

or, in space-phasor form,

$$\frac{\bar{u}_{i\psi s}}{R_r} = \frac{j u_{iy}}{R_r} = j\omega_{sl}(T_r - T'_r)|\bar{i}_{ms}|. \tag{4.2-39}$$

By considering Fig. 4.36, according to which the angle between the real axis of the stator-flux-oriented reference frame and the real axis of the reference frame fixed to the rotor is $(\rho_s - \theta_r)$, in the reference frame fixed to the rotor, $\bar{u}_{i\psi s}/R_r$ can be expressed as

$$\frac{\bar{u}_i}{R_r} = \bar{u}_{i\psi s}\frac{e^{j(\rho_s-\theta_r)}}{R_r} = j\omega_{sl}(T_r - T'_r)|\bar{i}_{ms}|\, e^{j(\rho_s-\theta_r)} = i_\alpha + j i_\beta, \tag{4.2-40}$$

where $\bar{u}_i$ is the space phasor of the induced voltages in the rotor reference frame, i_α and i_β are the required currents which have to be used as feedforward signals. Equation (4.2-40) also follows by considering the transformation of a space-phasor quantity from its natural reference frame into the general reference frame. Thus from eqn (2.1-143), the induced-voltage space phasor in the rotor reference frame $(\bar{u}_i)$ can be expressed in terms of the induced-voltage space phasor in the general reference frame as $\bar{u}_i = \bar{u}_{ig}\, e^{j(\theta_g-\theta_r)}$. Thus when, instead of the general reference frame, the stator-flux-oriented reference frame is used, $\bar{u}_{ig} = \bar{u}_{i\psi s}$, $\theta_g = \rho_s$, and finally $\bar{u}_i = \bar{u}_{i\psi s}\, e^{j(\rho_s-\theta_r)}$ is obtained, which agrees with the induced voltage space phasor used in eqn (4.2-40). Equation (4.2-40) will be utilized in the implementations discussed in the following two sections. When the currents i_α and i_β are added to the rotor currents $i_{r\alpha ref}$, $i_{r\beta ref}$, which are the reference values of the rotor currents expressed in the rotor reference frame, then the effects of the rotational voltages in the rotor windings due to the stator flux are cancelled (the stator flux has been assumed to be constant) and improved dynamic performance is obtained.

4.2.6 Control of the doubly fed induction machine with impressed rotor currents

First the stator-flux-oriented control of a doubly fed induction machine employing a cycloconverter in the rotor circuit is described. This is followed by a discussion on the stator-flux-oriented control of the doubly fed machine employing a converter cascade in the rotor.

Control of the induction machine with a current-controlled cycloconverter in the rotor circuit

In many countries, pumps, fans and blowers can be responsible for approximately half the electricity consumed by industry. Pump drives are used, for example, in sewage treatment, water supply applications, circulating water pumps, etc., and fans are used for mine ventilation, wind tunnel applications, etc. These can represent very large loads—over 10 MW being common—and there is an increasing trend to use variable-speed drives in an attempt to improve the efficiency. In these applications the load torque is proportional to the square of the speed and there is a restricted range of speed control. For such applications the doubly fed slip-ring induction machine with its stator connected to a constant-frequency power supply and its rotor connected to a static converter represents an ideal solution. It is one of the most important features of a slip recovery system, that if the application requires only limited range of speed control (in pump and fan drives only a speed variation of 10–30% is usually required), then the control system has to be rated only for slip power, which is only a fraction of the stator power. The resulting reduction in the converter costs makes the slip recovery systems very attractive in all applications where only a limited speed range is required and where energy saving is an important factor.

In this section a current-controlled cycloconverter is used in the rotor. The cycloconverter used is similar to that described in Section 3.2.1 (also used in Section 4.1.2), and it consists of three sets of antiparallel six-pulse thyristor bridges without circulating currents. The rotor currents of the slip-ring machine are separately controlled by fast current loops.

If sinusoidal stator currents and the decoupling of the active and reactive stator power is required, it is useful to employ stator-flux-oriented control which ensures the required decoupling by appropriate control of the amplitude and phase angle of the rotor-current space phasor. The application of stator-flux-oriented control also has the advantage of stabilizing the drive. It should be noted that another area of application for a doubly fed induction machine with stator-flux-oriented control can be in large wind power stations, where wind gusts can cause power and torque fluctuations which can result in unwanted mechanical stresses on the mechanical transmission system (shaft, gears) and can cause undesirable power fluctuations. When the doubly fed machine is used with a restricted speed range, the wind energy can be temporarily stored in the inertia of the rotor blades and the undesirable effects (power fluctuations, mechanical stresses) are reduced.

Figure 4.38 shows the schematic of the stator-flux-oriented control of an induction machine with a current-controlled cycloconverter in the rotor. The induction machine can operate as a motor or a generator in the supersynchronous and subsynchronous speed ranges with lagging or leading stator currents.

In Fig. 4.38 the input signals to the control system are the monitored stator and rotor currents, the monitored rotor angle (θ_r) and the active and reactive stator powers, respectively P_s and Q_s. The cycloconverter is connected to the grid via a three-phase transformer. The turns ratio of the transformer can be selected to minimize the voltage rating of the cycloconverter. The drive can be started in several ways. One possibility is to employ rotor resistances and to short-circuit them sequentially as the machine comes up to the operating

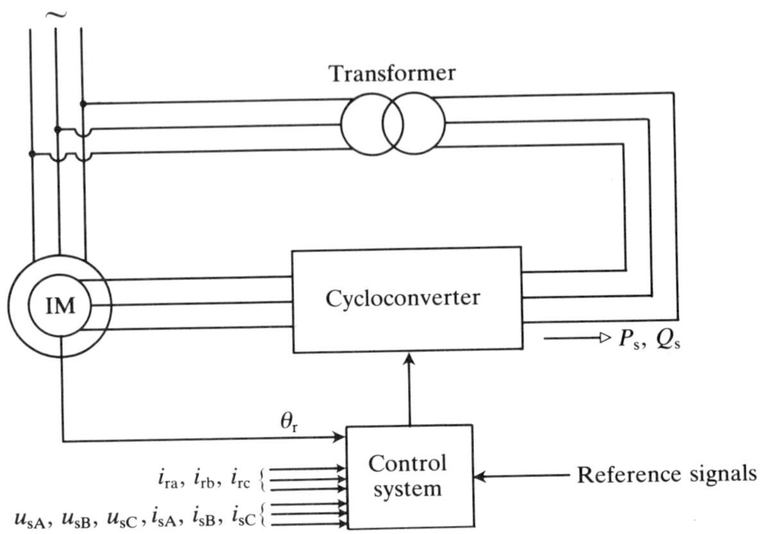

Fig. 4.38. Simplified schematic of the stator-flux-oriented control of an induction machine with a current-controlled cycloconverter in the rotor.

speed when the cycloconverter is switched on. It is also possible to short-circuit the stator and to supply the rotor from the cycloconverter (which is connected to the grid) and which operates as a variable-frequency power supply. This arrangement corresponds to an inverted induction machine. The machine will accelerate and at approximately half speed the stator is connected to the grid. The rotor frequency will be approximately one half of the supply frequency and the motor can accelerate further.

It will now be shown that in the stator-flux-oriented system, the real- and imaginary-axis components of the rotor currents (i_{rx}, i_{ry}) correspond directly to the two-axis stator currents (i_{sx}, i_{sy}) expressed in the stator-flux-oriented reference frame. Thus control of the i_{rx} current can be used to maintain the reactive stator current (i_{sx}) at a fixed value (or at a value dependent on the voltage) and the current i_{ry} can be used to control the electromagnetic torque.

By resolving eqn (4.2-24) into its two-axis components, it follows that the stator current components in the stator-flux-oriented reference frame (i_{sx}, i_{sy}) can be expressed in terms of the rotor current components established in the same reference frame (i_{rx}, i_{ry}) as follows:

$$i_{sy} = -i_{ry}\frac{L_m}{L_s}, \qquad (4.2\text{-}41)$$

which is the same as eqn (4.2-5) but is repeated here for convenience, and

$$i_{sx}=(|\bar{i}_{ms}|-i_{ry})\frac{L_m}{L_s}. \qquad (4.2\text{-}42)$$

It follows from eqn (4.2-41) that the quadrature-axis rotor current (i_{ry}) is proportional to the torque-producing (active) stator current component (i_{sy}). However, in eqn (4.2-42) the stator magnetizing-current space phasor is also present, and this depends on the stator voltage. This dependency will now be obtained. For this purpose, it is assumed that the stator resistance (R_s) of the induction machine can be neglected. This is usually justified in machines with a rating over 10 kW. Furthermore it is assumed that the frequency of the power supply on the stator is constant, ω_1=constant. In the steady state the modulus of the stator magnetizing current [which has been defined in eqn (4.2-2)] is constant ($|\bar{i}_{ms}|$=constant) and the speed of the stator magnetizing-current space phasor ($\omega_{ms}=d\rho_s/dt$, where ρ_s is the angle of the stator magnetizing current space phasor with respect to the real axis of the stationary reference frame and is shown in Fig. 4.36) is also constant.

Thus it follows from eqn (4.2-25) that

$$|\bar{u}_s|=\omega_1 L_m|\bar{i}_{ms}|, \qquad (4.2\text{-}43)$$

where $|\bar{u}_s|$ is the modulus of the stator-voltage space phasor and from eqn (4.2-30) $|\bar{u}_s|=\sqrt{2}U_s$ is obtained. It follows from eqn (4.2-43) that the modulus of the stator magnetizing current is

$$|\bar{i}_{ms}|=\frac{|\bar{u}_s|}{\omega_1 L_m}, \qquad (4.2\text{-}44)$$

which is an expected result since a steady-state has been assumed together with negligible stator resistance. It should be noted that the same result follows from eqns (4.2-27)–(4.2-32), which under the same assumptions yield

$$u_{sx}=0 \qquad u_{sy}=|\bar{u}_s| \qquad |\bar{i}_{ms}|=\frac{u_{sy}}{L_m\omega_1}. \qquad (4.2\text{-}45)$$

Thus from eqns (4.2-42) and (4.2-44)

$$i_{sx}=(|\bar{u}_s|/\omega_1-L_m i_{rx})/L_s, \qquad (4.2\text{-}46)$$

from which it follows that, since the machine parameters (L_s, L_m) and also $|\bar{u}_s|$ and ω_1 are constant, the direct-axis (reactive) stator currents (i_{sx}) can be controlled by the direct-axis rotor current (i_{rx}). These results will be used in the implementation of the stator-flux-oriented control of the doubly fed machine shown in Fig. 4.39.

As in eqn (2.1-87), the stator active power can be defined as

$$P_s=\tfrac{3}{2}\,\mathrm{Re}(\bar{u}_{s\psi s}\bar{i}^{*}_{s\psi s})=\tfrac{3}{2}(u_{sx}i_{sx}+u_{sy}i_{sy}) \qquad (4.2\text{-}47)$$

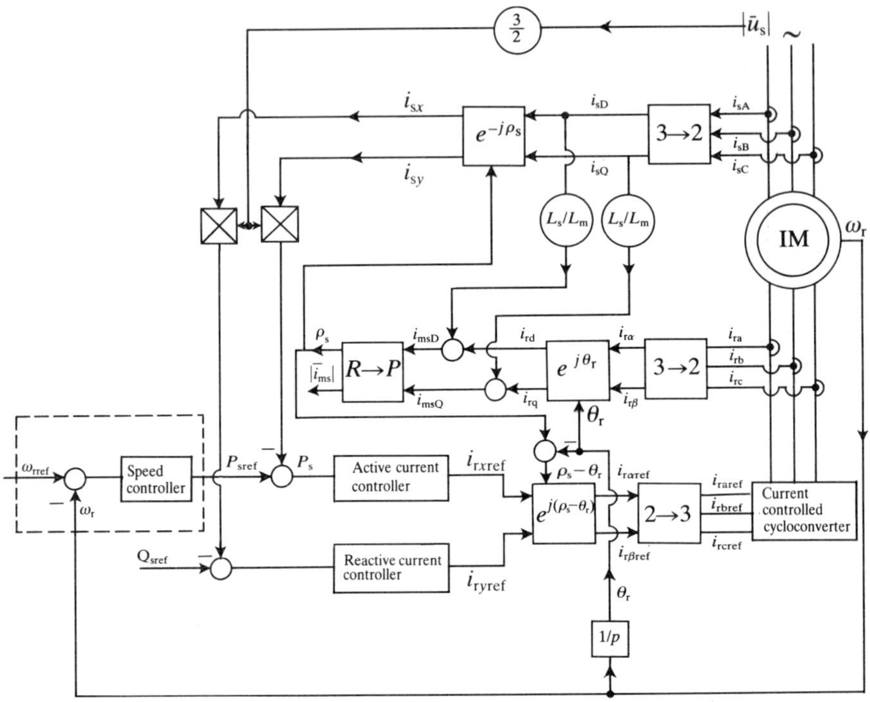

Fig. 4.39. Schematic of the stator-flux-oriented control of a doubly-fed induction machine with a current-controlled cycloconverter in the rotor.

and the stator reactive power as

$$Q_s=\tfrac{3}{2}\,\mathrm{Im}(\bar{u}_{s\psi s}\bar{i}^*_{s\psi s})=\tfrac{3}{2}(u_{sy}i_{sx}-u_{sx}i_{sy}). \tag{4.2-48}$$

From eqn (4.2-45) $u_{sx}=0$ and $\bar{u}_{sy}=|\bar{u}_s|$ and follows from eqns (4.2-47) and (4.12-48) that

$$P_s=\tfrac{3}{2}|\bar{u}_s|i_{sy} \tag{4.2-49}$$

and, since $|\bar{u}_s|$ is a constant, P_s is equal to a constant times the torque producing current, it also follows that

$$Q_s=\tfrac{3}{2}|\bar{u}_s|i_{sx} \tag{4.2-50}$$

and this is equal to a constant times the stator reactive current. Equations (4.2-49) and (4.2-50) are utilized in the implementation shown in Fig. 4.39. The modulus of the stator-voltage space phasor can be obtained from the monitored values of the stator voltages.

In Fig. 4.39 the space angle of the stator magnetizing-current space phasor (ρ_s) is obtained from the direct-axis (i_{msD}) and quadrature-axis (i_{msQ}) components of the stator magnetizing-current space phasor expressed in the stationary reference frame, by utilizing a rectangular-to-polar converter. This can be obtained by considering eqns (4.2-2), (4.2-3), and (4.2-4) and thus

$$\bar{\imath}_{ms}=i_{msD}+ji_{msQ}=\frac{L_s}{L_m}(i_{sD}+ji_{sQ})+(i_{r\alpha}+ji_{r\beta})e^{j\theta_r}$$

$$=\frac{L_s}{L_m}(i_{sD}+ji_{sQ})+(i_{rd}+ji_{rq})=|\bar{\imath}_{ms}|e^{j\rho_s}, \qquad (4.2\text{-}51)$$

where i_{sD}, i_{sQ} are the stator current components in the stationary reference frame and are obtained from the monitored three-phase stator currents (i_{sA}, i_{sB}, i_{sC}) by the application of the three-phase to two-phase transformation. The two-axis components of the rotor currents in the rotor reference frame ($i_{r\alpha}$, $i_{r\beta}$) are obtained from the monitored rotor currents (i_{ra}, i_{rb}, i_{rc}) by the application of the three-phase to two-phase transformation. The direct- and quadrature-axis rotor currents in the stator reference frame (i_{rd}, i_{rq}) are obtained from the rotor currents in the rotor reference frame ($i_{r\alpha}$, $i_{r\beta}$) by the application of the transformation $e^{j\theta_r}$ where θ_r is the rotor angle. It should be noted that eqn (4.2-51) can be obtained directly from eqn (2.1-128), which defines the stator flux-linkage space phasor in the stationary reference frame, if it is divided by the magnetizing inductance. It is also possible to obtain i_{msD}, i_{msQ} by using monitored stator voltages and currents. For this purpose $i_{msD}+ji_{msQ}=(\psi_{sD}+j\psi_{sQ})/L_m$ could bc utilized, where $\psi_{sD}+j\psi_{sQ}$ can be obtained by the integration of $(u_{sD}+ju_{sQ})-R_s(i_{sD}+ji_{sQ})$, where R_s is the stator resistance and u_{sD}, u_{sQ} are the two-axis components of the stator voltages in the stator reference frame.

In Fig. 4.39 the stator currents i_{sx}, i_{sy} are obtained from i_{sD}, i_{sQ} by utilizing eqn (4.2-3). The difference between the reference and actual values of the stator active power is the input to the active current controller, which is a PI controller and the output of which is the direct-axis reference rotor current component established in the stator-flux-oriented reference frame (i_{rxref}). Similarly, the reference value of the stator active power is compared with its actual value and their difference serves as the input to the reactive current controller, also a PI controller, and the output of which is i_{ryref}. The currents i_{rxref} and i_{ryref} are transformed into the two-axis rotor current references in the rotor reference frame $i_{r\alpha ref}$, $i_{r\beta ref}$ by using eqn (4.2-4). Finally by using the two-phase to three-phase transformation, these are transformed into the three-phase rotor current references (i_{raref}, i_{rbref}, i_{rcref}), which are required to control the cycloconverter.

It is also possible to implement a speed control loop, as shown in Fig. 4.39. In this case the output of the speed controller yields the referencc value of the stator active power. To achieve improved dynamic performance, the currents i_α and i_β, defined in eqn (4.2-40), should be added to the currents $i_{r\alpha ref}$ and $i_{r\beta ref}$ respectively, since then the effects of the voltage induced in the y-axis rotor winding due to the stator flux are cancelled. However, the same effect can be achieved by adding the current defined in eqn (4.2-38) to the reference current i_{ryref}.

Control of the induction machine with a static converter cascade in the rotor

It is possible to implement stator-flux-oriented control of a doubly fed induction machine with a static converter cascade in the rotor. This is similar to the control scheme described in Fig. 4.34, but there rotor-flux-oriented control was employed. Figure 4.40 shows the schematic of the drive system.

In the implementation shown in Fig. 4.40 a force-commutated thyristor converter is connected to the slip rings of the induction machine and this is connected via a d.c. link reactor to a naturally commutated controlled thyristor bridge converter (a special form of static Scherbius cascade). Thus slip power can flow in both directions and the speed of the induction machine can be controlled in the supersynchronous and subsynchronous speed ranges.

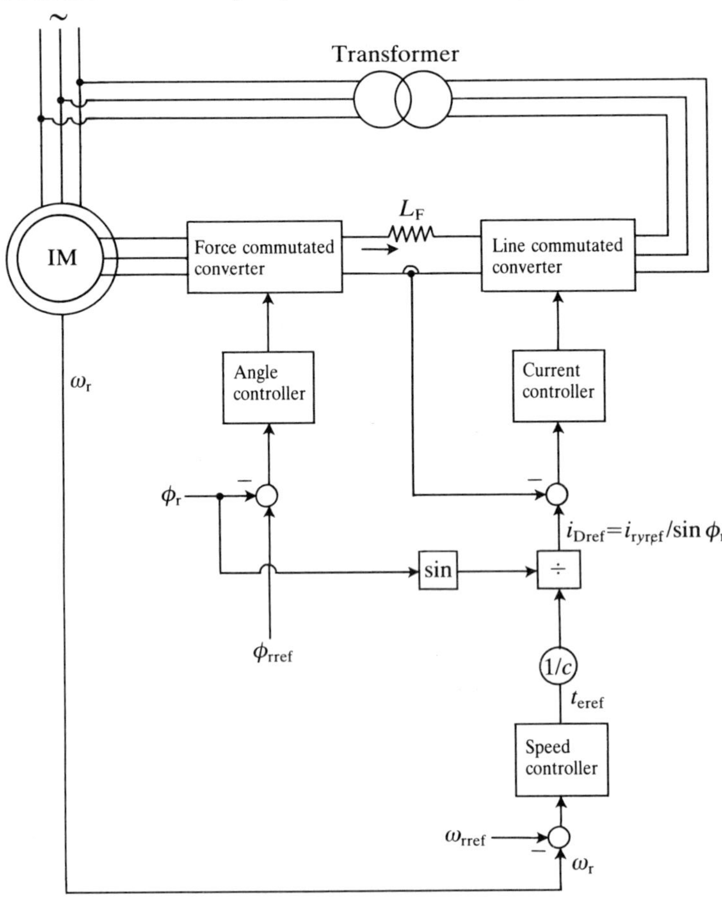

Fig. 4.40. Schematic of the stator-flux-oriented control of a doubly-fed induction machine with a static converter cascade in the rotor.

If the rotor-side converter was a line-commutated bridge converter, near to synchronous speed, when the a.c. voltage available to commutate this converter is small, the commutating ability of this converter would be lost. It is for this reason that a force-commutated rotor-side converter is used in the implementation shown in Fig. 4.40.

It follows from eqn (4.2-6) that in the stator-flux-oriented reference frame, if the stator magnetizing current $|\bar{\imath}_{ms}|$ is constant, the electromagnetic torque of the doubly fed induction machine is proportional to the quadrature-axis rotor current component (i_{ry}) and the torque can be controlled in the same way as for a separately excited d.c. machine where there is independent control of the excitation flux and the field current. For convenience eqn (4.2-6) is repeated here in a slightly different form,

$$t_e = c|\bar{\imath}_{ms}|i_{ry} = c|\bar{\imath}_{ms}||\bar{\imath}_r| \sin(\alpha_r + \theta_r - \rho_s), \tag{4.2-52}$$

where $c = -3L_m^2/(2L_s)$, $|\bar{\imath}_r|$ is the modulus of the rotor-current space phasor and the various angles are shown in Fig. 4.36. Thus the angle $\phi_r = (\alpha_r + \theta_r - \rho_s)$ is the angle of the rotor-current space phasor with respect to the real axis of the stator-flux-oriented reference frame (torque angle). In Fig. 4.40 the modulus of the rotor-current space phasor is controlled by the commutated converter and the angle ϕ_r is controlled by the machine-side converter.

In Fig. 4.40 the line-side converter is controlled by utilizing the closed-loop control of the d.c. link current, where the difference between the reference d.c. link current and the actual d.c. link current is fed into a PI controller. The torque reference (t_{eref}) is used to obtain the reference value of the d.c. link current (i_{Dref}) and the torque reference appears as the output of the speed controller, also a PI controller, and the input of which is the difference between the reference rotor speed (ω_{rref}) and the monitored rotor speed (ω_r). The reference value i_{Dref} is obtained as $i_{ryref}/\sin\phi_r$.

The rotor-side converter is controlled by utilizing the closed-loop control of the torque angle. For this purpose an angle controller is used, also a PI controller, the input of which is the difference between the reference (ϕ_{rref}) and the actual (ϕ_r) value of the torque angle.

It is possible to obtain the torque angle in several ways, and a method which does not utilize the stator voltages is now described. It has been shown that the torque angle is a function of three angles, $\phi_r = \alpha_r + \theta_r - \rho_s$. These can be obtained as follows. If eqn (4.2-51) is utilized, then the angle ρ_s can be obtained by monitoring the stator and rotor currents and the rotor angle, as shown in Fig. 4.39, and during the process of the determination of the angle ρ_s, and the rotor current components established in the rotor reference frame ($i_{r\alpha}$ and $i_{r\beta}$) are also obtained. Thus the angle α_r, which according to Fig. 4.36 is the angle of the rotor-current space phasor with respect to the real axis of the rotor reference frame, can be obtained if these two currents are inputs to a rectangular-to-polar converter. Finally, the rotor angle (θ_r) can be obtained by direct measurement.

4.3 MAGNETIZING-FLUX-ORIENTED CONTROL OF INDUCTION MACHINES

In this part of the chapter the magnetizing-flux-oriented control of induction machines will be discussed. The main assumptions are the same as those used in connection with rotor-flux-oriented control and stator-flux-oriented control of induction machines and in particular the effects of magnetic saturation are neglected. Because of the similarity between magnetizing-flux-oriented control and stator- and rotor-flux-oriented control, magnetizing-flux-oriented control will only be described briefly. However, it will be shown how this technique can be used in the case of a double-cage induction machine, whose space-phasor model will also be developed.

4.3.1 Expression for the electromagnetic torque

From the results presented in Section 2.1.8, it follows that the electromagnetic torque produced by an induction machine is proportional to the product of the modulus of the magnetizing-current space phasor ($|\bar{i}_{mm}|$) and the torque-producing stator current component (i_{sy}) and the torque can be controlled by the independent control of these two current components. It follows from Section 2.1.8 and also from Fig. 4.41 that in the magnetizing-flux-oriented reference frame $|\bar{i}_{mm}|=i_{mx}$, where i_{mx} is the direct-axis component of the magnetizing flux-linkage space phasor in the magnetizing-flux-oriented reference frame.

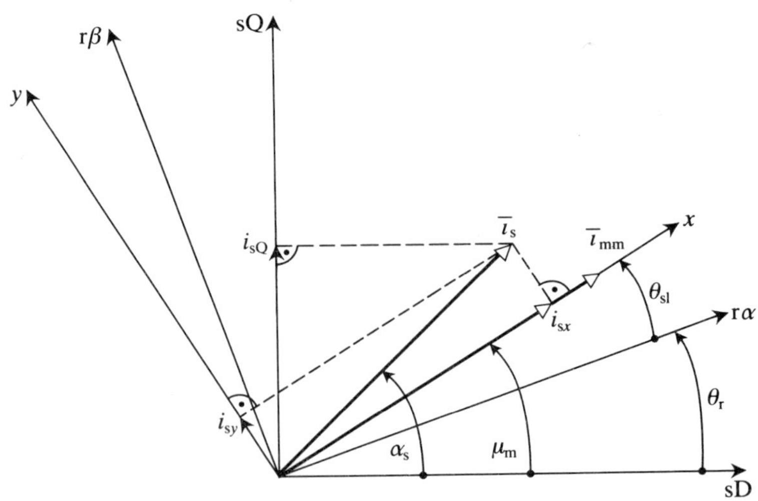

Fig. 4.41. Stator and magnetizing current space phasors.

Thus from eqns (2.1-186), and the relationship $|\bar{i}_{mm}| = i_{mx}$, it follows that the electromagnetic torque can be expressed as

$$t_e = \tfrac{3}{2}PL_m i_{mx} i_{sy} = \tfrac{3}{2}PL_m |\bar{i}_{mm}| i_{sy}, \tag{4.3-1}$$

where P is the number of pole pairs and L_m is the magnetizing inductance of the machine. In eqn (4.3-1) the magnetizing-current space phasor can be expressed in terms of the stator- and rotor-current space phasors in the magnetizing-flux-oriented reference frame $\bar{i}_{s\psi m}$, $\bar{i}_{r\psi m}$, which for simplicity are denoted by $\bar{i}_{sm}$ and $\bar{i}_{rm}$, as follows, if eqns (2.1-181), (2.1-184), and (2.1-185) are considered,

$$|\bar{i}_{mm}| = \bar{i}_{sm} + \bar{i}_{rm} = i_{sx} + ji_{sy} + i_{rx} + ji_{ry}, \tag{4.3-2}$$

where i_{sx} and i_{sy} are the direct- and quadrature-axis components of the stator currents in the magnetizing-flux-oriented reference frame and they are also shown in Fig. 4.41. i_{rx} and i_{ry} are the direct- and quadrature-axis components of the rotor currents in the magnetizing-flux-oriented reference frame. It should be noted that the space phasor of the stator currents in the magnetizing-flux-oriented reference frame ($\bar{i}_{sm}$) can be expressed as follows, if eqn (2.1-181) is considered,

$$\bar{i}_{sm} = \bar{i}_s e^{-j\mu_m} = (i_{sD} + ji_{sQ})e^{-j\mu_m} = i_{sx} + ji_{sy}, \tag{4.3-3}$$

where $\bar{i}_s$ is the space phasor of the stator currents in the stationary reference frame, i_{sD} and i_{sQ} are its two-axis components in the same reference frame and μ_m is the angle of the magnetizing-current (or magnetizing flux-linkage) space phasor with respect to the real axis of the stationary reference frame, as shown in Fig. 4.41. Similarly, the space phasor of the rotor currents in the magnetizing-flux-oriented reference frame can be expressed as follows, if eqn (2.1-185) is considered,

$$\bar{i}_{rm} = \bar{i}_r e^{-j(\mu_m - \theta_r)} = (i_{r\alpha} + ji_{r\beta})e^{-j(\mu_m - \theta_r)} = i_{rx} + ji_{ry}. \tag{4.3-4}$$

In eqn (4.3-4) $\bar{i}_r$ is the space phasor of the rotor currents in the reference frame fixed to the rotor, $i_{r\alpha}$ and $i_{r\beta}$ are its two-axis components in the rotor reference frame, and θ_r is the rotor angle, also shown in Fig. 4.41.

If the induction machine is supplied by impressed stator voltages or currents, eqn (4.3-1) is utilized when the magnetizing-flux-oriented control scheme is developed, but for a doubly fed induction machine, as in Section 4.2.1, the expression for the torque is put into a form which contains the rotor currents. Thus it follows from eqn (4.3-2) that

$$i_{sy} = -i_{ry} \tag{4.3-5}$$

and thus eqn (4.3-1) can be put into the form,

$$t_e = -\tfrac{3}{2}PL_m|\bar{i}_{mm}|i_{ry} = -\tfrac{3}{2}PL_m|\bar{i}_{mm}||\bar{i}_r| \sin(\alpha_r + \theta_r - \mu_m), \tag{4.3-6}$$

where i_{ry} has been obtained from eqn (4.3-4), and the Euler form of the space phasor of the rotor currents in the rotor reference frame, $\bar{\imath}_r = |\bar{\imath}_r| e^{j\alpha_r}$ (the angle α_r is the angle of $\bar{\imath}_r$ with respect to the rα-axis shown in Fig. 4.41) has been used. Equation (4.3-6) is similar to eqn (4.2-6), but as well as the differences in the constant parts, the modulus of the magnetizing-current space phasor is now present instead of the modulus of the stator magnetizing-current space phasor and the angle ρ_s is replaced by the angle μ_m. Equation (4.3-6) can be utilized for magnetizing-flux-oriented control of the doubly fed induction machine with impressed rotor currents.

4.3.2 Stator voltage equations

Here the space-phasor form of the stator voltage equation expressed in the magnetizing-flux-oriented reference frame will be obtained and this will be resolved into its two-axis components. For this purpose the voltage equations formulated in the general reference frame are utilized directly, these have been given in Section 2.1.7.

By considering eqns (2.1-148) and (2.1-150), which are the stator-voltage and flux-linkage space-phasor equations in the general reference frame, and by neglecting the effects of magnetic saturation, the space-phasor voltage equation of the stator will be as follows in the reference frame fixed to the magnetizing flux-linkage space phasor, which rotates at the speed $\omega_m = \mathrm{d}\mu_m/\mathrm{d}t$, where the angle μ_m has been defined above:

$$\bar{u}_{sm} = R_s \bar{\imath}_{sm} + L_s \frac{\mathrm{d}\bar{\imath}_{sm}}{\mathrm{d}t} + L_m \frac{\mathrm{d}\bar{\imath}_{rm}}{\mathrm{d}t} + j\omega_m (L_s \bar{\imath}_{sm} + L_m \bar{\imath}_{rm}). \qquad (4.3\text{-}7)$$

R_s and L_s are the stator resistance and self-inductance respectively and L_m is the magnetizing inductance and, as for the transformation of the stator currents given by eqn (4.1-27), $\bar{u}_{sm}$ is related to the space phasor of the stator voltages in the stationary reference frame ($\bar{u}_s$) by $\bar{u}_{sm} = \bar{u}_s e^{-j\mu_m} = (u_{sD} + ju_{sQ})e^{-j\mu_m} = u_{sx} + ju_{sy}$. By considering eqn (4.3-2), the rotor-current space phasor can be eliminated from eqn (4.3-7), since it follows that

$$\bar{\imath}_{rm} = |\bar{\imath}_{mm}| - \bar{\imath}_{sm} \qquad (4.3\text{-}8)$$

and when eqn (4.3-8) is substituted into eqn (4.3-7),

$$\bar{u}_{sm} = R_s \bar{\imath}_{sm} + L_{sl} \frac{\mathrm{d}\bar{\imath}_{sm}}{\mathrm{d}t} + L_m \frac{\mathrm{d}|\bar{\imath}_{mm}|}{\mathrm{d}t} + j\omega_m (L_{sl} \bar{\imath}_{sm} + L_m |\bar{\imath}_{mm}|), \qquad (4.3\text{-}9)$$

where L_{sl} is the leakage inductance of a stator winding, and $L_{sl} = L_s - L_m$ has been used. Physically it follows from eqn (4.3-9) that the stator-voltage space phasor is balanced by the stator ohmic drop ($R_s \bar{\imath}_{sm}$), a voltage drop across the stator leakage inductance ($L_{sl}\, \mathrm{d}\bar{\imath}_{sm}/\mathrm{d}t$), the magnetizing voltage drop ($L_m\, \mathrm{d}|\bar{\imath}_{mm}|/\mathrm{d}t$) and a rotational voltage component due to the rotation of the chosen reference frame: that last component is equal to $j\omega_m (L_{sl} \bar{\imath}_{sm} + L_m |\bar{\imath}_{mm}|)$, where $L_{sl} \bar{\imath}_{sm}$ is the stator leakage flux-linkage space phasor and $L_m |\bar{\imath}_{mm}|$ is the magnetizing flux-linkage space phasor and of course both are expressed in the magnetizing-flux-oriented reference frame.

Equation (4.3-9) is more complicated than eqn (4.2.10), which was expressed in the stator-flux-oriented reference frame and does not contain the derivative of the stator currents. Furthermore, in contrast to eqn (4.2-10), which does not contain a separate stator leakage voltage drop, in eqn (4.3-9) the stator leakage voltage drop is also present. Thus if the induction machine is supplied by impressed stator voltages, it is simpler to use the stator-flux-oriented control scheme than the magnetizing-flux-oriented control scheme. When compared with the corresponding stator voltage equation expressed in the rotor-flux-oriented reference frame (eqn 4.1-6), whilst in eqn (4.3-9) L_{sl} is present in the two corresponding terms, in eqn (4.1-6) the stator transient inductance (L_s') is present. Furthermore, in eqn (4.3-9) in two respective terms the modulus ($|\bar{\psi}_{mm}|=L_m|\bar{\imath}_{mm}|$) and the derivative of the magnetizing flux-linkage space phasor are present, whereas in the corresponding two terms of eqn (4.1-6) $|\bar{\imath}_{mr}|(L_s-L_s')=(L_m/L_r)|\bar{\psi}_{r\psi r}|$ and its derivative are present, where L_r is the self-inductance of the rotor and $|\bar{\psi}_{r\psi r}|$ is the modulus of the rotor flux-linkage space phasor.

Resolution of eqn (4.3-9) into its real- and imaginary-axis components yields

$$i_{sx}=\frac{u_{sx}}{R_s}-T_{sl}\frac{di_{sx}}{dt}-\frac{L_m}{R_s}\frac{d|\bar{\imath}_{mm}|}{dt}+\omega_m T_{sl} i_{sy} \tag{4.3-10}$$

$$i_{sy}=\frac{u_{sy}}{R_s}-T_{sl}\frac{di_{sy}}{dt}-\frac{\omega_m(L_{sl}i_{sx}+L_m|\bar{\imath}_{mm}|)}{R_s}, \tag{4.3-11}$$

where $T_{sl}=L_{sl}/R_s$ is the stator leakage time constant. Equations (4.3-10) and (4.3-11) must be considered for the magnetizing-flux-oriented control scheme of an induction machine supplied by impressed stator voltages. However, for this purpose, a decoupling circuit must be utilized, as discussed in the next section.

Decoupling circuit

It follows from eqns (4.3-10) and (4.3-11) that there are unwanted coupling terms and independent control of the flux- and torque-producing stator currents i_{sx} and i_{sy} is only possible if eqns (4.3-10) and (4.3-11) are decoupled and i_{sx} and i_{sy} are indirectly controlled by controlling the terminal voltages of the induction machine. For this purpose, as described previously, the following decoupling voltages are introduced, if it is assumed that the magnetizing flux is constant,

$$u_{dx}=-\omega_m L_{sl} i_{sy} \tag{4.3-12}$$

$$u_{dy}=\omega_m(L_{sl}i_{sx}+L_m|\bar{\imath}_{mm}|). \tag{4.3-13}$$

These equations are similar to eqns (4.1-9) and (4.1-10), which define the decoupling voltages in the case of rotor-flux-oriented control and the required decoupling circuit shown in Fig. 4.42 is similar to the one shown in Fig. 4.1.

Thus the flux- and torque-producing stator current components can be independently controlled if u_{dx} and u_{dy} are added to the outputs of the current controllers ($\hat{u}_{sx}$, $\hat{u}_{sy}$) which control i_{sx} and i_{sy} respectively, since $\hat{u}_{sx}+u_{dx}$ gives the direct-axis terminal voltage component in the magnetizing-flux-oriented reference frame and if the magnetizing flux is constant, it follows from eqn (4.3-10) that the voltage at the output of the direct-axis current controller is

$$\hat{u}_{sx}=R_s i_{sx}+L_{sl}\frac{di_{sx}}{dt}. \tag{4.3-14}$$

Thus $\hat{u}_{sx}$ directly controls i_{sx} through a simple delay element with the stator leakage time constant ($T_{sl}=L_{sl}/R_s$). Similarly, $\hat{u}_{sy}+u_{dy}$ gives the quadrature-axis terminal voltage in the same reference frame and it follows from eqn (4.3-11) that the voltage on the output of the quadrature-axis current controller is

$$\hat{u}_{sy}=R_s i_{sy}+L_{sl}\frac{di_{sy}}{dt} \tag{4.3-15}$$

and $\hat{u}_{sy}$ directly controls i_{sy} through a simple delay element with the time constant T_{sl}.

In the decoupling circuit shown in Fig. 4.42, the stator currents i_{sx} and i_{sy} can be obtained from the monitored three-phase stator currents by using eqn (4.3-3) and $|\bar{i}_{mm}|$ and ω_m (and μ_m) can be obtained either from a flux model which utilizes the rotor equations of the machine or by using direct measurements (e.g. Hall-sensors, special search coils, etc.). If the effect of inverter time delay is to be considered in the decoupling circuit, the method

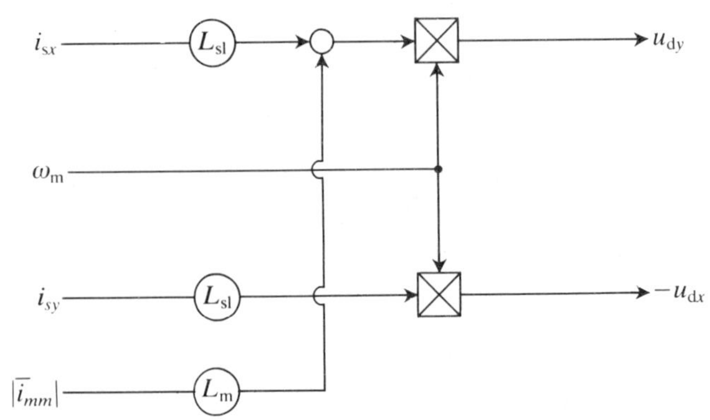

Fig. 4.42. Decoupling circuit to obtain the decoupling voltages u_{dx}, u_{dy}.

developed in Section 4.1.1 can be used and a decoupling circuit is obtained which is similar to that shown in Fig. 4.5.

When the machine is supplied by impressed stator currents, the stator voltage equations can be omitted from the equations of the drive.

4.3.3 Rotor voltage equations

The rotor voltage equations expressed in the magnetizing-flux-oriented reference frame can be used to obtain the modulus of the magnetizing flux-linkage space phasor ($|\bar{\psi}_{mm}|=L_m|\bar{i}_{mm}|$) and its speed ($\omega_m$) or its phase angle ($\mu_m$). For this purpose eqn (2.1-153) is used which gives the rotor voltage equation in a general reference frame. Thus by neglecting the effects of saturation, it follows from eqn (2.1-153) that in the magnetizing-flux-oriented reference frame ($\omega_g=\omega_m$, $\bar{i}_{rg}=\bar{i}_{rm}$, $\bar{i}_{sg}=\bar{i}_{sm}$) the rotor voltage equation is as follows if short-circuited rotor windings are assumed:

$$0=R_r\bar{i}_{rm}+L_r\frac{d\bar{i}_{rm}}{dt}+L_m\frac{d\bar{i}_{sm}}{dt}+j(\omega_m-\omega_r)(L_r\bar{i}_{rm}+L_m\bar{i}_{sm}), \quad (4.3\text{-}16)$$

where the stator- and rotor-current space phasors in the magnetizing-flux-oriented reference frame ($\bar{i}_{sm}$ and $\bar{i}_{rm}$) have been defined in eqns (4.3-3) and (4.3-4) respectively and ω_r is the rotor speed. Since eqn (4.3-16) contains the rotor-current space phasor, and the rotor currents cannot be directly measured in the case of a squirrel-cage induction machine, $\bar{i}_{rm}$ is eliminated by considering eqn (4.3-8) and

$$0=R_r(|\bar{i}_{mm}|-\bar{i}_{sm})+L_r\frac{d|\bar{i}_{mm}|}{dt}-L_{rl}\frac{d\bar{i}_{sm}}{dt}+j(\omega_m-\omega_r)(L_r|\bar{i}_{mm}|-L_{rl}\bar{i}_{sm}). \quad (4.3\text{-}17)$$

This equation is much more complicated than eqn (4.1-24), which defines the rotor voltage equation in the rotor-flux-oriented reference frame. Thus if the machine is supplied by impressed stator currents, magnetizing-flux-oriented control leads to more complicated implementation than the implementation based on rotor-flux-oriented control. When eqn (4.3-17) is compared with eqn (4.2-15), which gives the rotor voltage equation in the stator-flux-oriented reference frame, it can be seen that the two equations are similar.

Resolution of eqn (4.3-17) into its direct- and quadrature-axis components in the magnetizing-flux-oriented reference frame yields

$$(|\bar{i}_{mm}|+T_r\,d|\bar{i}_{mm}|/dt)/T_{rl}=\frac{di_{sx}}{dt}+\frac{i_{sx}}{T_{rl}}-\omega_{sl}i_{sy} \quad (4.3\text{-}18)$$

$$\omega_{sl}\left(|\bar{i}_{mm}|\frac{T_r}{T_{rl}}-i_{sx}\right)=\frac{di_{sy}}{dt}+\frac{i_{sy}}{T_{rl}}, \quad (4.3\text{-}19)$$

where $\omega_{sl}=\omega_m-\omega_r$ is the angular slip frequency and T_r and T_{rl} are the rotor time constant and rotor leakage time constant respectively ($T_r=L_r/R_r$, $T_{rl}=L_{rl}/R_r$). For an induction machine with impressed stator currents eqns (4.3-18) and (4.3-19) have to be considered. However, there is unwanted coupling between these two equations and as a consequence $|\bar{\imath}_{mm}|$ will be altered by a change in the torque-producing stator current component. This coupling can be removed by the application of a decoupling circuit, as obtained in the following section, where the magnetizing-flux-oriented control scheme of an induction machine supplied by a current-controlled PWM inverter is also described.

Decoupling circuit, control scheme of a current-controlled PWM inverter-fed induction machine

If the magnetizing current is controlled by a flux controller (which can be a PI controller) and the output of this controller is $\hat{\imath}_{sx}$, then the required decoupling current along the direct axis of the magnetizing-flux-oriented reference frame i_{dx} has to be added to $\hat{\imath}_{sx}$ to give the reference value i_{sxref}. Thus

$$i_{sx}=\hat{\imath}_{sx}+i_{dx} \tag{4.3-20}$$

and, when eqn (4.3-20) is substituted into eqn (4.3-18),

$$\frac{|\bar{\imath}_{mm}|+T_r\,\mathrm{d}|\bar{\imath}_{mm}|/\mathrm{d}t}{T_{rl}}=\frac{\mathrm{d}\hat{\imath}_{sx}}{\mathrm{d}t}+\frac{\hat{\imath}_{sx}}{T_{rl}}+\frac{\mathrm{d}i_{dx}}{\mathrm{d}t}+\frac{i_{dx}}{T_{rl}}-\omega_{sl}i_{sy}. \tag{4.3-21}$$

Thus $|\bar{\imath}_{mm}|$ can be decoupled from the torque-producing stator current if i_{dx} is selected in such a way that the term $\omega_{sl}i_{sy}$ should disappear in eqn (4.3-21). Thus it follows from eqn (4.3-20) that this can be satisfied if

$$i_{dx}=\frac{\omega_{sl}i_{sy}T_{rl}}{1+T_{rl}p} \tag{4.3-22}$$

where $p=\mathrm{d}/\mathrm{d}t$. The angular slip frequency in eqn (4.3-22) can be obtained from eqn (4.3-19) and it follows that it takes the form,

$$\omega_{sl}=\frac{(1+T_{rl}p)i_{sy}}{T_r|\bar{\imath}_{mm}|-T_{rl}i_{sx}}. \tag{4.3-23}$$

Equations (4.3-22) and (4.3-23) can be used to construct the required decoupling circuit. When this is used, the magnetizing-flux-oriented control scheme of an induction machine supplied by a current-controlled PWM inverter will be similar to that shown in Fig. 4.37, which uses stator-flux-oriented control. Naturally for the magnetizing-flux-oriented control the transformation $e^{j\rho_s}$ in Fig. 4.37 has to be replaced by the transformation $e^{j\mu_m}$ and $|\bar{\imath}_{ms}|$ and $|\bar{\imath}_{msref}|$ have to be replaced by $|\bar{\imath}_{mm}|$ and $|\bar{\imath}_{mmref}|$ respectively.

The modulus and space angle of the magnetizing flux-linkage space phasor can also be obtained from the stator voltages and currents. For this purpose the stator voltage equation is formulated in the stationary reference frame.

Thus by considering eqns (2.1-124) and (2.1-128), and also $L_s = L_{sl} + L_m$ and by neglecting the effects of magnetic saturation,

$$\bar{u}_s = R_s \bar{i}_s + L_{sl}\frac{d\bar{i}_s}{dt} + \frac{d\bar{\psi}_m}{dt} = u_{sD} + ju_{sQ}, \tag{4.3-24}$$

where $\bar{i}_s = i_{sD} + ji_{sQ}$ and $\bar{\psi}_m$ is the magnetizing flux-linkage space phasor in the stationary reference frame which has also been used in eqn (2.1-178),

$$\bar{\psi}_m = L_m(\bar{i}_s + \bar{i}_r e^{j\theta_r}) = L_m \bar{i}_m = L_m(i_{mD} + ji_{mQ}) = L_m|\bar{i}_{mm}|e^{j\mu_m}. \tag{4.3-25}$$

In eqn (4.3-25) $\bar{i}_m$ is the magnetizing flux-linkage space phasor in the stationary reference frame, which in polar form is $|\bar{i}_{mm}|e^{j\mu_m}$ (for simplicity, instead of $|\bar{i}_{mm}|$ the notation $|\bar{i}_m|$ can be used, since the modulus of the magnetizing-current space phasor in the stationary reference, $|\bar{i}_m|$, is equal to the modulus of the same quantity in the magnetizing-flux-oriented reference frame $|\bar{i}_{mm}|$). Thus it follows from eqns (4.3-24) and (4.3-25) that

$$\frac{di_{mD}}{dt} = u_{sD} - R_s i_{sD} - L_{sl}\frac{di_{sD}}{dt} \tag{4.3-26}$$

$$\frac{di_{mQ}}{dt} = u_{sQ} - R_s i_{sQ} - L_{sl}\frac{di_{sQ}}{dt} \tag{4.3-27}$$

and i_{mD} and i_{mQ} can be obtained by considering eqns (4.3-26) and (4.3-27) and by monitoring the terminal voltages and currents of the machine. Thus $i_{mD} + ji_{mQ} = |\bar{i}_{mm}|e^{j\mu_m}$ can be used to obtain the modulus and the space angle of the magnetizing current space phasor. In this case a flux model is obtained which is similar to that for an induction machine supplied by a voltage inverter. It is a disadvantage of the application of eqns (4.3-26) and (4.3-27) that they depend on the stator resistance and stator leakage inductance respectively and at low stator frequencies the stator ohmic drops will dominate. Thus prior to the integration of eqns (4.3-26) and (4.3-27), accurate stator voltage-drop compensation must be performed, which is a difficult task.

Equations (4.3-18) and (4.3-19) can be used to obtain the indirect magnetizing-flux-oriented control scheme of an induction machine with impressed stator currents. For this purpose the following expression for i_{sx} is obtained from eqn (4.3-18),

$$i_{sx} = \frac{(1 + T_r p)|\bar{i}_{mm}| + i_{sy} T_{rl}\omega_{sl}}{1 + T_{rl}p}, \tag{4.3-28}$$

where ω_{sl} has been defined in eqn (4.3-23).

Equations (4.3-28) and (4.3-23) are used in the indirect scheme shown in Fig. 4.43, where the induction machine is supplied by a current-controlled PWM inverter.

In Fig. 4.43(a) one of the inputs is the reference value of the modulus of the magnetizing current $|\bar{i}_{mmref}|$. The quadrature-axis stator current reference i_{syref}

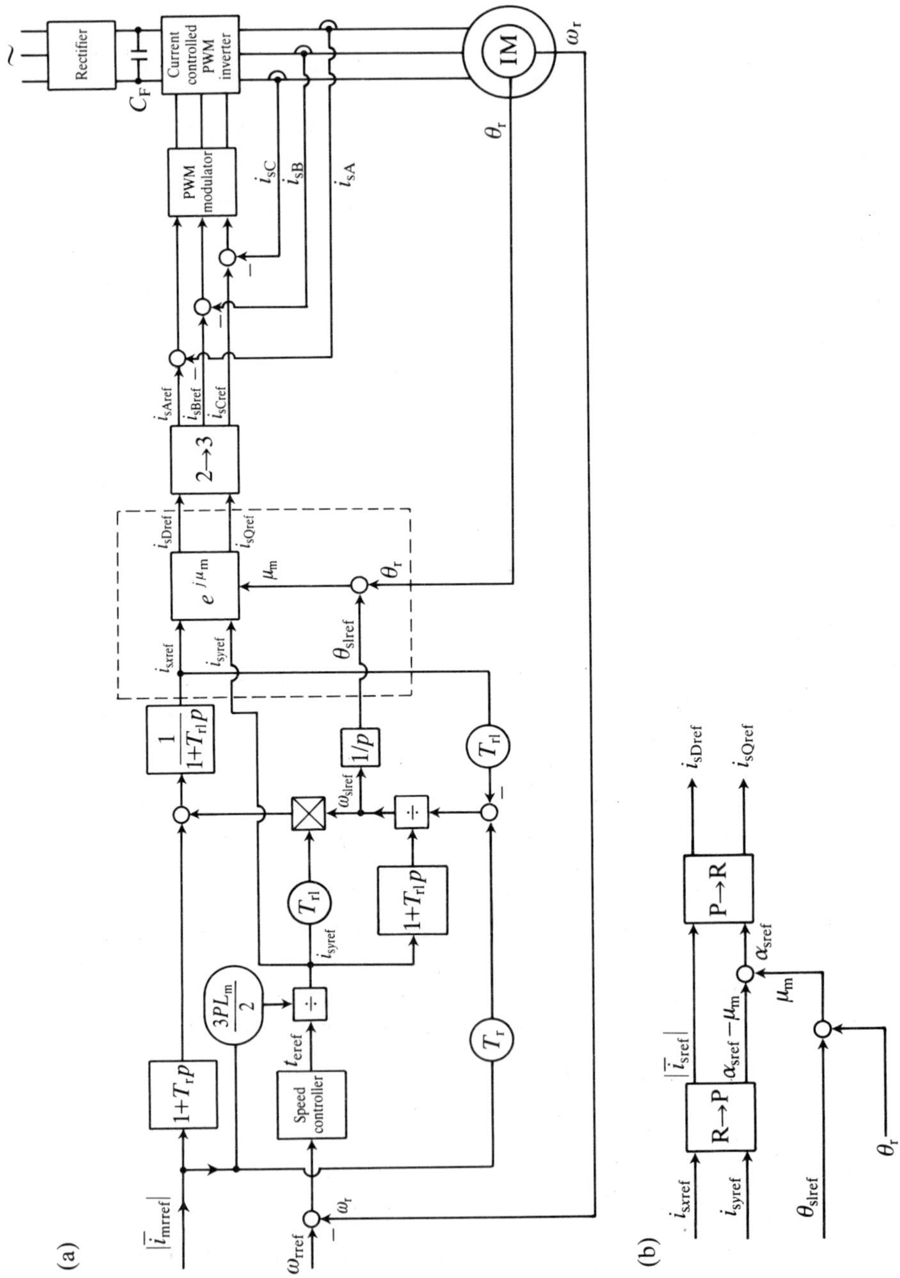

Fig. 4.43. Indirect magnetizing-flux-oriented control of a current-controlled PWM inverter-fed induction machine. (a) In Cartesian coordinates. (b) In polar coordinates.

is obtained from the torque reference (t_{eref}) by utilizing eqn (4.3-1) and the torque reference is obtained as the output of the speed controller. In Fig. 4.43(a), within the dashed lines, the two-axis stator current references expressed in the stationary reference frame, i_{sDref} and i_{sQref} are obtained by the application of the transformation $e^{j\mu_m}$ in accordance with eqn (4.3-3). These are then transformed into the three-phase reference values by the application of the three-phase to two-phase transformation. The angle μ_m is obtained in two steps; first the angular slip frequency is integrated and this gives the slip angle θ_{sl}, which is also shown in Fig. 4.41 and is the angle of the magnetizing-current space phasor with respect to the real axis of the rotor reference frame. During the second step, the monitored value of the rotor angle (θ_r) is added to the slip angle, and thus the angle of the magnetizing-current space phasor with respect to the direct-axis of the stationary reference frame (μ_m) is obtained, as shown in Fig. 4.41.

In Fig.4.43(a) the part shown within the dashed lines can be replaced by the part shown in Fig. 4.43(b). In this case, i_{sxref} and i_{syref} are inputs to a rectangular-to-polar converter, the inputs of which are the reference value of the modulus of the stator-current space phasor ($|\bar{i}_{sref}|$) and the space angle of the stator-current space phasor with respect to the direct axis of the magnetizing-flux-oriented reference frame ($\alpha_{sref}-\mu_m$) (see Fig. 4.41). From Fig. 4.41 the sum of the slip angle and the rotor angle give μ_m and when this is added to ($\alpha_{sref}-\mu_m$), α_{sref} is obtained, which is the angle of the stator-current space phasor with respect to the direct axis of the stationary reference frame. Finally, a polar-to-rectangular converter is used to obtain i_{sDref} and i_{sQref}.

When the implementation given in Fig. 4.43(a) is compared with the implementation shown in Fig. 4.27(a), which corresponds to indirect rotor-flux-oriented control of the induction machine supplied by a current-controlled PWM inverter, it can be seen that it is much simpler to implement the rotor-flux-oriented control.

Furthermore, in the steady state, with constant-magnetizing-flux control, the electromagnetic torque has a pull-out value (t_{emax}) and this corresponds to the pull-out value of the angular slip frequency (ω_{slmax}). Beyond the pull-out value of the slip frequency, the induction machine can have a static stability limit. However, in contrast to this, in the steady state, with constant-rotor-flux control, there is no pull-out torque (if there is no limitation on the stator currents) and there is no static stability limit. With constant-stator-flux control, in the steady state there is also a pull-out torque and thus a static stability limit. These conclusions can be proved by considering the equations for the torque in the steady state. This will be discussed in the next section and the stability limits will be derived.

4.3.4 Steady-state stability limit

For magnetizing-flux-oriented control it follows from eqn (4.3-17) that in the steady state (when all the derivative terms are zero), the stator-current space phasor in the magnetizing-flux-oriented reference frame can be expressed as

$$\bar{i}_{sm} = |\bar{i}_{mm}| \frac{R_r + j\omega_{sl}L_r}{R_r + j\omega_{sl}L_{rl}} = i_{sx} + ji_{sy}. \tag{4.3-29}$$

The torque producing stator current (i_{sy}) is equal to the imaginary part of eqn (4.3-29) and thus

$$i_{sy} = \frac{\omega_{sl}R_r L_m|\bar{i}_{mm}|}{R_r^2 + \omega_{sl}^2 L_{rl}^2} = \frac{\omega_{sl}R_r|\bar{\psi}_{mm}|}{R_r^2 + \omega_{sl}^2 L_{rl}^2}, \tag{4.3-30}$$

where $|\bar{\psi}_{mm}|$ is the modulus of the magnetizing flux-linkage space phasor. Thus by the substitution of eqn (4.3-30) into eqn (4.3-1), the following expression is obtained for the electromagnetic torque:

$$t_e = \frac{\frac{3}{2}PR_r|\bar{\psi}_{mm}|^2\omega_{sl}}{R_r^2 + \omega_{sl}^2 L_{rl}^2}. \tag{4.3-31}$$

This (and the torque-producing stator current) will have their maxima at the following value of the angular slip frequency,

$$\omega_{slmax} = \pm\frac{R_r}{L_{rl}} = \pm\frac{1}{T_{rl}}, \tag{4.3-32}$$

where T_{rl} is the leakage time constant of the rotor. Equation (4.3-32) can also be proved by differentiation of the electromagnetic torque (or the torque-producing stator current) with respect to the angular slip frequency, and setting this first derivative to zero. Substitution of eqn (4.3-32) into eqn (4.3-31) yields

$$t_{emax} = \pm\frac{\frac{3}{4}P|\bar{\psi}_{mm}|^2}{L_{rl}} = \pm\frac{\frac{3}{4}PL_m^2|\bar{i}_{mm}|^2}{L_{rl}} = \tfrac{3}{2}P|\bar{\psi}_{mm}|i_{symax}, \tag{4.3-33}$$

where i_{symax} is the maximal value of the torque-producing stator current. When the angular slip frequency is higher than ω_{slmax}, static instability arises. It follows from eqns (4.3-32) and (4.3-33) that the pull-out angular slip frequency depends only on the rotor leakage time constant and does not depend on the magnetizing flux. However, the pull-out torque is proportional to the square of the modulus of the magnetizing flux-linkage space phasor and thus a small increase of the magnetizing flux will cause a significant increase in the electromagnetic torque.

It is also possible to obtain the limit of static stability in terms of the torque-producing stator current. From eqn (4.3-33)

$$i_{symax} = \frac{\pm L_m|\bar{i}_{mm}|}{2L_{rl}} \tag{4.3-34}$$

and it follows that unstable operation arises if the torque-producing stator current is larger than $L_m|\bar{i}_{mm}|/(2L_{rl})$. If $L_{rl}/L_m = 1/30$, which is a typical value, instability only arises if $i_{sy} > 15|\bar{\imath}_{mm}|$, which is an unusually high value under steady-state conditions.

The static-stability limit in the case of stator-flux-oriented control with constant stator flux can be similarly determined by utilizing eqns (4.2-1) and (4.2-15). It follows from eqn (4.2-15), that the stator-current space phasor in the steady state is

$$\bar{\imath}_{s\psi s} = \frac{L_m}{L_s} \frac{|\bar{\imath}_{ms}|(R_r + j\omega_{sl} L_r)}{R_r + j\omega_{sl} L'_r} \tag{4.3-35}$$

and the torque-producing stator current (i_{sy}) is equal to its imaginary-axis component. The maximum of this is at

$$\omega_{slmax} = \pm \frac{1}{T'_r}, \tag{4.3-36}$$

where T'_r is the rotor transient time constant and thus i_{symax} can also be obtained. By the substitution of i_{symax} into eqn (4.2-1), the pull-out torque is obtained as

$$t_{emax} = \pm \frac{3}{4} P \left(\frac{L_m}{L_s}\right)^2 L_m^2 |\bar{\imath}_{ms}|^2 / L'_r = \pm \frac{3}{4} P \left(\frac{L_m}{L_s}\right)^2 |\bar{\psi}_s|^2 / L'_r$$
$$= \tfrac{3}{2} P L_m |\bar{\imath}_{ms}| i_{symax}, \tag{4.3-37}$$

where $|\bar{\psi}_s|$ is the modulus of the stator flux-linkage space phasor. If the angular slip frequency is larger than ω_{slmax}, static instability arises. It follows from eqn (4.3-36) that this critical value of the angular slip frequency depends only on the rotor transient time constant and is not determined by the flux level. However, the pull-out torque depends on the square of the modulus of the stator linkage space phasor. It follows from eqn (4.3-37) that

$$i_{symax} = \pm \frac{L_m^3 |\bar{\imath}_{ms}|}{2 L_s^2 L'_r} \tag{4.3-38}$$

and static instability arises if $i_{sy} > L_m^3 |\bar{\imath}_{ms}| / (2 L_s^2 L'_r)$. By assuming equal stator and rotor leakages ($L_{sl} = L_{rl}$), the following inequality is obtained

$$i_{sy} > \frac{(1-\sigma)|\bar{\imath}_{ms}|}{2\sigma(1 + L_{rl}/L_m)},$$

where σ is the resultant leakage constant, $\sigma = 1 - L_m^2/(L_s L_r)$. Thus when typical values of σ and L_{rl}/L_m are chosen, e.g. $\sigma = 1/15$ and $L_{rl}/L_m = 1/30$, it follows that the instability only arises if the torque-producing stator current is greater than $6.78|\bar{\imath}_{ms}|$. However, in the steady state it is unusual to have such a high torque-producing current and thus this instability does not cause a serious practical problem. It is also possible to obtain eqn (4.3-37) or eqn(4.3-38) by considering the general steady-state equivalent circuit of the induction machine shown in Fig. 4.10(b) and by utilizing that the general turns ratio is equal to $a = L_s/L_m$.

Finally for the case of rotor-flux-oriented control it follows from eqn (4.1-24), that in the steady state the torque-producing stator current can be expressed as

$$i_{sy} = \omega_{sl} T_r |\bar{i}_{mr}|. \tag{4.3-39}$$

This varies linearly with ω_{sl} (when $|\bar{i}_{mr}|$ = constant and T_r = constant), and thus the torque will also vary linearly with the angular slip frequency and there is no pull-out slip and no pull-out torque. This has also been discussed in Section 4.1.1.

4.3.5 Magnetizing-flux-oriented control of double-cage induction machines

In all the previous sections, the squirrel-cage induction machine has been represented by its single-cage model and the effects of deep bars have not been considered. However, during transient operation, in an induction machine with deep bars, the deep-bar effect can significantly influence the rotor time constants of the machine, which are important parameters during vector control. A similar situation exists in a squirrel-cage machine with double-cage rotor. If the rotor time constants are inaccurate, the angular slip frequency (ω_{sl}) will be inaccurate, and this will result in an inaccurate slip angle (θ_{sl}). Thus the angle of the magnetizing flux-linkage space phasor (μ_m), which is equal to the sum of the slip angle and the monitored rotor angle (θ_r) (see Fig. 4.41), will also be inaccurate. Therefore the desired decoupling of flux- and torque-producing stator current components (i_{sx}, i_{sy}) cannot be achieved and this can lead to unwanted oscillations and a degradation of the performance of the induction machine. Compensation for the deep bar effects can improve the performance of the drive. For this purpose the angular slip frequency has to be calculated in such a way that it contains the effects of the deep bars (or double cage). In this section, the space-phasor model of the double-cage induction machine is developed and magnetizing-flux-oriented control of the double-cage machine with impressed stator currents is discussed in detail.

In practice, squirrel-cage rotors can have deep-bar rotors (with narrow and deep bars) or double-cage rotors with an outer and inner cage. In the case of the deep-bar rotor, it is possible to consider a bar with a large number of layers, and the leakage inductance of the bottom layer is greater than that of the top layer since the bottom layer links a greater amount of leakage flux. Thus the leakage inductance of the outer layer will be low and the leakage inductance of the inner layer will be high. Since all the layers can be considered to be electrically connected in parallel, when a.c. current flows in the cage (e.g. when the machine is started and the frequency of the rotor currents is high and equal to the stator frequency), the current in the upper layer (with low leakage reactance) will be larger than the current in a lower layer (with high leakage reactance). Thus the current will mainly flow in the top layers and the uneven rotor current distribution will result in an increase of effective rotor resistance and thus high starting torque is produced. When the machine accelerates, the rotor frequency decreases, so that at nominal speed the rotor frequency is very low and thus the effective rotor resistance is almost equal to the d.c. value. Thus in the induction machine with deep-bar rotor, high starting torque (resulting from high effective rotor resistance) and good running performance (resulting from low effective rotor resistance) are automatically ensured. As mentioned above, a rotor bar can be divided into several layers. In the dynamic state, to achieve high accuracy, several layers must be considered, but even the application of a model where a rotor bar is divided into two equal sections can give a substantial improvement in the computed characteristics, when compared with the application of a model where the rotor bar is not divided into layers. When a rotor bar is divided into two equal layers, the resistances of the two layers will be the same.

Similar conditions can be achieved by the application of a double-cage winding. In this case, the outer bars have smaller cross section, and thus they have a larger resistance than that of the inner bars. Furthermore, the leakage inductance of the inner bars is larger than that of the outer bars, since the inner bars link more leakage flux. During starting, most of the rotor current will flow in the outer bars which have high resistance and small leakage reactance. Thus high starting torque is produced. At nominal speed, the rotor frequency is so low that the reactance of the inner cage is considerably lower than its resistance and the effective rotor resistance will be low and will be approximately equal to the d.c. resistance of the two cages in parallel.

In the following sections the dynamic model of the double-cage induction machine is established by the utilization of space-phasor theory. Although the goal is to obtain the dynamic model in the magnetizing-flux-oriented reference frame, which could be directly obtained by considering the voltage equations in the general reference frame, for better understanding of the various quantities (space phasors of various flux linkages, space phasors of rotor currents in the upper and lower cages etc.) first the space-phasor voltage equations of the stator and rotor are established in the stationary reference frame. This will be followed by transformation of these into the magnetizing-flux-oriented reference frame.

Stator- and rotor-voltage space-phasor equations in the stationary reference frame

The main assumptions are those used previously, in particular, the effects of magnetic saturation are neglected and the rotor is assumed to consist of two cages. Figure 4.44 shows the construction of the rotor. It is assumed that there is no mutual leakage between the stator and the upper cage (outer cage). Furthermore, in accordance with the discussion presented above, the leakage

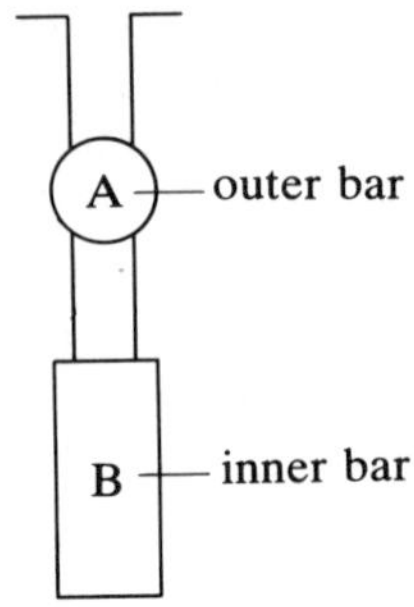

Fig. 4.44. Schematic of a double-cage.

inductance of the upper cage is neglected. It is assumed that an upper bar and a lower bar occupy the same slot.

As in eqn (2.1-124), the space-phasor form of the stator voltage equation, in the stationary reference frame is

$$\bar{u}_s = R_s \bar{i}_s + \frac{d\bar{\psi}_s}{dt}, \tag{4.3-40}$$

where $\bar{u}_s$ and $\bar{i}_s$ are the space phasors of the stator voltages and currents respectively in the stationary reference frame, R_s is the stator resistance and $\bar{\psi}_s$ is the space phasor of the stator flux linkages in the stationary reference frame. This contains three components. The first component is a flux linkage due to the stator currents, $L_s \bar{i}_s$, where L_s is the self-inductance of the stator, which is the sum of the stator leakage inductance L_{sl} and the magnetizing inductance L_m ($L_s = L_{sl} + L_m$). The second component is a mutual flux linkage due to the rotor currents in the upper bars (denoted by A in Fig. 4.44), $L_{mA} \bar{i}'_{rA}$, where L_{mA} is the mutual inductance between the stator and the upper cage and $\bar{i}'_{rA}$ is the space phasor of the rotor currents in the upper bars, but expressed in the stationary reference frame. Since the mutual leakage inductance between the stator and the upper cage has been neglected, it follows that $L_{mA} = L_m$. Finally the third component is a mutual flux-linkage component which is due to the rotor currents in the lower bars (denoted by B in Fig. 4.44), $L_{mB} \bar{i}'_{rB}$, where L_{mB} is the mutual inductance between the stator and the bottom cage and $\bar{i}'_{rB}$ is the space phasor of the rotor currents in the bottom bars, but expressed in the stationary reference frame. From the assumptions made above $L_{mA} = L_{mB} = L_m$. Thus $\bar{\psi}_s$ can be expressed as

$$\bar{\psi}_s = L_{sl} \bar{i}_s + L_m(\bar{i}_s + \bar{i}'_{rA} + \bar{i}'_{rB}) = L_{sl} \bar{i}_s + \bar{\psi}'_m, \tag{4.3-41}$$

where $\bar{\psi}'_m$ is the magnetizing flux-linkage space phasor in the stationary reference frame,

$$\bar{\psi}'_m = L_m(\bar{i}_s + \bar{i}'_{rA} + \bar{i}'_{rB}) = L_m \bar{i}'_m = L_m |\bar{i}_{mm}| e^{j\mu_m} \tag{4.3-42}$$

and in eqn (4.3-42) $\bar{i}'_m$ is the magnetizing current in the stationary reference frame, $|\bar{i}_{mm}|$ is its modulus and μ_m is its angle with respect to the real axis of the stationary reference frame, as shown in Fig. 4.41.

The space-phasor forms of the rotor voltage equations corresponding to the upper and lower cages, when expressed in the stationary reference frame, are similar to eqn (2.1-125),

$$0 = R_{rA}\bar{i}'_{rA} + \frac{d\bar{\psi}'_{rA}}{dt} - j\omega_r\bar{\psi}'_{rA} \qquad (4.3\text{-}43)$$

and

$$0 = R_{rB}\bar{i}'_{rB} + \frac{d\bar{\psi}'_{rB}}{dt} - j\omega_r\bar{\psi}'_{rB}. \qquad (4.3\text{-}44)$$

In eqn (4.3-43), $\bar{\psi}'_{rA}$ is the space phasor of the rotor flux linkages which link the upper cage and it is expressed in the stationary reference frame. $\bar{\psi}'_{rB}$ is the space phasor of the rotor flux linkages which link the bottom cage and it is also expressed in the stationary reference frame. From the assumptions made above, $\bar{\psi}'_{rA}$ will contain three components, two of which are leakage components, $L_{AB}\bar{i}'_{rA}$ and $L_{AB}\bar{i}'_{rB}$, where L_{AB} is the mutual leakage inductance between the upper and lower cages. The third component is equal to the magnetizing flux-linkage space phasor $\bar{\psi}'_m$ defined by eqn (4.3-42). Thus

$$\bar{\psi}'_{rA} = L_{AB}(\bar{i}'_{rA} + \bar{i}'_{rB}) + \bar{\psi}'_m. \qquad (4.3\text{-}45)$$

In eqn (4.3-45) it has also been assumed that the leakage inductance of the upper cage is neglected. However, the leakage inductance of the lower cage $(L_{rBl} - L_{rl})$ is not neglected and thus $\bar{\psi}'_{rB}$ in eqn (4.3-44) can be expressed in a similar way to eqn (4.3-45), but it also has to include the leakage flux component $L_{rl}\bar{i}'_{rB}$. Thus

$$\bar{\psi}'_{rB} = L_{rl}\bar{i}'_{rB} + L_{AB}(\bar{i}'_{rA} + \bar{i}'_{rB}) + \bar{\psi}'_m. \qquad (4.3\text{-}46)$$

Equations (4.3-40)–(4.3-46) define the voltage equations of the double-cage machine under dynamic conditions in the stationary reference frame. In the following section these equations will be expressed in the magnetizing-flux-oriented reference frame. It should be noted that, since there are two rotor flux-linkage space phasors (corresponding to the upper and lower cages respectively), in the case of the double-cage machine, it is not possible to establish a rotor-flux-oriented reference frame, where the stator currents can be decoupled into flux- and torque-producing components. This is in contrast to the single-cage induction machine, where it has been possible to use the rotor-flux-oriented reference frame and the stator currents have been decoupled into flux- and torque-producing components.

Stator- and rotor-voltage space-phasor equations in the magnetizing-flux-oriented reference frame

The voltage equations of the double-cage machine, expressed in the magnetizing-flux-oriented reference frame can be obtained directly from the corresponding voltage equations expressed in the stationary reference frame by utilizing the appropriate transformations between these two reference frames. These will be described in the present section. The stator voltage equations expressed in the magnetizing-flux-oriented reference frame are required for the magnetizing-flux-oriented control scheme of the double-cage machine supplied by impressed stator voltages, but are not required when the same machine is supplied by impressed stator currents. In the latter case only the rotor voltage equations are required.

Stator voltage equations: By considering eqns (4.3-40) and (4.3-41), together with the required transformations of the stator voltage, current and flux-linkage space phasors [see eqns (4.3-48), (4.3-49), and (4.3-50)], it follows that in the special reference frame fixed to the magnetizing flux-linkage space phasor, which rotates at the speed of ω_m, the stator-voltage space-phasor equation will be

$$\bar{u}_{sm} = R_s \bar{i}_{sm} + \frac{d\bar{\psi}_{sm}}{dt} + j\omega_m \bar{\psi}_{sm}. \tag{4.3-47}$$

This is similar to eqn (2.1-148), which gives the stator-voltage space-phasor equation in the general reference frame. The required transformations used for the derivation of eqn (4.3-47) are as follows. The space phasor of the stator currents in the magnetizing-flux-oriented reference frame ($\bar{i}_{sm}$) has been defined in eqn (2.1-181), but for convenience it is repeated here,

$$\bar{i}_{sm} = \bar{i}_s e^{-j\mu_m} = (i_{sD} + j i_{sQ}) e^{-j\mu_m} = i_{sx} + j i_{sy}, \tag{4.3-48}$$

and the space phasor of the stator voltages in the same reference frame can similarly be defined as

$$\bar{u}_{sm} = \bar{u}_s e^{-j\mu_m} = (u_{sD} + j i_{sQ}) e^{-j\mu_m} = u_{sx} + j u_{sy}. \tag{4.3-49}$$

In these equations the components with the subscripts D and Q are direct- and quadrature-axis components in the stationary reference frame and the components with subscripts x and y are the direct- and quadrature-axis components in the magnetizing-flux-oriented reference frame (see Fig. 4.41). In eqn (4.3-47), $\bar{\psi}_{sm}$ is the space phasor of the stator flux linkages in the magnetizing-flux-oriented reference frame and the following transformation has been utilized during the derivation of eqn (4.3-47):

$$\bar{\psi}_{sm} = \bar{\psi}_s e^{-j\mu_m} = (\psi_{sD} + j\psi_{sQ}) e^{-j\mu_m} = \psi_{sx} + j\psi_{sy}, \tag{4.3-50}$$

which is similar to the transformation defined by eqns (4.3-48) and (4.3-49). By substitution of eqn (4.3-41) into eqn (4.3-50) and by utilizing the definition of $\bar{i}_{sm}$ given in eqn (4.3-48) and the definition of $\bar{\psi}'_m$ given in eqn (4.3-42),

$$\begin{aligned} \bar{\psi}_{sm} &= \bar{\psi}_s e^{-j\mu_m} = (L_{sl}\bar{i}_s + \bar{\psi}'_m) e^{-j\mu_m} = L_{sl}\bar{i}_s e^{-j\mu_m} + \bar{\psi}'_m e^{-j\mu_m} \\ &= L_{sl}\bar{i}_{sm} + L_m |\bar{i}_{mm}| = \psi_{sx} + j\psi_{sy}, \end{aligned} \tag{4.3-51}$$

where ψ_{sx} and ψ_{sy} are the direct- and quadrature-axis components of the stator flux-linkage space phasor in the magnetizing-flux-oriented reference frame and it follows from eqn (4.3-51) that they can be expressed as

$$\psi_{sx}=L_{sl}i_{sx}+L_m|\bar{i}_{mm}|=L_{sl}i_{sx}+|\bar{\psi}_{mm}| \tag{4.3-52}$$

$$\psi_{sy}=L_{sl}i_{sy}, \tag{4.3-53}$$

where $|\bar{\psi}_{mm}|$ is the modulus of the magnetizing flux-linkage space phasor and is equal to the space phasor of the magnetizing flux linkages in the magnetizing-flux-oriented reference frame. Thus the two-axis forms of the stator voltage equation in the magnetizing-flux-oriented reference frame are obtained by resolving eqn (4.3-47) into its real- and imaginary-axis components,

$$u_{sx}=R_si_{sx}+\frac{d\psi_{sx}}{dt}-\omega_m\psi_{sy} \tag{4.3-54}$$

$$u_{sy}=R_si_{sy}+\frac{d\psi_{sy}}{dt}+\omega_m\psi_{sx} \tag{4.3-55}$$

where ψ_{sx} and ψ_{sy} have been defined in eqns (4.3-52) and (4.3-53) respectively.

By multiplying eqn (4.3-42) by $e^{-j\mu_m}$, it follows that the space phasor of the magnetizing flux linkages in the magnetizing-flux-oriented reference frame is

$$\begin{aligned}\bar{\psi}'_m e^{-j\mu_m}&=L_m(\bar{i}_s e^{-j\mu_m}+\bar{i}'_{rA}e^{-j\mu_m}+\bar{i}'_{rB}e^{-j\mu_m})\\&=L_m|\bar{i}_{mm}|=|\bar{\psi}_{mm}|=\psi_{mx}+j\psi_{my},\end{aligned} \tag{4.3-56}$$

where according to eqn (4.3-48) $\bar{i}_s e^{-j\mu_m}=\bar{i}_{sm}$ is the stator-current space phasor in the magnetizing-flux-oriented reference frame and the rotor-current phasors $\bar{i}'_{rA}e^{-j\mu_m}$ and $\bar{i}'_{rB}e^{-j\mu_m}$ are the space phasors of the rotor currents in the upper and lower cages respectively, but expressed in the magnetizing-flux-oriented reference frame,

$$\bar{i}_{rAm}=\bar{i}'_{rA}e^{-j\mu_m}=i_{rAx}+ji_{rAy} \tag{4.3-57}$$

$$\bar{i}_{rBm}=\bar{i}'_{rB}e^{-j\mu_m}=i_{rBx}+ji_{rBy} \tag{4.3-58}$$

It should be noted that $\bar{i}'_{rA}$ and $\bar{i}'_{rB}$ are the space phasors of the rotor currents in the upper and lower cages respectively, but expressed in the stationary reference frame. Equations (4.3-57) and (4.3-58) are similar to eqn (4.3-4), which defines the space phasor of the rotor currents in the magnetizing-flux-oriented reference frame for a single-cage machine. In eqns (4.3-57) and (4.3-58) i_{rAx}, i_{rAy}, and i_{rBx}, i_{rBy} are the two-axis rotor currents of the upper and lower cages respectively in the magnetizing-flux-oriented reference frame. Thus by resolving eqn (4.3-56) into its real- and imaginary-axis components,

$$\psi_{mx}=|\bar{\psi}_{mm}|=L_m(i_{sx}+i_{rAx}+i_{rBx}) \tag{4.3-59}$$

$$\psi_{my}=0=L_m(i_{sy}+i_{rAy}+i_{rBy}). \tag{4.3-60}$$

These two equations will be used below, when the state-variable equations of the double-cage induction machine with impressed stator currents are obtained, to express $i_{rAx}+i_{rBx}$ in terms of $|\psi_{mm}|$ and i_{sx}, and to express $(i_{rAy}+i_{rBy})$ in terms of i_{sy}.

Equations (4.3-54), (4.3-55), (4.3-52), and (4.3-53) have to be considered if the double-cage induction machine is supplied by impressed stator voltages. They can be put into a form which is similar to eqns (4.3-10) and (4.3-11), for the single-cage machine. Thus the required decoupling circuit can be constructed in a similar way to that shown in Fig. 4.42 and it should be noted that $|\bar{\psi}_{mm}|=L_m|\bar{i}_{mm}|$. In such a decoupling circuit, the stator currents i_{sx} and i_{sy} can be obtained from the monitored three-phase stator currents by utilizing eqn (4.3-48), and $|\bar{i}_{mm}|$ and ω_m (and μ_m) can be obtained either from a flux-model which uses the rotor equations of the machine expressed in the magnetizing-flux-oriented reference frame or by using direct measurements (e.g. Hall-sensors, special search coils, etc.). The rotor voltage equations will be obtained in the following section. They will also be used to obtain the magnetizing-flux-oriented control scheme of the double-cage machine with impressed stator currents.

Rotor voltage equations: By utilizing the required transformations [see eqns (4.3-57), (4.3-58), (4.3-63), and (4.3-64)], it follows from eqns (4.3-43) and (4.3-44) that the rotor-voltage space-phasor equations in the magnetizing-flux-oriented reference frame will take the form,

$$0=R_{rA}\bar{i}_{rAm}+\frac{d\bar{\psi}_{rAm}}{dt}+j(\omega_m-\omega_r)\bar{\psi}_{rAm} \tag{4.3-61}$$

$$0=R_{rB}\bar{i}_{rBm}+\frac{d\bar{\psi}_{rBm}}{dt}+j(\omega_m-\omega_r)\bar{\psi}_{rBm} \tag{4.3-62}$$

These are similar to the rotor voltage equation of a single-cage machine expressed in the same reference frame, which can be obtained from eqn (2.1-149). For the derivation of eqns (4.3-61) and (4.3-62) eqns (4.3-57) and (4.3-58) have also been utilized; they give the space-phasor forms of the rotor currents of the upper and lower cages respectively in the magnetizing-flux-oriented reference frame ($\bar{i}_{rAm}$, $\bar{i}_{rBm}$). A similar definition has been used for the space phasors $\bar{\psi}_{rAm}$ and $\bar{\psi}_{rBm}$, which are the space phasors of the rotor flux linkages for the upper and lower cages, expressed in the magnetizing-flux-oriented reference frame,

$$\bar{\psi}_{rAm}=\bar{\psi}'_{rA}e^{-j\mu_m}=\psi_{rAx}+j\psi_{rAy} \tag{4.3-63}$$

$$\bar{\psi}_{rBm}=\bar{\psi}'_{rB}e^{-j\mu_m}=\psi_{rBx}+j\psi_{rBy} \tag{4.3-64}$$

where $\bar{\psi}'_{rA}$ and $\bar{\psi}'_{rB}$ have been defined in eqns (4.3-45) and (4.3-46) respectively.

By substituting eqn (4.3-45) into eqn (4.3-63) and considering eqns (4.3-57), (4.3-58), and (4.3-56),

$$\bar{\psi}_{\mathrm{rAm}} = [L_{\mathrm{AB}}(\bar{i}'_{\mathrm{rA}} + \bar{i}'_{\mathrm{rB}}) + \bar{\psi}'_{\mathrm{m}}]e^{-j\mu_{\mathrm{m}}} = L_{\mathrm{AB}}(\bar{i}_{\mathrm{rAm}} + \bar{i}_{\mathrm{rBm}}) + |\bar{\psi}_{\mathrm{mm}}|. \quad (4.3\text{-}65)$$

Similarly, substituting eqn (4.3-46) into eqn (4.3-64) and considering eqns (4.3-57), (4.3-58), and (4.3-56) gives

$$\bar{\psi}_{\mathrm{rBm}} = [L_{\mathrm{rl}}\bar{i}'_{\mathrm{rB}} + L_{\mathrm{AB}}(\bar{i}'_{\mathrm{rA}} + \bar{i}'_{\mathrm{rB}}) + \bar{\psi}'_{\mathrm{m}}]e^{-j\mu_{\mathrm{m}}}$$

$$= L_{\mathrm{rl}}\bar{i}_{\mathrm{rBm}} + L_{\mathrm{AB}}(\bar{i}_{\mathrm{rAm}} + \bar{i}_{\mathrm{rBm}}) + |\bar{\psi}_{\mathrm{mm}}|. \quad (4.3\text{-}66)$$

Resolution of eqns (4.3-61), (4.3-62), (4.3-65), and (4.3-66) into their real- and imaginary-axis components give

$$0 = R_{\mathrm{rA}}i_{\mathrm{rA}x} + \frac{\mathrm{d}\psi_{\mathrm{rA}x}}{\mathrm{d}t} - (\omega_{\mathrm{m}} - \omega_{\mathrm{r}})\psi_{\mathrm{rA}y} \quad (4.3\text{-}67)$$

$$0 = R_{\mathrm{rA}}i_{\mathrm{rA}y} + \frac{\mathrm{d}\psi_{\mathrm{rA}y}}{\mathrm{d}t} + (\omega_{\mathrm{m}} - \omega_{\mathrm{r}})\psi_{\mathrm{rA}x} \quad (4.3\text{-}68)$$

$$0 = R_{\mathrm{rB}}i_{\mathrm{rB}x} + \frac{\mathrm{d}\psi_{\mathrm{rB}x}}{\mathrm{d}t} - (\omega_{\mathrm{m}} - \omega_{\mathrm{r}})\psi_{\mathrm{rB}y} \quad (4.3\text{-}69)$$

$$0 = R_{\mathrm{rB}}i_{\mathrm{rB}y} + \frac{\mathrm{d}\psi_{\mathrm{rB}y}}{\mathrm{d}t} + (\omega_{\mathrm{m}} - \omega_{\mathrm{r}})\psi_{\mathrm{rB}x} \quad (4.3\text{-}70)$$

where

$$\psi_{\mathrm{rA}x} = L_{\mathrm{AB}}(i_{\mathrm{rA}x} + i_{\mathrm{rB}x}) + |\bar{\psi}_{\mathrm{mm}}| \quad (4.3\text{-}71)$$

$$\psi_{\mathrm{rA}y} = L_{\mathrm{AB}}(i_{\mathrm{rA}y} + i_{\mathrm{rB}y}) \quad (4.3\text{-}72)$$

$$\psi_{\mathrm{rB}x} = L_{\mathrm{rl}}i_{\mathrm{rB}x} + L_{\mathrm{AB}}(i_{\mathrm{rA}x} + i_{\mathrm{rB}x}) + |\bar{\psi}_{\mathrm{mm}}| \quad (4.3\text{-}73)$$

$$\psi_{\mathrm{rB}y} = L_{\mathrm{rl}}i_{\mathrm{rB}y} + L_{\mathrm{AB}}(i_{\mathrm{rA}y} + i_{\mathrm{rB}y}), \quad (4.3\text{-}74)$$

where $|\bar{\psi}_{\mathrm{mm}}|$ has been defined in eqn (4.3-59). The stator voltage equations, eqns (4.3-54) and (4.3-55), which use the expressions for the stator flux linkages, eqns (4.3-52) and (4.3-53) together with the rotor voltage equations, eqns (4.3-67)–(4.3-70), which use the rotor flux-linkage equations, eqns (4.3-71)–(4.3-74), describe the dynamic behaviour of the double-cage induction machine if the equation of motion is also considered. The expression for the electromagnetic torque will be obtained in the next section.

The electromagnetic torque: The electromagnetic torque can be obtained in a similar way to that shown in eqn (4.3-1) for the single-cage machine. Physically this must be the case since it has been shown earlier that the electromagnetic torque can be considered to be produced by the interaction of the stator flux linkages and the stator currents and from this point of view, it is irrelevant how many cages are there on the rotor. Mathematically, it is possible to prove in several ways that the expression for the electromagnetic torque in the double-cage machine is similar to that given by eqn (4.3-1), but, of course, the modulus of the magnetizing flux-linkage space phasor ($|\bar{\psi}_{mm}|$) is different for the double-cage machine from that for the single-cage machine, since as well as the stator currents, the rotor currents of both rotor cages contribute to the magnetization current. Furthermore the torque-producing stator current (i_{sy}) is different in the double-cage machine from that of the single-cage machine. It is very convenient to utilize a form of the torque equation which contains the stator flux-linkage space phasor, rather than other forms where two rotor flux-linkage space phasors have to be used.

In eqn (2.1-167) the electromagnetic torque has been defined by utilizing the space phasors of the stator flux linkages and the stator current in the general reference frame. Thus when the magnetizing-flux-oriented reference frame is used, it follows from eqn (2.1-167) that

$$t_e = \tfrac{3}{2}P\bar{\psi}_{sm} \times \bar{i}_{sm}, \tag{4.3-75}$$

where $\bar{\psi}_{sm}$ and $\bar{i}_{sm}$ have been defined in eqns (4.3-48) and (4.3-51) respectively. By performing the vector product, the following expression is obtained for the electromagnetic torque, if the stator flux-linkage components are replaced by the expressions given in eqns (4.3-52) and (4.3-53)

$$\begin{aligned} t_e &= \tfrac{3}{2}P(\psi_{sx}i_{sy} - \psi_{sy}i_{sx}) \\ &= \tfrac{3}{2}P[(L_{sl}i_{sx} + |\bar{\psi}_{mm}|)i_{sy} - L_{sl}i_{sy}i_{sx}] = \tfrac{3}{2}P|\bar{\psi}_{mm}|)i_{sy}, \end{aligned} \tag{4.3-76}$$

which as expected, agrees with the form of eqn (4.3-1). In eqn (4.3-76) the modulus of the magnetizing flux-linkage space phasor, ($|\bar{\psi}_{mm}|$), is defined in eqn (4.3-59) and the torque-producing stator current (i_{sy}) is related to the direct-axis and quadrature-axis stator currents in the stationary reference frame by eqn (4.3-48).

Magnetizing-flux-oriented control of the double-cage induction machine with impressed stator currents

In this chapter the indirect magnetizing-flux-oriented control scheme of the double-cage induction machine with impressed stator currents is developed. For this purpose the rotor equations derived in the stationary reference frame are utilized together with the expression for the electromagnetic torque given in eqn (4.3-76). However, the rotor voltage equations are reformulated in such a way that the stator currents i_{sx}, i_{sy} are retained, since their reference values transformed into the stationary reference frame command the current-fed converter. Furthermore, the modulus of the magnetizing flux-linkage space phasor, and its space angle ($|\bar{\psi}_{mm}|$, μ_m) are selected as state variables since

$|\bar{\psi}_{mm}|$ will be an input command and μ_m will be controlled indirectly. Since there are four rotor voltage equations, two other state variables also have to be selected. These are chosen to be the real- and imaginary-axis components (ψ_{rBlx}, ψ_{rBly}) of the space phasor of the rotor leakage flux linkages in the lower cage. This space phasor has been defined in eqn (4.3-66) as $\bar{\psi}_{rBl}=L_{rl}\bar{i}_{rBm}$ and thus its two-axis components in the magnetizing-flux-oriented reference frame are

$$\psi_{rBlx}=L_{rl}i_{rBx} \tag{4.3-77}$$

$$\psi_{rBly}=L_{rl}i_{rBy}. \tag{4.3-78}$$

These are selected as state variables since they are essential for the description of the double-cage (deep-bar) effect. The direct-axis stator current (i_{sx}) and the angular slip frequency (ω_{sl}) are strongly dependent on these. It should be noted that L_{rl} is the leakage inductance of the bottom cage, and in the transient state, the non-uniform current distribution between the upper and lower cages depends on the energy stored in this inductance.

The four rotor voltage equations, eqns (4.3-67)–(4.3-70), are now rearranged to contain the required four state variables. For this purpose, first the rotor flux-linkage components given by eqns (4.3-71)–(4.3-74) are substituted into eqns (4.3-67)–(4.3-70). The resulting equations will thus contain the four rotor current components (i_{rAx}, i_{rAy}, i_{rBx}, i_{rBy}):

$$0=R_{rA}i_{rAx}+\frac{d|\bar{\psi}_{mm}|}{dt}+L_{AB}\frac{d(i_{rAx}+i_{rBx})}{dt}-\omega_{sl}L_{AB}(i_{rAy}+i_{rBy}) \tag{4.3-79}$$

$$0=R_{rA}i_{rAy}+L_{AB}\frac{d(i_{rAy}+i_{rBy})}{dt}+\omega_{sl}[|\bar{\psi}_{mm}|+L_{AB}(i_{rAx}+i_{rBx})] \tag{4.3-80}$$

$$0=R_{rB}i_{rBx}+\frac{d|\bar{\psi}_{mm}|}{dt}+\frac{d(L_{rl}i_{rBx})}{dt}+L_{AB}\frac{d(i_{rAx}+i_{rBx})}{dt}$$
$$-\omega_{sl}[L_{rl}i_{rBy}+L_{AB}(i_{rAy}+i_{rBy})] \tag{4.3-81}$$

$$0=R_{rB}i_{rBy}+\frac{d(L_{rl}i_{rBy})}{dt}+L_{AB}\frac{d(i_{rAy}+i_{rBy})}{dt}$$
$$+\omega_{sl}[L_{rl}i_{rBx}+L_{AB}(i_{rAx}+i_{rBx})+|\bar{\psi}_{mm}|], \tag{4.3-82}$$

where $\omega_{sl}=(\omega_m-\omega_r)$ is the angular slip frequency.

However, the rotor current components can be eliminated by expressing them in terms of the four state-variables, since it follows from eqns (4.3-77) and (4.3-78) that

$$i_{rBx}=\frac{\psi_{rBlx}}{L_{rl}} \tag{4.3-83}$$

$$i_{\mathrm{rB}y}=\frac{\psi_{\mathrm{rBl}y}}{L_{\mathrm{rl}}},\tag{4.3-84}$$

and from eqns (4.3-59) and (4.3-60) that

$$i_{\mathrm{rA}x}=\frac{|\bar{\psi}_{\mathrm{mm}}|}{L_{\mathrm{m}}}-i_{\mathrm{s}x}-i_{\mathrm{rB}x}=\frac{\bar{\psi}_{\mathrm{mm}}}{L_{\mathrm{m}}}-i_{\mathrm{s}x}-\frac{\psi_{\mathrm{rBl}x}}{L_{\mathrm{rl}}}\tag{4.3-85}$$

$$i_{\mathrm{rA}y}=-i_{\mathrm{s}y}-i_{\mathrm{rB}y}=-i_{\mathrm{s}y}-\frac{\psi_{\mathrm{rBl}y}}{L_{\mathrm{rl}}},\tag{4.3-86}$$

where $i_{\mathrm{rB}x}$ and $i_{\mathrm{rB}y}$ have been substituted by the expressions given by eqns (4.3-83) and (4.3-84) respectively. Thus when eqns (4.3-83)–(4.3-86) are substituted into the rotor voltage equations, eqns (4.3-79)–(4.3-82), the following four equations are obtained. The direct-axis rotor voltage equation of the upper cage yields

$$\frac{\mathrm{d}|\bar{\psi}_{\mathrm{mm}}|}{\mathrm{d}t}=-\frac{R_{\mathrm{rA}}|\bar{\psi}_{\mathrm{mm}}|}{L_{\mathrm{m}}+L_{\mathrm{AB}}}+\frac{L_{\mathrm{m}}}{L_{\mathrm{m}}+L_{\mathrm{AB}}}\left(R_{\mathrm{rA}}i_{\mathrm{s}x}+L_{\mathrm{AB}}\frac{\mathrm{d}i_{\mathrm{s}x}}{\mathrm{d}t}+\frac{R_{\mathrm{rA}}\psi_{\mathrm{rBl}x}}{L_{\mathrm{rl}}}-L_{\mathrm{AB}}\omega_{\mathrm{sl}}i_{\mathrm{s}y}\right).\tag{4.3-87}$$

The quadrature-axis rotor voltage equation of the upper cage yields

$$\omega_{\mathrm{sl}}=\left[R_{\mathrm{rA}}\left(i_{\mathrm{s}y}+\frac{\psi_{\mathrm{rBl}y}}{L_{\mathrm{rl}}}\right)+L_{\mathrm{AB}}\frac{\mathrm{d}i_{\mathrm{s}y}}{\mathrm{d}t}\right]\Big/\psi_{\mathrm{rA}x},\tag{4.3-88}$$

where in the denominator,

$$\psi_{\mathrm{rA}x}=|\bar{\psi}_{\mathrm{mm}}|\frac{(L_{\mathrm{m}}+L_{\mathrm{AB}})}{L_{\mathrm{m}}}-L_{\mathrm{AB}}i_{\mathrm{s}x}\tag{4.3-89}$$

is the direct-axis flux-linkage component of the upper cage. It follows from the direct-axis rotor voltage equation of the lower cage that

$$\frac{\mathrm{d}\psi_{\mathrm{rBl}x}}{\mathrm{d}t}=-\frac{R_{\mathrm{rB}}\psi_{\mathrm{rBl}x}}{L_{\mathrm{rl}}}-\frac{L_{\mathrm{AB}}+L_{\mathrm{m}}}{L_{\mathrm{m}}}\frac{\mathrm{d}|\bar{\psi}_{\mathrm{mm}}|}{\mathrm{d}t}+L_{\mathrm{AB}}\frac{\mathrm{d}i_{\mathrm{s}x}}{\mathrm{d}t}-\omega_{\mathrm{sl}}L_{\mathrm{AB}}i_{\mathrm{s}y}+\omega_{\mathrm{sl}}\psi_{\mathrm{rBl}y}.\tag{4.3-90}$$

Finally, the following equation is obtained from the quadrature-axis rotor voltage equation of the lower cage,

$$\frac{\mathrm{d}\psi_{\mathrm{rBl}y}}{\mathrm{d}t}=-\frac{R_{\mathrm{rB}}\psi_{\mathrm{rBl}y}}{L_{\mathrm{rl}}}-\omega_{\mathrm{sl}}\psi_{\mathrm{rBl}x}+L_{\mathrm{AB}}\frac{\mathrm{d}i_{\mathrm{s}y}}{\mathrm{d}t}-\omega_{\mathrm{sl}}\psi_{\mathrm{rA}x},\tag{4.3-91}$$

where $\psi_{\mathrm{rA}x}$ has been defined in eqn (4.3-89). However, eqn (4.3-91) contains the first derivative $\mathrm{d}i_{\mathrm{s}y}/\mathrm{d}t$ and this can be eliminated by using eqn (4.3-88).

Thus from eqn (4.3-88) ψ_{rAx} is expressed in terms of ω_{sl}, i_{sy}, and ψ_{rBly} and, when this is substituted into eqn (4.3-91),

$$\frac{d\psi_{rBly}}{dt}=-\frac{R_{rA}+R_{rB}}{L_{rl}}\psi_{rBly}-R_{rA}i_{sy}-\omega_{sl}\psi_{rBlx}. \tag{4.3-92}$$

Thus eqns (4.3-87), (4.3-88), (4.3-89), (4.3-90), and (4.3-92), together with the expression of the electromagnetic torque, eqn (4.3-76), can be used to obtain the indirect magnetizing-flux-oriented control scheme of the double-cage induction machine supplied by impressed stator currents. For this purpose it should be considered that if the reference value of the modulus of the magnetizing flux-linkage space phasor ($|\bar{\psi}_{mmref}|$ and the reference value of the torque-producing stator current (i_{syref}) are known, in the indirect control scheme i_{sxref} and ω_{slref} have to be determined. Thus these quantities have to be expressed in terms of ($|\bar{\psi}_{mmref}|$) and (i_{syref}), but in accordance with eqns (4.3-87)–(4.3-90) and eqn (4.3-92), they will depend on the leakage flux components ψ_{rBlx} and ψ_{rBly}. Therefore to obtain the control scheme, the equations given above have to be arranged for i_{sx} and ω_{sl}, and the leakage fluxes of the bottom cages (ψ_{rBlx} and ψ_{rBly}) have to be expressed in terms of $|\bar{\imath}_{mm}|$, i_{sx}, and i_{sy}. Thus when eqn (4.3-87) is rearranged, the following equation is obtained for the direct-axis stator current

$$(1+T_{AB}p)i_{sx}=(1+T_{mAB}p)|\bar{\imath}_{mm}|+T_{AB}\omega_{sl}i_{sy}-\frac{\psi_{rBlx}}{L_{rl}}, \tag{4.3-93}$$

where $p=d/dt$ and the mutual time constants associated with the upper cage are

$$T_{AB}=\frac{L_{AB}}{R_{rA}},\qquad T_{mAB}=T_m+T_{AB},\quad \text{and}\quad T_m=\frac{L_m}{R_{rA}}.$$

Equation (4.3-88) yields the following expression for the angular slip frequency

$$\omega_{sl}=\frac{i_{sy}(1+T_{AB}p)+\psi_{rBly}/L_{rl})}{T_{mAB}|\bar{\imath}_{mm}|-T_{AB}i_{sx}}. \tag{4.3-94}$$

In eqns (4.3-93) and (4.3-94) the rotor leakage-flux linkages are also present. These are obtained from eqns (4.3-89) and (4.3-90) as

$$(p+T_{Bl})\psi_{rBlx}=-pR_{rA}(T_{mAB}|\bar{\imath}_{mm}|-T_{AB}i_{sx})-\omega_{sl}L_{AB}i_{sy}+\omega_{sl}\psi_{rBly}, \tag{4.3-95}$$

where $T_{Bl}=R_{rB}/L_{rl}$ is the leakage time constant of the lower cage,

$$(p+T_{ABl})\psi_{rBly}=-R_{rA}i_{sy}-\omega_{sl}\psi_{rBlx} \tag{4.3-96}$$

and T_{ABl} is the total leakage time constant of the upper and lower cages,

$$T_{ABl}=T_{Bl}+T_{Al}\quad \text{where } T_{Al}=\frac{R_{rA}}{L_{rl}}.$$

Equations (4.3-93)–(4.3-96) are used in Fig. 4.45, which shows the

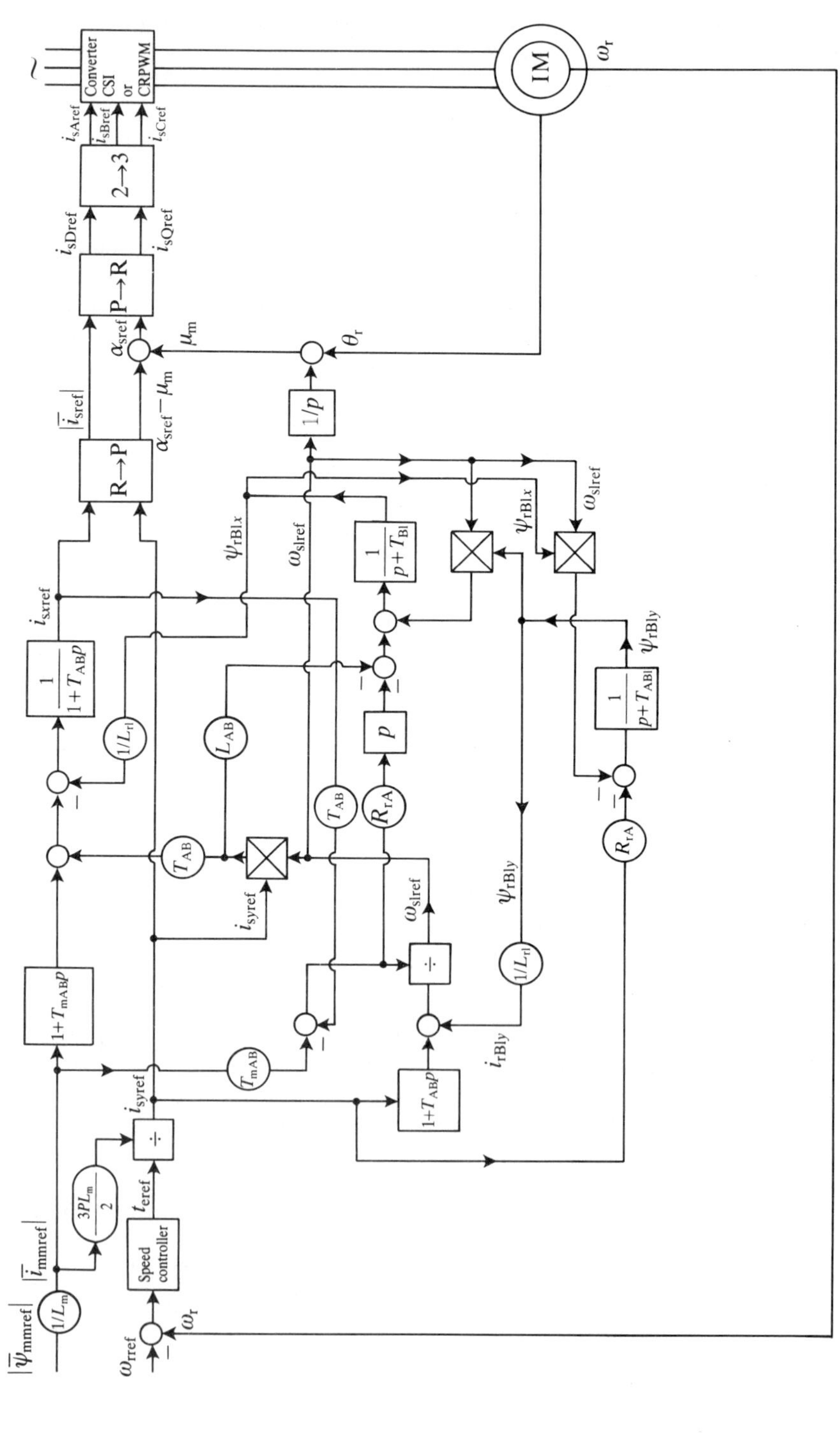

Fig. 4.45. Schematic of the indirect magnetizing-flux-oriented control of a double-cage induction machine with impressed stator currents utilizing an asymmetrical rotor flux-linkage estimator.

magnetizing-flux-oriented indirect control scheme of the double-cage induction machine with impressed stator currents. The converter which supplies the induction machine can be either a CSI or a current-controlled PWM inverter.

In Fig.4.45 the required reference values of the stator currents are obtained in a way similar to that shown in Fig.4.43 for the single-cage machine. Thus by using the reference values of the direct- and quadrature-axis stator current components expressed in the magnetizing-flux-oriented reference frame (i_{sxref}, i_{syref}), it is possible to obtain the reference value of the space phasor of the stator currents in the magnetizing-flux-oriented reference frame,

$$(i_{sxref}+ji_{syref})=|\bar{i}_{sref}|e^{j(\alpha_{sref}-\mu_m)} \tag{4.3-97}$$

(see Fig. 4.41). Equation (4.3-97) can be used to obtain the reference values of the direct- and quadrature-axis stator currents in the stationary reference frame (i_{sDref}, i_{sQref}), by either of the methods shown in Fig. 4.43(a) or Fig. 4.43(b). If the method of Fig. 4.43(b) is used, $|\bar{i}_{sref}|$ and $(\alpha_{sref}-\mu_m)$ are obtained by polar-to-rectangular conversion from i_{sxref} and i_{syref}, since eqn (4.3-97) is utilized. When the angular slip frequency (ω_{sl}), obtained from eqn (4.3-94), is integrated, the slip angle (θ_{sl}) is obtained and when the monitored rotor angle (θ_r) is added to this, the angle of the magnetizing flux-linkage space phasor with respect to the direct-axis of the magnetizing oriented reference frame (μ_m) is obtained (this also follows from Fig. 4.41) as

$$\mu_m=\theta_{slref}+\theta_r. \tag{4.3-98}$$

Thus when μ_m is added to $(\alpha_{sref}-\mu_m)$, α_{sref} is obtained, which is the space angle of the stator-current space phasor with respect to the direct axis of the stationary reference frame (see Fig. 4.41).

In Fig. 4.45, the circuit which produces the required leakage-flux components of the bottom cage (ψ_{rBlx}, ψ_{rBly}), contains a differentiator (indicated by the block containing the operator p) and furthermore, it is asymmetrical. However, it is possible to construct a circuit which produces the required rotor leakage fluxes, where differentiation is not required and which is symmetrical. For this purpose, eqn (4.3-95) is rearranged to resemble the simple structure of eqn (4.3-96), but eqn (4.3-96) itself will not be changed.

Thus when the expression for the voltage component $R_{rA}|\bar{\psi}_{mm}|/L_{rl}$ is taken from eqn (4.3-93) [or it may be easier to use eqn (4.3-87) directly] and the expression is added to the expression for the rotational voltage, $\omega_{sl}\psi_{rBly}$, which can be obtained from eqn (4.3-95) [or eqn (4.3-90)], the following voltage equation is obtained,

$$(p+T_{ABl})\psi_{rBlx}=R_{rA}(|\bar{i}_{mm}|-i_{sx})+\omega_{sl}\psi_{rBly}. \tag{4.3-99}$$

The structure of eqn (4.3-99) resembles that of eqn (4.3-96) and thus eqns (4.3-96) and (4.3-99) describe a rotor flux-linkage estimator circuit which is symmetrical and does not contain a differentiator. Figure 4.46 shows the schematic of the indirect magnetizing-flux-oriented control of the double-cage

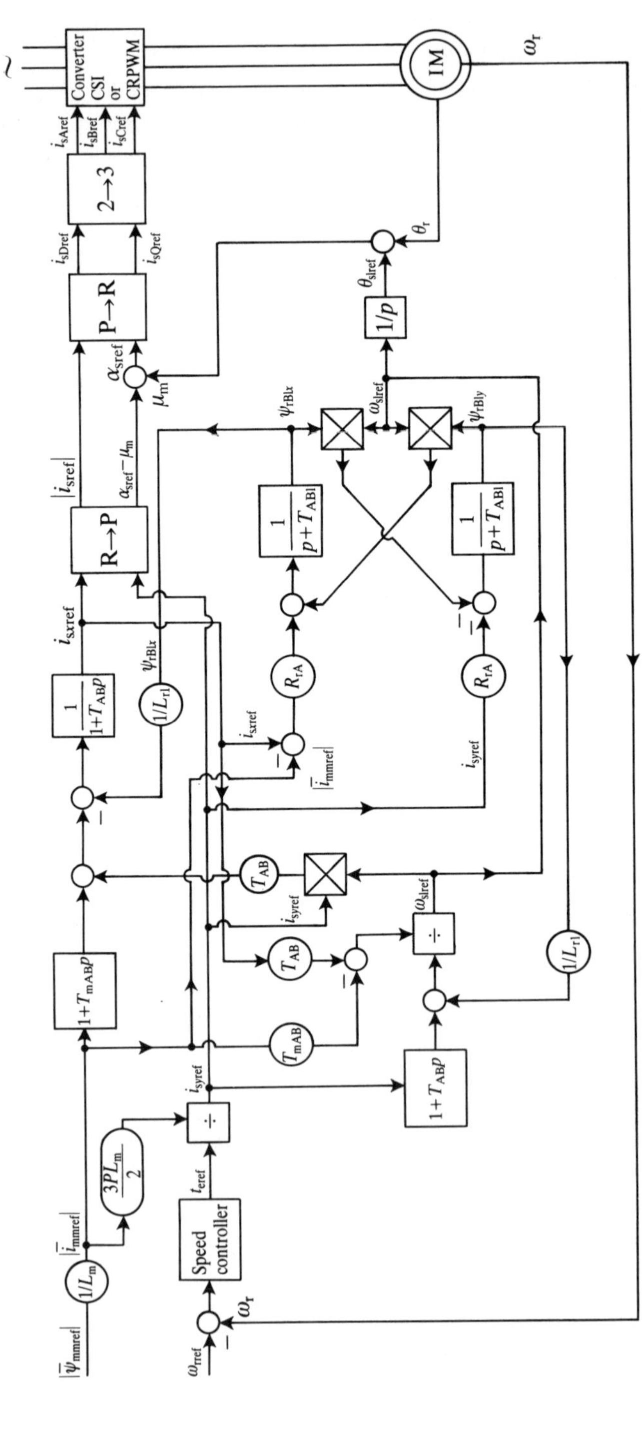

Fig. 4.46. Schematic of the indirect magnetizing-flux-oriented control of a double-cage induction machine with impressed stator currents utilizing a symmetrical rotor flux-linkage estimator.

induction machine with impressed stator currents where the rotor flux-linkage estimator circuit is symmetrical.

In the implementations shown in Fig. 4.45 and Fig. 4.46, the quadrature-axis stator current reference (i_{syref}) is obtained from the torque reference (t_{eref}), by using eqn (4.3-76). The torque reference is obtained as the output of the speed controller, which can be a PI controller. When the vector-control schemes shown in Fig. 4.45 and Fig. 4.46 are compared with the scheme shown in Fig. 4.43, which corresponds to the single-cage machine, there is a resemblance, although as a result of the double-cage, extra parts are present in Fig. 4.45 and Fig. 4.46. The main difference between the magnetizing-flux-oriented control schemes corresponding to the single-cage and double-cage machines is that in the scheme for the double-cage machine, an estimation circuit is present, which derives the direct- and quadrature-axis leakage flux-linkage components of the bottom cage. However, the required flux linkages can be obtained in real time at very high speed by using digital signal processors.

For indirect vector control of both the single-cage machine and the double-cage machine, it has been necessary to calculate the angular slip frequency, by, for example, the circuits shown in Fig. 4.43(a) and Fig. 4.46 respectively. It follows from eqn (4.3-94), which defines ω_{sl}, by considering eqns (4.3-93), (4.3-95), and (4.3-96), and by setting all the derivative terms to be zero that the expression for ω_{sl} is obtained in the steady state. This gives a fourth-order equation for ω_{sl}. It follows from this equation that if $i_{sx}/|\bar{i}_{mm}|$ and i_{sy}/i_{sx} are small and ω_{sl} is also small, the angular slip frequency of the double-cage machine takes the same form as the angular slip frequency of the single-cage machine. However, while for the single-cage machine the angular slip frequency is proportional to the rotor resistance (R_r), for the double-cage machine it is proportional to a resultant rotor resistance, which is obtained by connecting in parallel the resistors of the upper and lower cages, giving $R_{rA}R_{rB}/(R_{rA}+R_{rB})$. Usually the rotor resistance of the single-cage machine R_r and the resultant rotor resistance of the double-cage machine $R_{rA}R_{rB}/(R_{rA}+R_{rB})$ are different and when a double-cage machine is subjected to vector control and ω_{sl} is obtained by utilizing the equations of the double-cage machine, since the resistances of the upper and lower cages are independent of the rotor slip frequency, more accurate values of ω_{sl} are obtained than by utilizing the single-cage equations. This also holds at low values of the torque-producing current. It can also be shown, by considering the steady-state expression for ω_{sl}, that when ω_{sl} is calculated as a function of $i_{sx}/|\bar{i}_{mm}|$ or of i_{sy}/i_{sx}, it will have a maximum value. In the steady state this limits the stability region of the double-cage induction machine subjected to magnetizing-flux-oriented control.

When eqn (4.3-94) is used to obtain the expression for the angular slip frequency in the transient state, by also considering eqns (4.3-93), (4.3-95), and (4.3-96) it is possible to show that in the transient state, ω_{sl} can be expressed in terms of i_{sy}, i_{sx} and the various inductance and resistance parameters of the machine. If ω_{sl} is small, the rotor time constant for slow and fast variations of i_{sx} decreases from $T_{r1}=L_m/[R_{rA}R_{rB}/(R_{rA}+R_{rB})]$ to $T_{r2}=L_{rl}/(R_{rA}+R_{rB})$. If, say, a machine with deep rotor bars is considered and it is assumed that a rotor bar is divided into two equal sections, it follows that $R_{rA}=R_{rB}=R_r$. If $L_m/L_{rl}=30$, the ratio of T_{r1} to T_{r2} is equal to 120, while if $L_m/L_{rl}=10$, it is equal to 40. This shows that there can be a significant change in the rotor time constant. Thus to obtain fast torque response for high-performance a.c. servo applications, it is very important to incorporate the effects of the double-cage in the vector control scheme.

BIBLIOGRAPHY

Acarnley, P. P. and Finch, J. W. (1987). Review of control techniques for field-orientation in a.c. drives. 22 Universities Power Engineering Conference, Sunderland, paper 9.08.

Akamatsu, M., Ikeda, K., Tomei, H., and Yano, S. (1982). High performance IM drive by coordinate control using a controlled current inverter. *IEEE Transactions on Industry Applications* **IA-18**, 382–92.

Albrecht, P. and Vollstedt, W. (1982). Microcomputer control of a variable speed double-fed induction generator operating on the fixed frequency grid. *Conference on Microelectronics in Power Electronics and Electrical Drives, Darmstadt*, pp. 363–8.

Bassi, E., Benzi, F., Bolognani, S., and Buja, G. S. (1987). Torque-angle control scheme for field-oriented CSIM drives. *Beijing International Conference on Electrical Machines, Beijing*, pp. 368–70.

Bausch, H. and Hontheim, H. (1985). Transient performance of induction machines with field-oriented control. Second International Conference on Electrical Machines—Design and Applications, IEE, London, pp. 199–203.

Bausch, H., Hontheim, H., and Kolletschke, D. (1986). The influence of decoupling methods on the dynamic behaviour of a field-oriented controlled induction machine. In *Proceedings of International Conference on Electrical Machines, München*, pp. 648–51.

Bayer, K. H. and Blaschke, F. (1977). Stability problems with the control of induction machines using the method of field orientation. *IFAC Symposium on Control in Power Electronics and Drives, Düsseldorf*, pp. 483–92.

Bedford, B. D. and Hoft, R. G. (1964). *Principles of inverter circuits*. Wiley, New York.

Bellini, A., Figalli, G., and Ulivi, G. (1986). A microcomputer based direct field oriented control of induction motors. In *Proceedings of International Conference on Electrical Machines, München*, pp. 652–65.

Blaschke, F. (1972). The principle of field-orientation as applied to the new Transvektor closed-loop control system for rotating-field machines. *Siemens Review* **34**, 217–20.

Boldeau, I. and Nasar, S. A. (1986). Electric machine dynamics. Macmillan, New York.

Bolognani, S. and Buja, G. S. (1988). Parameter variation and computation error effects in indirect field-oriented induction motor drives. In *Proceedings of International Conference on Electrical Machines, Pisa*, pp. 545–9.

Bolognani, S. and Buja, G. S. (1986). Dc link curent control for high performance CSIM drive systems, *IEEE IAS Annual Meeting*, pp. 112–16.

Bose, B. K. (1986). *Power electronics and drives*. Prentice Hall, Englewood Cliffs, New Jersey.

Bowes, S. R. (1984). Steady-state performance of P.W.M. inverter drives. *Institute of Electrical Engineers Proc.* **B130**, 229–44.

Deng, D. and Lipo, T. A. (1986). A modified control method for fast response current source inverter drives, *IEEE Transactions on Industry Applications* **IA-22**, 653–65.

De Doncker, R., Vandenput, A., and Geysen, W. (1986). A digital field-oriented controller using the double-cage induction motor model. *IEEE Power Electronics Specialists Conference, Vancouver*, pp. 502–9.

De Doncker, R. (1987). Field oriented controllers with rotor deep bar compensation circuits. *IEEE IAS Annual Meeting, Atlanta*, pp. 142–9.

De Doncker, R. and Novotny, D. W. (1988). The universal field-oriented controller. *IEEE IAS Annual Meeting, Pittsburgh*, pp. 450–6.

Dewan, S. B., Slemon, G. R., and Straughen, A. (1984). *Power semiconductor drives*. Wiley, New York.

Feller, P. (1983). Speed control of an a.c. motor by state variables feedback with decoupling. *IFAC Symposium on Control in Power Electronics and Drives, Lausanne, Switzerland*, pp. 87–93.

Ferraris, P., Fratta, A., Vagati, A., and Villata, F. (1986). About the vector control of induction motors for special applications without speed sensor. *Int. Conference on the evolution and modern aspects of induction machines, Torino*, pp. 444–50.

Finney, D. (1988). *Variable-frequency a.c. motor drive systems*. Peter Peregrinus Ltd., London.

Fratta, A., Vagati, A., and Villata, F. (1988). Vector control of induction motors without transducers. *IEEE Power Electronics Specialists Conference, Kyoto*, pp. 839–46.

Garces, L. J. (1980). Parameter adaptation of the speed-controlled static a.c. drive with a squirrel-cage induction motor. *IEEE Transactions on Industry Applications* **IA-16**, 173–8.

Harashima, F., Kondo, S., Ohnishi, K., Kajita, M., and Susono, M. (1985). Multimicroprocessor-based control system for quick response induction motor drive. *IEEE Transactions on Industry Applications* **IA-21**, 602–9.

Hasse, K. (1977). Control of cycloconverters for feeding asynchronous machines. *IFAC Conference on Control in Power Electronics and Drives, Dusseldorf*, pp. 537–46.

Ho, E. Y. Y. and Sen, P. C. S. Decoupling control of induction motor drives. *IEEE Transactions on Industrial Electronics* **35**, 253–62.

Holtz, J. and Stadtfeld, S. (1983). Field-oriented control by forced motor currents in a voltage-fed induction motor drive. *IFAC Symposium on Control in Power Electronics and Electrical Drives, Lausanne*, pp. 103–10.

Irisa, T., Takata, S., Ueda, R., and Sonoda, T. (1984). Effect of machine structure identification of spatial position and its magnitude of rotor flux in induction motor vector control, *IEEE Conference on Power Electronics and Variable Speed Drives, London*, pp. 352–6.

Islam, S. M. and Somouah, C. B. (1989). An efficient high performance voltage decoupled induction motor drive with excitation control. *IEEE Transactions on Energy Conversion* **4**, 109–17.

Joetten, R. and Maeder, G. (1982). Control methods for good dynamic performance induction motor drives based on current and voltage as measured quantities. *International Semiconductor Power Converter Conference, IEEE/IAS, Orlando*, pp. 397–407.

Kawamura, A. and Hoft, R. (1983). An analysis of induction motor for field oriented or vector control. *IEEE Power Electronics Specialists Conference, Albuquerque, New Mexico*, pp. 91–100.

Kazmierkowski, M. P. and Köpcke, H. J. (1983). Comparison of dynamic behaviour of frequency converter fed induction machine drives. *IFAC Symposium on Control in Power Electronics and Drives, Lausanne*, pp. 313–20.

Krishnan, R., Doran, F. C., and Latos, T. S. (1987). Identification of thermally safe load cycles for an induction motor position servo. *IEEE Transactions on Industry Applications* **23**, 636–43.

Krishnan, R. and Doran, F.C. (1987). Study of parameter sensitivity in high-performance inverter-fed induction motor drive systems. *IEEE Transactions on Industry Applications* **IA-23**, 623–35.

Landau, Y. D. (1986). *Adaptive control, the model reference approach*. Marcel Dekker, New York.

Leonhard, W. (1985). *Control of electrical drives*. Springer-Verlag, Berlin.

Leonhard, W. (1986). Microcomputer control of high dynamic performance a.c. drives—a survey, *Automatica* **22**, 1–19.

Leonhard, W. (1988). Adjustable-speed a.c. drives. *Proceedings of the IEEE* **76**, 455–70.

Leonhard, W. (1988). Field-orientation for controlling a.c. machines—Principle and application. *3rd International Conference on Power Electronics and Variable Speed Drives, London*, pp. 277–82. IEE, London.

Lorenz, R. D., Lucas, M. O., and Lawson, D. B. (1986). Synthesis of a state-variable motion controller for high-performance field oriented induction machine drives. *IEEE IAS Annual Meeting*, Denver, 80–5.

Murphy, J. M.D. and Turnbull, F. G. (1987). *Power electronic control of a.c. motors*. Pergamon Press, Oxford.

Novotny, D. W. and Lorenz, R. D. (1985). Introduction to field orientation and high performance a.c. drives. Tutorial Course. IAS Annual Meeting, IEEE, 6–7 October. Toronto.

Ohnishi, K. and Miyachi, K. (1982). Principles of constant magnitude regulation of secondary flux based on slip frequency control in induction motor drive. In Proceedings of International Conference on Electrical Machines, Budapest, pp. 203–6.

Ohnishi, K., Suzuki, H., Miyachi, K. and Terashima, M. (1985). Decoupling control of secondary flux and secondary current in induction motor drive with controlled voltage source and its comparison with volts/hertz control. *IEEE Transactions on Industry Applications* **IA-21**, 241–7.

Ohnishi, K., Ueda, Y., and Miyachi, K. (1986). Model reference adaptive system against rotor resistance variation in induction motor drive. *IEEE Transactions on Industrial Electronics* **IA-33**, 217–23.

Okuyama, T., Fujimoto, N., Matsui, T., and Kubota, Y. (1986). A high performance speed control scheme of induction motor without speed and voltage sensors. *IEEE IAS Annual Meeting, Denver*, pp. 106–10.

Plunkett, A. B., Kliman, G. B., and Boyle, M. J. (1985). Digital techniques in evaluation of high-efficiency induction motors for inverter drives. *IEEE Transactions on Industry Applications* **IA-21**, 456–63.

Polmann, A. (1986). Software pulsewidth modulation for microprocessor control of a.c. drives. *IEEE Transactions on Industry Applications* **IA-22**, 691–6.

Rowan, T. M., Kerkman, R. J., and Lipo, T. A. (1987). Operation of naturally sampled current regulators in transition mode. *IEEE Transactions on Industry Applications* **IA-23**, 586–96.

Saito, K., Kamiyama, K., and Sukegawa, T. (1986). A multi-processor based, fully digital, a.c. drive system for rolling mills. *IEEE IAS Annual Meeting, Denver*, pp. 36–41.

Sathikumar, S. and Vithayathil, J. (1984). Digital simulation of field-oriented control of induction motor. *IEEE Transactions on Industrial Electronics* **IE-31**, 141–8.

Schauder, C. D., Choo, F. H., and Roberts, M. T. (1983). High-performance torque-controlled induction motor drives. *IEEE Transactions on Industry Applications* **IA-19**, 349–55.

Stepina, J. (1979). Non-transformational matrix analysis of electrical machinery. *Electrical Machines and Electromechanics* **4**, 255–68.

Sugimoto, H. and Tamai, S. (1987). Secondary resistance identification of an induction motor applied model reference adaptive system and its characteristics. *IEEE Transactions on Industry Applications* **IA-23**, 296–303.

Svensson, T. (1983). Current reference estimator for inverter-fed a.c. machines. *IFAC Symposium on Control in Power Electronics and Electrical Drives, Lausanne*, pp. 539–46.

Tamai, S., Sugimoto, H., and Yano, M. (1987). Speed sensorless vector control of induction motor with model reference adaptive system. *IEEE IAS Annual Meeting, Atlanta*, pp. 189–95.

Tsuji, M., Yamada, E., Izumi, K., and Oyama, J. (1984). Stability analysis of a current-source inverter-fed induction motor under vector control. In *Conference Proceedings of International Conference on Electrical Machines, Lausanne*, pp. 867–74.

Vagati, A., Villata, F. (1984). A.c. servo system for position control. In *Conference Proceedings of International Conference on Electrical Machines, Lausanne*, pp. 871–4.

Vagati, A. and Villata, F. (1985). Field-oriented control, a particular implementation. In *Conference Record of Motor-Con, Hannover*, pp. 160–72.

Vas, P. and Alaküla, M. (1989) Field-oriented control of saturated induction machines. *IEEE Transactions on Energy Conversion*, IEEE Power Engineering Society Summer Meeting, Long Beach, California, paper 89SM 738-6EC.

Vas, P., Brown, J. E., and Hallenius, K. E. (1987). Small-signal analysis of smooth-air-gap machines using space vector theory. *Electric Energy Conference, Adelaide*, pp. 210–4.

Verghese, G. C. and Sanders, S. R. (1985). Observers for faster flux estimation in induction machines. *IEEE Power Electronics Specialists Conference.*

Wu, Z. K. and Strangas, E. G. (1988). Feedforward field orientation control of an induction motor using a PWM voltage source inverter and standardized single-board computers. *IEEE Transactions on Industrial Electronics* **35**, 75–9.

Xu, X., De Doncker, R., and Novotny, D. W. (1988). Stator flux-orientation control of induction machines in the field-weakening region. *IEEE IAS Annual Meeting, Pittsburgh*, pp. 437–43.

Xu, X., De Doncker, R., and Novotny, D. W. (1988). A stator flux oriented induction machine drive, *IEE Power Electronics Specialists Conference, Kyoto*, pp. 870–6.

Yamamura, S. (1986). *A.c. motors for high-performance applications*. Marcel Dekker, New York.

Zhang, J., Thiagaranjan, V., Grant, T., and Barton, T. H. (1988). New approach to field orientation control of a CSI induction motor drive. *IEE Proc. B*. **135**, 1–7.

Zhang, J. and Barton, T. H. (1987). Microprocessor based primary current control for cage induction motor drive. *IEE Power Electronics Specialists Conference, Blacksburg*, pp. 616–24.

5 Effects of magnetic saturation

In this section the effects of main flux saturation the equations required for the various implementations of vector control (e.g. rotor-flux-oriented control, magnetizing-flux-oriented control, stator-flux-oriented control) are discussed for both smooth-air-gap and salient-pole machines. However, only the derivation and a brief description of the required equations is given, since these can be applied similarly to the corresponding equations derived in Chapters 3 and 4, which, however, are only valid under linear magnetic conditions. It will be shown that as a result of saturation, the linear equations are modified and to obtain the required decoupling between the flux and torque-producing currents of the machine under consideration, it is these equations which have to be considered when the machine operates under saturated conditions.

As a result of saturation of the main flux paths, the magnetizing inductance and thus also the stator and rotor inductances (and the stator and rotor transient time constants) are not constant, but vary with saturation. The variation of the magnetizing inductance will be incorporated into the voltage equations. However, it will be shown that the voltage equations which are valid under saturated conditions differ from the equations which can be derived from their linear forms given in Chapters 3 and 4, where the currents have been used as state variables. The new voltage equations are not identical to the voltage equations which could be obtained by simply substituting the variable magnetizing inductance into the voltage equations which are valid under linear magnetic conditions, but they also contain new and/or modified terms.

The equations to be derived will be based on two-axis (or space-phasor) theory. Saturation of the main flux paths distorts the flux density distributions, which in the case of the linear theory (which neglects saturation) are sinusoidal in space, (see Chapter 2) if sinusoidal m.m.f. distributions are assumed. The resulting space harmonics can, however, be neglected if sinusoidally distributed windings are assumed since only the fundamental sinusoidal component of a flux wave can produce flux linkages with sinusoidally distributed windings, and thus it is possible to use two-axis or space-phasor theories.

5.1 VECTOR CONTROL OF SMOOTH-AIR-GAP MACHINES BY CONSIDERING THE EFFECTS OF MAIN FLUX SATURATION

The effects of main flux saturation are briefly discussed for vector-controlled induction machines with single-cage rotor. The various expressions for the electromagnetic torque will not be derived for the saturated machine, since formally they are the same as in the case of the unsaturated machine, since saturation does not introduce new terms in the expression for the electromagnetic torque, and the assumptions used in the previous section hold. Of course, as a result of saturation, the saturation-dependent machine parameters which are present in the different expressions for the torque, e.g. the magnetizing inductance, or the rotor self-inductance or the stator self-inductance, will be different to their unsaturated values and are variables which depend on the machine currents.

5.1.1 Rotor-flux-oriented control

In this section the stator and rotor voltage equations for an induction machine are described in the rotor-flux-oriented reference frame. However, because of saturation of the main flux paths, the magnetizing inductance and thus the stator and rotor self-inductances are not constant but vary with saturation. There are several ways to consider this variation with saturation and two approaches will be discussed. First the magnetizing inductance will be expressed as a function of the modulus of the rotor magnetizing-current space phasor and secondly it will be expressed in terms of the modulus of the magnetizing-current space phasor. It will be shown that when the second method is used, more complicated equations are obtained than in the first case, since in contrast to the rotor magnetizing-current space phasor, which is coaxial with the direct axis of the rotor-flux-oriented reference frame, the magnetizing flux-linkage space phasor is not coaxial with the direct axis of the rotor-flux-oriented reference frame and under saturated conditions owing to the effects of cross-magnetization (cross-saturation), extra terms will be obtained in the equations. Furthermore, it will be shown that the equations derived in this section differ from the equations which can be derived from their linear forms given in Section 4.1 by simply replacing the constant magnetizing inductance by the variable magnetizing inductance.

In Section 4.1.1 the stator and rotor voltage equations in the rotor-flux-oriented reference frame were given for an induction machine and the effects of magnetic saturation were neglected. In this section it is assumed that the effects of leakage flux saturation can be neglected but the effects of main flux saturation are incorporated in the analysis. All other assumptions are the same as those used in Section 4.1.

Rotor-flux-oriented control, expressing the magnetizing inductance as a function of the rotor magnetizing-current space phasor

In this section the stator and rotor voltage equations are obtained which contain the effects of main-flux saturation and they are formulated in the reference frame fixed to the rotor-flux linkage space phasor. The method followed here is similar to that used in Section 4.1.1, where the space-phasor forms of the voltage equations formulated in the general reference frame were utilized directly for the derivation of the required forms of the voltage equations. The rotor-current space phasor is expressed in terms of the rotor magnetizing-current space phasor and therefore the resulting voltage equations will contain the modulus and space angle of the rotor magnetizing-current space phasor. These quantities are necessary to implement rotor-flux-oriented control.

Stator voltage equations: By considering eqns (2.1-148) and (2.1-150), which are the stator space-phasor voltage and stator flux-linkage space-phasor equations in the general reference frame, the following stator-voltage space-phasor equation is obtained in the reference frame fixed to the rotor flux-linkage space phasor, which rotates at the speed ω_{mr}, where it follows from eqn (2.1-188) that this speed is equal to the first time derivative of the space angle ρ_r of the rotor magnetizing-current space phasor with respect to the direct axis of the stationary reference frame,

$$\bar{u}_{s\psi r} = R_s \bar{\imath}_{s\psi r} + \frac{d(L_s \bar{\imath}_{s\psi r})}{dt} + \frac{d(L_m \bar{\imath}_{r\psi r})}{dt} + j\omega_{mr}(L_s \bar{\imath}_{s\psi r} + L_m \bar{\imath}_{r\psi r}). \quad (5.1\text{-}1)$$

L_s is the self-inductance of a stator winding and it is not constant since $L_s = L_{sl} + L_m$, where L_{sl} is the leakage inductance of a stator winding, which is assumed to be constant, but L_m varies with saturation. It is assumed that the magnetizing inductance is a non-linear function of the rotor magnetizing current $\bar{\imath}_{mr}(t)$, which in general varies with time, thus $L_m = L_m(|\bar{\imath}_{mr}|)$. It follows from eqn (4.1-28) that the magnetizing inductance can be defined as $L_m = \bar{\psi}'_r / \bar{\imath}_{mr} = |\bar{\psi}_r| / |\bar{\imath}_{mr}|$. However, it should be noted that in the theory of electrical machines it is usually defined as $L_m = |\bar{\psi}_m| / |\bar{\imath}_{mm}|$, which follows from eqn (2.1-184), where $|\bar{\psi}_m|$ and $|\bar{\imath}_{mm}|$ are the modulus of the magnetizing flux-linkage space phasor and magnetizing-current space phasor respectively ($|\bar{\imath}_{mm}| = |\bar{\imath}_m|$). It follows from eqn (2.1-184) that the magnetizing-current space phasor is equal to the sum of the stator-current and rotor-current space phasors.

Substitution of eqn (4.1-5) into eqn. (5.1-1) yields the following stator voltage equation

$$\bar{u}_{s\psi r} = R_s \bar{\imath}_{s\psi r} + \frac{d(L'_s \bar{\imath}_{s\psi r})}{dt} + \frac{d(L_m^2 |\bar{\imath}_{mr}| / L_r)}{dt} + j\omega_{mr}\left(L'_s \bar{\imath}_{s\psi r} + \frac{L_m^2 |\bar{\imath}_{mr}|}{L_r}\right). \quad (5.1\text{-}2)$$

L'_s is the stator transient inductance, $L'_s = (L_s - L_m^2/L_r)$, L_r is the self-inductance of the rotor and $L_r = L_{rl} + L_m$, where L_{rl} is the leakage inductance of the rotor, which is assumed to be constant. There are two derivatives in eqn (5.1-2) which contain L_m, which varies with saturation, these are

$$\frac{d(L'_s \bar{\imath}_{s\psi r})}{dt} = \frac{d[(L_s - L_m^2/L_r)\bar{\imath}_{s\psi r}]}{dt}$$

and

$$\frac{\mathrm{d}(L_{\mathrm{m}}^2|\bar{\imath}_{\mathrm{mr}}|/L_{\mathrm{r}})}{\mathrm{d}t}=\frac{\mathrm{d}}{\mathrm{d}t}\left[\frac{L_{\mathrm{m}}^2|\bar{\imath}_{\mathrm{mr}}|}{L_{\mathrm{rl}}+L_{\mathrm{m}}}\right].$$

By utilizing the chain differentiation rule, these can be expanded as

$$\frac{\mathrm{d}(L_{\mathrm{s}}'\bar{\imath}_{\mathrm{s}\psi\mathrm{r}})}{\mathrm{d}t}=\frac{\mathrm{d}[(L_{\mathrm{s}}-L_{\mathrm{m}}^2/L_{\mathrm{r}})\bar{\imath}_{\mathrm{s}\psi\mathrm{r}}]}{\mathrm{d}t}=L_{\mathrm{s}}\frac{\mathrm{d}\bar{\imath}_{\mathrm{s}\psi\mathrm{r}}}{\mathrm{d}t}-\frac{L_{\mathrm{m}}^2}{L_{\mathrm{r}}}\frac{\mathrm{d}\bar{\imath}_{\mathrm{s}\psi\mathrm{r}}}{\mathrm{d}t}$$

$$+\bar{\imath}_{\mathrm{s}\psi\mathrm{r}}\frac{\mathrm{d}(L_{\mathrm{sl}}+L_{\mathrm{m}})}{\mathrm{d}t}-\bar{\imath}_{\mathrm{s}\psi\mathrm{r}}\frac{\mathrm{d}[L_{\mathrm{m}}^2/(L_{\mathrm{rl}}+L_{\mathrm{m}})]}{\mathrm{d}t}$$

$$=L_{\mathrm{s}}'\frac{\mathrm{d}\bar{\imath}_{\mathrm{s}\psi\mathrm{r}}}{\mathrm{d}t}+\bar{\imath}_{\mathrm{s}\psi\mathrm{r}}\frac{\mathrm{d}L_{\mathrm{m}}}{\mathrm{d}t}\left[1-\frac{L_{\mathrm{m}}}{L_{\mathrm{r}}^2}(2L_{\mathrm{r}}-L_{\mathrm{m}})\right]$$

$$=L_{\mathrm{s}}'\frac{\mathrm{d}\bar{\imath}_{\mathrm{s}\psi\mathrm{r}}}{\mathrm{d}t}+\bar{\imath}_{\mathrm{s}\psi\mathrm{r}}\left(\frac{L_{\mathrm{rl}}}{L_{\mathrm{r}}}\right)^2\frac{\mathrm{d}L_{\mathrm{m}}}{\mathrm{d}t} \qquad (5.1\text{-}3)$$

and

$$\frac{\mathrm{d}(L_{\mathrm{m}}^2|\bar{\imath}_{\mathrm{mr}}|/L_{\mathrm{r}})}{\mathrm{d}t}=\frac{L_{\mathrm{m}}^2}{L_{\mathrm{r}}}\frac{\mathrm{d}|\bar{\imath}_{\mathrm{mr}}|}{\mathrm{d}t}+|\bar{\imath}_{\mathrm{mr}}|\frac{\mathrm{d}[L_{\mathrm{m}}^2/(L_{\mathrm{rl}}+L_{\mathrm{m}})]}{\mathrm{d}t}$$

$$=\frac{L_{\mathrm{m}}^2}{L_{\mathrm{r}}}\frac{\mathrm{d}|\bar{\imath}_{\mathrm{mr}}|}{\mathrm{d}t}+|\bar{\imath}_{\mathrm{mr}}|\left(\frac{\mathrm{d}L_{\mathrm{m}}}{\mathrm{d}t}\right)\left(1-\frac{L_{\mathrm{rl}}^2}{L_{\mathrm{r}}^2}\right). \qquad (5.1\text{-}4)$$

In eqns (5.1-3) and (5.1-4) the term $\mathrm{d}L_{\mathrm{m}}/\mathrm{d}t$ is also present. However, this can be expanded as

$$\frac{\mathrm{d}L_{\mathrm{m}}}{\mathrm{d}t}=\frac{\mathrm{d}L_{\mathrm{m}}}{\mathrm{d}|\bar{\imath}_{\mathrm{mr}}|}\frac{\mathrm{d}|\bar{\imath}_{\mathrm{mr}}|}{\mathrm{d}t}=\frac{L-L_{\mathrm{m}}}{|\bar{\imath}_{\mathrm{mr}}|}\frac{\mathrm{d}|\bar{\imath}_{\mathrm{mr}}|}{\mathrm{d}t}, \qquad (5.1\text{-}5)$$

where

$$L=\frac{\mathrm{d}\bar{\psi}_{\mathrm{r}}'}{\mathrm{d}\bar{\imath}_{\mathrm{mr}}}=\frac{\mathrm{d}|\bar{\psi}_{\mathrm{r}}|}{\mathrm{d}|\bar{\imath}_{\mathrm{mr}}|}. \qquad (5.1\text{-}6)$$

In eqn (5.1-6) L is a dynamic (tangent-slope or incremental) inductance, and it is equal to the derivative of the modulus of the rotor flux-linkage space phasor with respect to the modulus of the rotor magnetizing-current space phasor. It should be noted that in contrast to this, L_{m} is the static (chord slope) inductance, and it follows from the considerations given above that it can also be expressed as

$$L_{\mathrm{m}}=\frac{|\bar{\psi}_{\mathrm{r}}|}{|\bar{\imath}_{\mathrm{mr}}|}. \qquad (5.1\text{-}7)$$

Fig. 5.1 shows the variation of L and it should be noted that according to eqn (5.1-5) L and L_{m} are related by $L=L_{\mathrm{m}}+|\bar{\imath}_{\mathrm{mr}}|\mathrm{d}L_{\mathrm{m}}/\mathrm{d}|\bar{\imath}_{\mathrm{mr}}|$ and in Fig. 5.1 the variation of $\mathrm{d}L_{\mathrm{m}}/\mathrm{d}|\bar{\imath}_{\mathrm{mr}}|$ is also shown.

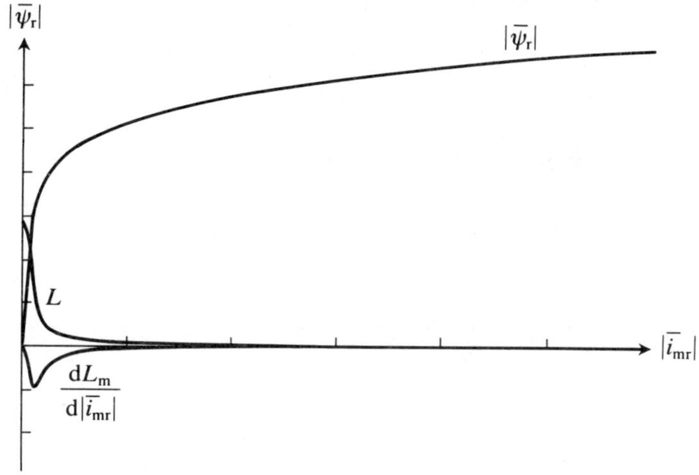

Fig. 5.1. Rotor magnetization curve. Variations of the dynamic inductance and the derivative of the magnetizing inductance.

It follows from eqns (5.1-5), (5.1-6) and (5.1-7) that under linear magnetic conditions $L=L_m$ and $dL_m/dt=0$ as expected, since in this case $L_m=$constant. Thus substitution of eqns (5.1-3) and (5.1-4) into eqn (5.1-2) gives the following equation, after the substitution of eqn (5.1-5),

$$\bar{u}_{s\psi r}=R_s\bar{i}_{s\psi r}+L'_s\frac{d\bar{i}_{s\psi r}}{dt}+\bar{i}_{s\psi r}\left(\frac{L_{rl}}{L_r}\right)^2\frac{L-L_m}{|\bar{i}_{mr}|}\frac{d|\bar{i}_{mr}|}{dt}$$
$$+\left[\frac{L_m^2}{L_r}+(L-L_m)\left(\frac{1-L_{rl}^2}{L_r^2}\right)\right]\frac{d|\bar{i}_{mr}|}{dt}+j\omega_{mr}\left(\frac{L'_s\bar{i}_{s\psi r}+L_m^2|\bar{i}_{mr}|}{L_r}\right). \quad (5.1\text{-}8)$$

When eqn (5.1-8) is divided by the stator resistance and is deliberately arranged into a form which is similar to eqn (4.1-6), finally the following space-phasor equation is obtained

$$T'_s\frac{d\bar{i}_{s\psi r}}{dt}+\bar{i}_{s\psi r}=\frac{\bar{u}_{s\psi r}}{R_s}-\left\{\frac{L-L_m}{R_s}\left[1-\frac{L_{rl}^2}{L_r^2(1-\bar{i}_{s\psi r}/|\bar{i}_{mr}|)}\right]\right.$$
$$\left.+\frac{L_m^2}{L_rR_s}\right\}\frac{d|\bar{i}_{mr}|}{dt}-j\omega_{mr}\left[T'_s\bar{i}_{s\psi r}+\frac{L_m^2}{L_rR_s|\bar{i}_{mr}|}\right]. \quad (5.1\text{-}9)$$

It follows that as a result of the saturation of the main flux paths, the extra term

$$\Delta\bar{i}_s=\left\{(L_m-L)\left[1-\frac{L_{rl}^2}{L_r^2(1-\bar{i}_{s\psi r}/|\bar{i}_{mr}|)}\right]\Big/R_s\right\}\frac{d|\bar{i}_{mr}|}{dt} \quad (5.1\text{-}10)$$

is also present in the voltage equation, but under linear magnetic conditions, when $L=L_m$, this term is zero. Furthermore, it follows from eqn (5.1-9) that the effects of changing saturation are also present in T'_s, since it depends on the changing L_m. By considering that $T_s-T'_s=(L_s-L'_s)/R_s=L_m^2/(L_rR_s)$, it can be seen that under linear magnetic conditions eqn (5.1-9) yields eqn (4.1-6).

By resolving eqn (5.1-9) into its real (x-axis) and imaginary (y-axis) components, the following two-axis voltage equations are obtained

$$T'_s\frac{\mathrm{d}i_{sx}}{\mathrm{d}t}+i_{sx}=\frac{u_{sx}}{R_s}+\omega_{mr}T'_si_{sy}-(T_s-T'_s)\frac{\mathrm{d}|\bar{\imath}_{mr}|}{\mathrm{d}t}+\Delta i_{sx} \tag{5.1-11}$$

$$T'_s\frac{\mathrm{d}i_{sy}}{\mathrm{d}t}+i_{sy}=\frac{u_{sy}}{R_s}-\omega_{mr}T'_si_{sx}-(T_s-T'_s)\omega_{mr}|\bar{\imath}_{mr}|+\Delta i_{sy} \tag{5.1-12}$$

where

$$\Delta i_{sx}=\left\{(L_m-L)\frac{[1-(L_{rl}^2/L_r^2)(1-i_{sx}/|\bar{\imath}_{mr}|)]}{R_s}\right\}\frac{\mathrm{d}|\bar{\imath}_{mr}|}{\mathrm{d}t} \tag{5.1-13}$$

$$\Delta i_{sy}=\frac{(L_m-L)(L_{rl}^2i_{sy})}{L_r^2R_s|\bar{\imath}_{mr}|}\frac{\mathrm{d}|\bar{\imath}_{mr}|}{\mathrm{d}t}. \tag{5.1-14}$$

Comparison with eqns (4.1-7) and (4.1-8) shows that under linear magnetic conditions eqns (5.1-13) and (5.1-14) agree with eqns (4.1-7) and (4.1-8). However, as a result of saturation, extra transformer terms arise along both the direct and quadrature axes and these result in the extra current terms Δi_{sx} and Δi_{sy} respectively. Both of these extra current components are proportional to the rate of change of the rotor magnetizing current. It should be noted that it is possible to obtain many other similar forms of these extra terms. For example, it follows from eqn (5.1-5) that it is possible to replace the factor (L_m-L) by

$$-|\bar{\imath}_{mr}|\,\mathrm{d}L_m/\mathrm{d}|\bar{\imath}_{mr}| \text{ and } \Delta i_{sy}=-\frac{L_{rl}^2}{R_sL_r^2}\frac{\mathrm{d}L_m}{\mathrm{d}|\bar{\imath}_{mr}|}\frac{\mathrm{d}|\bar{\imath}_{mr}|}{\mathrm{d}t}$$

$$=-\frac{L_{rl}^2}{R_sL_r^2}\frac{\mathrm{d}L_m}{\mathrm{d}t}.$$

If the rotor leakage inductance is assumed to be zero, then $\Delta i_{sy}=0$, or in other words, there is no quadrature-axis extra induced e.m.f. due to saturation. This is an expected result, since in this case the rotor flux-linkage space phasor, which has only a direct-axis component in the rotor-flux-oriented reference frame, is equal to the magnetizing flux-linkage space phasor, which will also have a direct-axis component only, and therefore there cannot result a quadrature-axis transformer e.m.f. due to the change of this zero quadrature-axis magnetizing flux-linkage space phasor. Furthermore, if $L_{rl}=0$, $\Delta i_{sx}\neq 0$, but $\Delta i_{sx}=[(L_m-L)/R_s]\,\mathrm{d}|\bar{\imath}_{mr}|/\mathrm{d}t$ or, in other words, in the direct-axis stator winding (x) there will be an extra direct-axis induced transformer e.m.f. due to saturation, which is equal to $-[(L_m - L)]\,d|\bar{\imath}_{mr}|/dt$. This is also an expected result, since when $L_{rl}=0$, the direct-axis rotor flux-linkage component ($\psi_{rx}=|\bar{\psi}_r|$) is equal to the direct-axis magnetizing flux-linkage component $\psi_{mx}=|\bar{\psi}_r|=L_m|\bar{\imath}_{mr}|$ and its rate of change under linear magnetic conditions is equal to $d|\bar{\psi}_r|/dt=L_m\,d|\bar{\imath}_{mr}|/dt$. This latter voltage component plus the extra component voltage due to saturation gives the total voltage $L_m\,d|\bar{\imath}_{mr}|/dt-(L_m-L)\,d|\bar{\imath}_{mr}|/dt=L\,d|\bar{\imath}_{mr}|/dt$ and this is the correct result since when $L_{rl}=0$, the total transformer e.m.f. in the x-axis stator winding must be equal to the rate of change of $|\bar{\psi}_r|$, which is indeed equal to $d|\bar{\psi}_r|/dt=(d|\psi_r|/d|\bar{\imath}_{mr}|)\,d|\bar{\imath}_{mr}|/dt)=L\,d|\bar{\imath}_{mr}|/dt$.

By analogy it follows that if the stator voltage equations are formulated in the magnetizing-flux-oriented reference frame, where the direct-axis component of the magnetizing flux-linkage space phasor is equal to its modulus and its quadrature-axis component is zero, then under saturated conditions, in the direct-axis stator voltage equation there must be a transformer e.m.f., which is equal to the rate of change of the modulus of the magnetizing flux-linkage space phasor and in the quadrature-axis stator winding there will be no induced e.m.f. due to the rate of change of the quadrature-axis magnetizing flux-linkage component (the details of this will be discussed in Section 5.1.2). This is an advantage over the formulation of the voltage equations of a saturated machine in the stationary reference frame (fixed to the stator), where the magnetizing flux-linkage space phasor must contain both the direct-axis and also the quadrature-axis components and therefore their rate of change is not zero and in general there will exist cross-magnetization coupling due to saturation (cross-saturation) between the windings in space quadrature. The phenomenon of cross-saturation due to cross-magnetization in a smooth-air-gap machine is similar to the phenomenon of the demagnetizing effect of the cross-magnetizing armature reaction in a d.c. machine. Its physical existence can be proved experimentally and also a simple explanation of its existence can be given by utilizing space-phasor theory. This will be described below.

The relationship between the stationary-axis voltage components (u_{sD}, u_{sQ}), the stator current components (i_{sD}, i_{sQ}) and the corresponding voltage (u_{sx}, u_{sy}) and current (i_{sx}, i_{sy}) components can be obtained by utilizing the transformations defined in eqns (4.1-3) and (4.1-4) respectively. It follows from eqns (5.1-11)–(5.1-14) that with respect to the stator currents i_{sx} and i_{sy}, the induction machine behaves as a first-order time delay element, whose time constant is equal to the stator transient time constant of the machine and whose gain is equal to the inverse of the stator resistance. However, because of saturation, the stator transient time constant is not a real constant, since it depends on the magnetizing inductance, which changes with saturation. Furthermore, it can be seen that there is an unwanted coupling between the stator circuits in the two axes. For the purposes of rotor-flux-oriented-control, it is the direct-axis stator current i_{sx} (rotor flux-producing component) and the quadrature-axis stator current i_{sy} (torque-producing component) which must be independently controlled. It will now be assumed that the induction machine is supplied by impressed stator voltages. Since the voltage equations are coupled and the coupling term in u_{sx} also depends on i_{sy} and the coupling term in u_{sy} also depends on i_{sx}, u_{sx} and u_{sy} cannot be considered as decoupled control variables for the rotor flux and electromagnetic torque and the stator currents i_{sx} and i_{sy} can only be independently controlled (decoupled control) if the stator voltage equations are decoupled and the stator current components i_{sx} and i_{sy} are indirectly controlled by controlling the terminal voltages of the induction machine. The required decoupling circuits can be obtained from the equations given above in the same way as described in Chapter 4.

If an induction machine with impressed stator currents is assumed, then the stator voltage equations can be omitted from the dynamic model of the drive and only the rotor equations have to be considered. These are obtained in the next section, where the effects of main-flux saturation are incorporated in the rotor voltage equations.

Rotor voltage equations: In this section the rotor voltage equations of the induction machine are obtained in the rotor-flux-oriented reference frame if the effects of main-flux saturation are incorporated in the equations. It follows from eqns (4.1-5), (4.1-22) and (4.1-23) that when the magnetizing inductance is changing due to saturation, the space-phasor form of the rotor voltage equations is

$$0=R_r L_m \frac{|\bar{\imath}_{mr}|-\bar{\imath}_{s\psi r}}{L_r}+\frac{d(L_m|\bar{\imath}_{mr}|)}{dt}+j(\omega_{mr}-\omega_r)L_m|\bar{\imath}_{mr}|, \qquad (5.1\text{-}15)$$

where $L_r=L_{rl}+L_m$ is the self-inductance of the rotor and changes with the saturation. In eqn (5.1-15) the derivative $d(L_m|\bar{\imath}_{mr}|)/dt$ is also present and this contains the non-linear element L_m. This can be expressed as

$$\frac{d(L_m|\bar{\imath}_{mr}|)}{dt}=\frac{d|\bar{\psi}_r|}{dt}=\frac{d|\bar{\psi}_r|}{d|\bar{\imath}_{mr}|}\frac{d|\bar{\imath}_{mr}|}{dt}=L\frac{d|\bar{\imath}_{mr}|}{dt}, \qquad (5.1\text{-}16)$$

where $|\bar{\psi}_r|$ is the modulus of the rotor flux-linkage space phasor, and it should be noted that of course the same result follows from the application of the chain differentiation rule

$$\begin{aligned}\frac{d(L_m|\bar{\imath}_{mr}|)}{dt}&=L_m\frac{d|\bar{\imath}_{mr}|}{dt}+|\bar{\imath}_{mr}|\frac{dL_m}{dt}\\&=L_m\frac{d|\bar{\imath}_{mr}|}{dt}+|\bar{\imath}_{mr}|\frac{dL_m}{d|\bar{\imath}_{mr}|}\frac{d|\bar{\imath}_{mr}|}{dt}\\&=\left(L_m+\frac{dL_m}{d|\bar{\imath}_{mr}|}\right)\frac{d|\bar{\imath}_{mr}|}{dt}=L\frac{d|\bar{\imath}_{mr}|}{dt}.\end{aligned}$$

This result is in agreement with that stated in the previous section. Substitution of eqn (5.1-16) into eqn (5.1-15) yields

$$T_r\left(\frac{L}{L_m}\right)\frac{d|\bar{\imath}_{mr}|}{dt}+|\bar{\imath}_{mr}|=\bar{\imath}_{s\psi r}-j(\omega_{mr}-\omega_r)T_r|\bar{\imath}_{mr}|, \qquad (5.1\text{-}17)$$

where $T_r=L_r/R_r=(L_{rl}+L_m)/R_r$ is the rotor time constant, which because of saturation, is not constant. In this expression L_{rl} is the leakage inductance of the rotor, which has been assumed to be constant. Under linear magnetic conditions $L=L_m$, and eqn (5.1-17) yields eqn (4.1-24) as expected. On the left-hand side of eqn (5.1-17) the modified rotor time constant appears, and this can be put into the form,

$$T_r^*=T_r(L/L_m)=(L_{rl}+L_m)\,(L/L_m)/R_r=T_{rl}L/L_m+L/R_r$$

where T_{rl} is the rotor leakage time constant, $T_{rl}=L_{rl}/R_r$. Thus the modified rotor time constant is equal to the sum of two rotor time constants. The first component is equal to the reduced value of the rotor leakage time constant and the reduction factor is equal to the ratio of the dynamic and static inductances, which ratio is equal to 1 under linear conditions. The second component is equal to the ratio of the dynamic inductance and the rotor resistance.

It follows from eqn (5.1-17) that it can be expressed in terms of T_r^* and T_r and therefore when eqn (5.1-17) is resolved into its real- (x-axis) and imaginary-axis (y-axis) components, the direct-axis rotor voltage equation will contain the modified rotor time constant T_r^* and the quadrature-axis rotor voltage equation will contain T_r. When these equations are deliberately put into a similar form to eqns (4.1-25) and (4.1-26), the following equations are obtained

$$T_r^*\frac{d|\bar{\imath}_{mr}|}{dt}+|\bar{\imath}_{mr}|=i_{sx} \qquad (5.1\text{-}18)$$

$$\omega_{mr}=\omega_r+\frac{i_{sy}}{T_r|\bar{\imath}_{mr}|}. \qquad (5.1\text{-}19)$$

Under linear magnetic conditions these agree with eqns (4.1-24) and (4.1-25). However, it is is important to note that owing to saturation both T_r and T_r^* are changing parameters and depend on the magnetizing inductance, which changes with $|\bar{\imath}_{mr}|$. By utilizing eqn (5.1-5), it is possible to put T_r^* into the form,

$$T_r^*=L\left(\frac{T_{rl}}{L_m}+\frac{1}{R_r}\right)=\left(L_m+|\bar{\imath}_{mr}|\frac{dL_m}{d|\bar{\imath}_{mr}|}\right)\left(\frac{T_{rl}}{L_m}+\frac{1}{R_r}\right)$$

and the dependence on the magnetizing inductance is clearer.

On the basis of eqns (5.1-18) and 5.1-19) it is possible to obtain the flux models of the saturated induction machine in the rotor-oriented reference frame, which will be similar to those shown in Fig. 4.6(a) and Fig. 4.6(b), but will now contain both T_r^* and T_r.

Rotor-flux oriented control, expressing the magnetizing inductance as a function of the magnetizing-current space phasor

In this section the voltage equations required for the implementation of the various forms of rotor-flux-oriented control of induction machines are derived, but it is assumed that the magnetizing inductance is a non-linear function of the modulus of the magnetizing-current space phasor.

Rotor magnetizing current: It follows from eqn (4.1-5), by considering that the rotor self-inductance is the sum of the leakage and the magnetizing inductances ($L_r = L_{rl} + L_m$), that the modulus of the rotor magnetizing-current space phasor can be expressed as

$$|\bar{\imath}_{mr}| = \frac{L_r}{L_m}\bar{\imath}_{r\psi r} + \bar{\imath}_{s\psi r} = \frac{L_{rl}}{L_m}\bar{\imath}_{r\psi r} + (\bar{\imath}_{r\psi r} + \bar{\imath}_{s\psi r}) = \frac{L_{rl}}{L_m}\bar{\imath}_{r\psi r} + \bar{\imath}_{m\psi r}, \qquad (5.1\text{-}20)$$

where

$$|\bar{\imath}_{m\psi r}| = \bar{\imath}_{s\psi r} + \bar{\imath}_{r\psi r} = \bar{\imath}_s e^{-j\rho_r} + \bar{\imath}_r' e^{-j\rho_r} = (\bar{\imath}_s + \bar{\imath}_r')e^{-j\rho_r} = \bar{\imath}_m e^{-j\rho_r} = |\bar{\imath}_m| e^{j(\mu_m - \rho_r)} = |\bar{\imath}_m| e^{j\mu} \qquad (5.1\text{-}21)$$

is the rotor magnetizing-current space phasor in the rotor-flux-oriented reference frame. Furthermore, $\bar{\imath}_m$ is the space phasor of the magnetizing currents in the stationary reference frame, $|\bar{\imath}_m|$ is its modulus, and μ_m is the space angle of $\bar{\imath}_m$ with respect to the real axis of the stationary reference frame. However, the angle $\mu = \mu_m - \rho_r$ is the angle of $\bar{\imath}_m$ with respect to the real axis of the rotor-flux-oriented reference frame. In Fig. 5.2 $\bar{\imath}_m$ and $\bar{\imath}_{mr}$ are shown, together with the angles μ_m and μ.

Since in general the leakage inductance is not zero, $L_{rl} \neq 0$, the magnetizing-current space phasor is not coaxial with the rotor flux-linkage space phasor and thus $\bar{\imath}_m$ is not coaxial with $\bar{\imath}_{mr}$.

An alternative method to that used previously is now presented, where instead of expressing the stator and rotor voltage equations in terms of $|\bar{\imath}_{mr}|$, they are expressed in terms of $|\bar{\imath}_m|$. Thus by considering eqns (5.1-20) and (5.1-21), the relationship of $|\bar{\imath}_{mr}|$ and $|\bar{\imath}_m|$ is

$$|\bar{\imath}_{mr}| = (L_{rl}/L_m)\bar{\imath}_{r\psi r} + |\bar{\imath}_m| e^{j\mu}. \qquad (5.1\text{-}22)$$

When this is differentiated with respect to time, under saturated conditions cross-coupling terms must arise in the direct- and quadrature-axis stator and rotor equations expressed in the rotor-flux-oriented reference, frame and the resulting equations will be more complicated than eqns (5.1-11), (5.1-12), (5.1-18) and (5.1-19). This will now be proved and the resulting equations will be derived. To obtain these equations, eqn (5.1-2) or (5.1-8) and eqn (5.1-15)

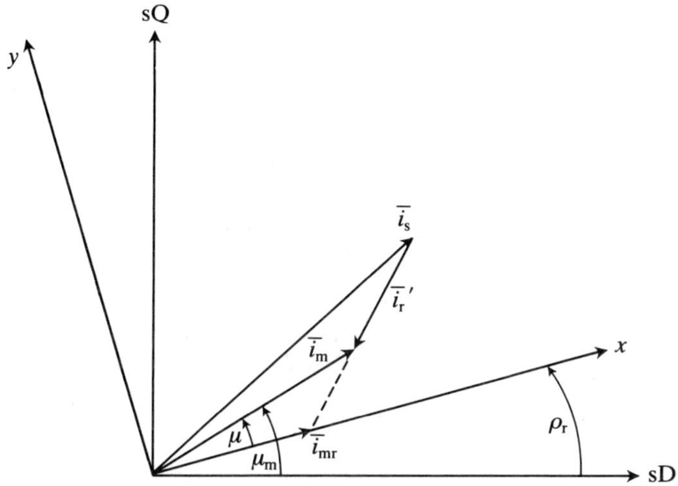

Fig. 5.2. Space phasor diagram of the rotor magnetizing current and the magnetizing current.

or (5.1-17) can be used directly, together with eqn (5.1-22), but it is more convenient to use a different approach as shown in the following section.

Stator voltage equations: It follows from eqns (5.1-1) and (5.1-21) that since the stator self-inductance (L_s) is equal to the sum of the stator leakage inductance (L_{sl}) and the magnetizing inductance (L_m), and it is assumed that L_{sl}=constant and only the magnetizing inductance varies with saturation,

$$\bar{u}_{s\psi r}=R_s\bar{i}_{s\psi r}+L_{sl}\frac{d\bar{i}_{s\psi r}}{dt}+\frac{d(L_m|\bar{i}_m|e^{j\mu})}{dt}+j\omega_{mr}(L_{sl}\bar{i}_{s\psi r}+L_m|\bar{i}_m|e^{j\mu}). \tag{5.1-23}$$

In eqn (5.1-23) L_m can be obtained from the no-load curve of the induction machine, since $L_m=|\bar{\psi}_m|/|\bar{i}_m|$, where $|\bar{\psi}_m|$ and $|\bar{i}_m|$ are the moduli of the space phasors of the magnetizing flux-linkage space phasor and magnetizing-current space phasor respectively. It should be noted that $L_m=|\bar{\psi}_r|/|\bar{i}_{mr}|$ also holds, [see eqn (5.1-7)], where $|\bar{\psi}_r|$ and $|\bar{i}_{mr}|$ are the moduli of the space phasors of the rotor flux linkage and rotor magnetizing current respectively. In eqn (5.1-23)

$$L_m|\bar{i}_m|e^{j\mu}=\bar{\psi}_{m\psi r}=L_m(\bar{i}_s+\bar{i}_r')e^{-j\rho_r}=\psi_{mx}+j\psi_{my} \tag{5.1-24}$$

is the magnetizing flux-linkage space phasor expressed in the rotor-flux-oriented reference frame. Since in general this is not coaxial with the rotor flux (it is only coaxial when $\mu=0$, i.e. when $\mu_m=\rho_r$) in eqn (5.1-23), under saturated conditions, the derivative $d\bar{\psi}_{m\psi r}/dt$ will lead to cross-coupling between the direct- and quadrature-axis stator voltage equations. This will now be discussed in detail.

Resolution of eqn (5.1-23) into its real and imaginary axes components yields

$$u_{sx}=R_s i_{sx}+L_{sl}\frac{di_{sx}}{dt}+\frac{d\psi_{mx}}{dt}-\omega_{mr}(L_{sl}i_{sy}+\psi_{my}) \tag{5.1-25}$$

$$u_{sy}=R_s i_{sy}+L_{sl}\frac{di_{sy}}{dt}+\frac{d\psi_{my}}{dt}+\omega_{mr}(L_{sl}i_{sx}+\psi_{mx}), \tag{5.1-26}$$

where the direct- and quadrature-axis components of the magnetizing flux-linkage space phasor in the rotor-flux-oriented reference frame are

$$\psi_{mx}=L_m(i_{sx}+i_{rx})=L_m i_{mx} \tag{5.1-27}$$

$$\psi_{my}=L_m(i_{sy}+i_{ry})=L_m i_{my}. \tag{5.1-28}$$

In eqns (5.1-27) and (5.1-28) i_{mx} and i_{my} are respectively the direct-axis and imaginary-axis components of the magnetizing-current space phasor in the rotor-flux- oriented reference frame and in eqns (5.1-25) and (5.1-26) their first time derivatives ($d\psi_{mx}/dt$ and $d\psi_{my}/dt$) are present, which can be expanded as follows by utilizing the chain differentiation rule:

$$\frac{d\psi_{mx}}{dt}=\frac{d(L_m i_{mx})}{dt}=L_m\frac{di_{mx}}{dt}+i_{mx}\frac{dL_m}{dt} \tag{5.1-29}$$

$$\frac{d\psi_{my}}{dt}=\frac{d(L_m i_{my})}{dt}=L_m\frac{di_{my}}{dt}+i_{my}\frac{dL_m}{dt}. \tag{5.1-30}$$

In these equations the first time derivative of the magnetizing inductance is present, which owing to saturation is not zero. By the application of the chain differentiation rule and also by considering that $L_m=|\bar{\psi}_m|/|\bar{i}_m|$,

$$\frac{dL_m}{dt}=\frac{dL_m}{d|\bar{i}_m|}\frac{d|\bar{i}_m|}{dt}=\frac{d|\bar{\psi}_m|}{d|\bar{i}_m|}\frac{d|\bar{i}_m|}{dt}=\frac{L_d-L_m}{|\bar{i}_m|}\frac{d|\bar{i}_m|}{dt}, \tag{5.1-31}$$

where $L_d=d|\bar{\psi}_m|/d|\bar{i}_m|$ is a dynamic (tangent slope) inductance, which under saturated conditions is not zero, but under linear magnetic conditions is equal to the (chord slope) magnetizing inductance L_m. In eqn (5.1-31) the first time derivative of the modulus of the magnetizing-current space phasor is present, and since this depends on both the direct- and quadrature-axis magnetizing currents, which however, contain the direct- and quadrature-axis stator and rotor currents respectively [see eqns (5.1-27) and (5.1-28)], it follows that dL_m/dt will contribute to the cross-saturation coupling terms. Mathematically, this can be proved by considering that

$$\frac{d|\bar{\imath}_m|}{dt}=\frac{d(i_{mx}^2+i_{my}^2)^{1/2}}{dt}=\frac{i_{mx}}{|\bar{\imath}_m|}\frac{di_{mx}}{dt}+\frac{i_{my}}{|\bar{\imath}_m|}\frac{di_{my}}{dt}=\cos\mu\frac{di_{mx}}{dt}+\sin\mu\frac{di_{my}}{dt},$$

(5.1-32)

where μ has been defined above (see Fig. 5.2, $\mu=\mu_m-\rho_r$), i_{mx} and i_{my} have been defined in eqns (5.1-27) and (5.1-28) respectively and from eqns (5.1-24) and (5.1-27), (5.1-28) (or from Fig. 5.2)

$$i_{mx}=|\bar{\imath}_m|\cos\mu$$
$$i_{my}=|\bar{\imath}_m|\sin\mu. \qquad (5.1\text{-}33)$$

Thus by considering eqns (5.1-29), (5.1-30), (5.1-31), (5.1-32) and (5.1-33), finally

$$\frac{d\psi_{mx}}{dt}=L_{mx}\frac{di_{mx}}{dt}+L_{xy}\frac{di_{my}}{dt} \qquad (5.1\text{-}34)$$

$$\frac{d\psi_{my}}{dt}=L_{my}\frac{di_{my}}{dt}+L_{xy}\frac{di_{mx}}{dt}, \qquad (5.1\text{-}35)$$

where L_{mx} and L_{my} are the magnetizing inductances along the direct (x) and quadrature (y) axes respectively,

$$L_{mx}=L_d\cos^2\mu+L_m\sin^2\mu=L_d\cos^2(\mu_m-\rho_r)+L_m\sin^2(\mu_m-\rho_r) \qquad (5.1\text{-}36)$$

$$L_{my}=L_d\sin^2\mu+L_m\cos^2\mu=L_d\sin^2(\mu_m-\rho_r)+L_m\cos^2(\mu_m-\rho_r), \qquad (5.1\text{-}37)$$

and L_{xy} is the cross-coupling inductance between the x and y axes,

$$L_{xy}=\tfrac{1}{2}[(L_d-L_m)\sin(2\mu)]=\tfrac{1}{2}(L_d-L_m)\sin[2(\mu_m-\rho_r)]. \qquad (5.1\text{-}38)$$

Under linear magnetic conditions $L_d=L_m$ and $L_{mx}=L_{my}=L_m$ and $L_{xy}=0$. However, owing to saturation of the main flux paths, in general the magnetizing inductances in the two axes are not equal, $L_{mx}\neq L_{my}$ and the cross-coupling inductance is not zero, $L_{xy}\neq 0$. For the special selection of $\mu=\mu_m-\rho_r=0$, it follows that $L_{mx}=L_d$, $L_{my}=0$ and $L_{xy}=0$, which are expected results since in this case it follows from eqn (5.1-33) or Fig. 5.2 that $i_{mx}=|\bar{\imath}_m|$, $i_{my}=0$ and thus $\psi_{mx}=L_m i_{mx}=L_m|\bar{\imath}_m|$, and $d\psi_{mx}/dt=(d\psi_{mx}/d|\bar{\imath}_m|)\cdot(d|\bar{\imath}_m|/dt)=L_d(di_{mx}/dt)$ and $\psi_{my}=0$; thus $d\psi_{my}/dt=0$. The same result also follows from eqns (5.1-33)–(5.1-38). Substitution of eqns (5.1-34), (5.1-35), (5.1-27) and (5.1-28) into eqns (5.1-25) and (5.1-26) yields the following stator voltage equations:

$$u_{sx}=R_s i_{sx}+L_{sx}\frac{di_{sx}}{dt}+L_{mx}\frac{di_{rx}}{dt}+L_{xy}\left(\frac{di_{sy}}{dt}+\frac{di_{ry}}{dt}\right)$$
$$-\omega_{mr}[L_{sl}i_{sy}+L_m(i_{sy}+i_{ry})] \qquad (5.1\text{-}39)$$

$$u_{sy}=R_s i_{sy}+L_{sy}\frac{di_{sy}}{dt}+L_{my}\frac{di_{ry}}{dt}+L_{xy}\left(\frac{di_{sx}}{dt}+\frac{di_{rx}}{dt}\right)$$
$$+\omega_{mr}[L_{sl}i_{sx}+L_m(i_{sx}+i_{rx})], \qquad (5.1\text{-}40)$$

where L_{mx}, L_{my} and L_{xy} have been defined in eqns (5.1-36), (5.1-37) and (5.1-38) respectively, and

$$L_{sx}=L_{sl}+L_{mx}$$
$$L_{sy}=L_{sl}+L_{my} \qquad (5.1\text{-}41)$$

are the self-inductances of the stator winding along the direct- and quadrature-axes of the rotor-flux-oriented reference frame and, as a result of saturation, anisotropy (asymmetry) exists, $L_{sx}\neq L_{sy}$. Under linear magnetic conditions, however, $L_{sx}=L_{sy}=L_{sl}+L_m$, as expected. It follows from eqns (5.1-39) and (5.1-40) that even when the speed of the rotor-flux-oriented reference frame is zero ($\omega_{mr}=0$), the two stator voltage equations are coupled as a result of saturation.

Under linear magnetic conditions, $L_{sx}=L_{sy}=L_{sl}+L_m$ and it follows from the resolution of eqn (5.1-20) into its direct- and quadrature-axis components that

$$i_{rx}=(|\bar{i}_{mr}|-i_{sx})\frac{L_m}{L_r} \qquad (5.1\text{-}42)$$

$$i_{ry}=-i_{sy}\frac{L_m}{L_r} \qquad (5.1\text{-}43)$$

and when eqns (5.1-42), (5.1-43) are substituted into eqns (5.1-39) and (5.1-40), the stator equations derived in Section 4.1, eqns (4.1-7) and (4.1-8), are obtained. Comparison of the equations which are valid under linear and saturated conditions show that with saturation extra and modified terms are present. If $\rho_r=0$, the rotor-flux-oriented reference frame coincides with the stationary reference frame (see Fig. 5.2) and by considering $\omega_{mr}=d\rho_r/dt=0$, the voltage equations in the stationary reference frame are obtained. Thus it follows from eqns. (5.1-39), (5.1-40) and (5.1-36), (5.1-37), (5.1-38) that

$$u_{sD}=R_s i_{sD}+L_{sl}\frac{di_{sD}}{dt}+\frac{d\psi_{mD}}{dt} \qquad (5.1\text{-}44)$$

$$u_{sQ}=R_s i_{sQ}+L_{sl}\frac{di_{sQ}}{dt}+\frac{d\psi_{mQ}}{dt}, \qquad (5.1\text{-}45)$$

where ψ_{mD} and ψ_{mQ} are respectively the direct- and quadrature-axis components of the magnetizing flux-linkage space phasor in the stationary reference frame,

$$\psi_{mD}=L_m i_{mD}=L_m(i_{sD}+i_{rd}) \tag{5.1-46}$$

$$\psi_{mQ}=L_m i_{mQ}=L_m(i_{sQ}+i_{rq}) \tag{5.1-47}$$

$$\frac{d\psi_{mD}}{dt}=L_{mD}\frac{di_{mD}}{dt}+L_{DQ}\frac{di_{mQ}}{dt} \tag{5.1-48}$$

$$\frac{d\psi_{mQ}}{dt}=L_{mQ}\frac{di_{mQ}}{dt}+L_{DQ}\frac{di_{mD}}{dt}, \tag{5.1-49}$$

where i_{mD}, i_{sD}, i_{rd} and i_{mQ}, i_{sQ}, i_{rq} are the direct- and quadrature-axis components of the magnetizing current, stator-current and rotor-current space phasor respectively in the stationary reference frame. L_{mD} and L_{mQ} are the magnetizing inductances along the direct and quadrature axes of the stationary reference frame, and L_{DQ} is the cross-coupling inductance between the stator windings sD and sQ, which are in space quadrature, but the same cross-coupling exists between all the windings of the stationary-axis model which are in space quadrature, e.g. between windings sD and rq, sQ and rd etc.:

$$L_{mD}=L_d\cos^2\mu_m+L_m\sin^2\mu_m \tag{5.1-50}$$

$$L_{mQ}=L_d\sin^2\mu_m+L_m\cos^2\mu_m \tag{5.1-51}$$

$$L_{DQ}=(L_d-L_m)\sin\mu_m\cos\mu_m. \tag{5.1-52}$$

Eqns (5.1-44)–(5.1-52) can be put into a more compact form:

$$u_{sD}=R_s i_{sD}+L_{sD}\frac{di_{sD}}{dt}+L_{mD}\frac{di_{rd}}{dt}+L_{DQ}\left(\frac{di_{sQ}}{dt}+\frac{di_{rq}}{dt}\right)$$
$$-\omega_r[L_{sl}i_{sQ}+L_m(i_{sQ}+i_{rq})] \tag{5.1-53}$$

$$u_{sQ}=R_s i_{sQ}+L_{sQ}\frac{di_{sQ}}{dt}+L_{mQ}\frac{di_{rQ}}{dt}+L_{DQ}\left(\frac{di_{sD}}{dt}+\frac{di_{rd}}{dt}\right)$$
$$+\omega_r[L_{sl}i_{sD}+L_m(i_{sD}+i_{rd})], \tag{5.1-54}$$

where $L_{sD}=L_{sl}+L_{mD}$ and $L_{sQ}=L_{sl}+L_{mQ}$ are the self-inductances of the stator winding along the direct and quadrature axes of the stationary reference frame and, in general, as a result of saturation $L_{sD}\neq L_{sQ}$.

A very simple physical explanation can be given for the existence of the phenomenon of cross-saturation (cross-magnetization coupling due to saturation of the main flux paths). This is now described. For simplicity a symmetrical quadrature-phase machine is assumed and it is also assumed that the direct- and quadrature-axis m.m.f. distributions around the periphery of the machine (spatial distributions) are sinusoidal, in agreement with the discussion given above (see introduction to Chapter 5).

In Fig. 5.3(a) $\bar{f}_D = F_D$ is the m.m.f. space phasor corresponding to the case where there is only direct-axis excitation. By using the magnetization curve of the machine, which is assumed to be non-linear, the space phasor of the flux-linkage space phasor $\bar{\psi}_m$ is obtained. However, when both the direct-axis and quadrature-axis windings are excited, the quadrature-axis m.m.f. component F_Q is also present and the resultant m.m.f. space phasor will take the form $\bar{f} = F_D + jF_Q = |\bar{f}|e^{j\mu_m}$, where μ_m is the space angle of the m.m.f. space phasor with respect to the direct axis of the stationary reference frame a shown in Fig. 5.3(a). By using the same magnetization curve as before, the corresponding space phasor of the magnetizing flux linkages $\bar{\psi}'_m$ is obtained $\bar{\psi}'_m = |\bar{\psi}'_m|e^{j\mu_m}$, and this is also shown in Fig. 5.3(a). It follows from Fig. 5.3(a) that the flux linkage along the direct axis has changed from $|\bar{\psi}_m|$ to $|\bar{\psi}'_m|\cos \mu_m$ and the reduction of the magnetizing flux linkage is $\Delta\bar{\psi}_m = |\bar{\psi}_m| - |\bar{\psi}'_m| \cos \mu_m$.

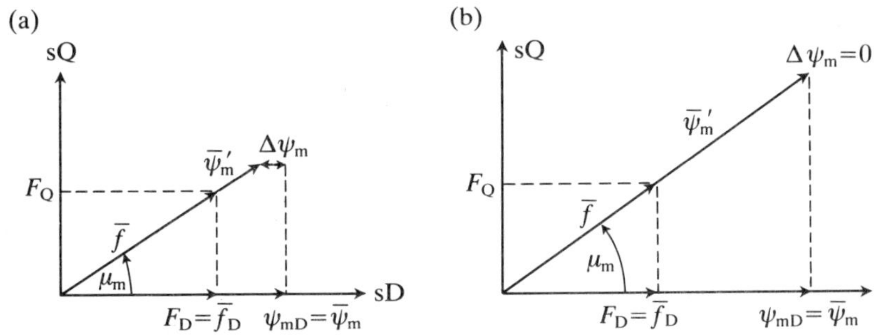

Fig. 5.3. Space-phasor diagram to prove the existence of cross-saturation. (a) Non-linear case. (b) Linear case.

However, under linear magnetic conditions (by using the linear part of the previously used magnetizing curve), the space-phasor diagram shown in Fig. 5.3(b) is obtained. If the slope of the linear section of the magnetization curve is $\tan \alpha$ and only the direct axis is excited, the flux linkage along the direct axis is $\psi_{mD} = |\bar{f}| \cos \mu_m \tan \alpha$. However, if the quadrature-axis winding is also excited, the space phasor of the resultant flux linkages will be $\bar{\psi}'_m$, and its direct-axis component is equal to $\psi_{mD} = |\bar{\psi}'_m| \cos \mu_m$. As a result of linear magnetic conditions $|\bar{\psi}'_m| = |\bar{f}| \tan \alpha$, and thus $\psi'_{mD} = |\bar{f}| \cos \mu_m \tan \alpha$ and it follows that ψ'_{mD} is equal to ψ_{mD}, which is an expected result. Thus $\Delta\psi_m = \psi'_{mD} - \psi_{mD} = 0$. It is important to note that under linear magnetic conditions the magnetizing flux-linkage components correspond exactly (in per-unit terms) to the m.m.f. components, and angular displacement and a change of the resultant magnetizing flux-linkage space phasor does not lead to an induced e.m.f. in the direct-axis winding, since the direct-axis component of the magnetizing flux-linkage space phasor is unchanged. However, under saturated magnetic conditions, the change in the magnitude of the resultant m.m.f. space phasor leads to a non-proportional change in the direct-axis magnetizing flux-linkage component and an induced e.m.f. in the direct-axis winding. It should be emphasized that an angular displacement of the magnetizing flux-linkage space phasor (relative to a stationary winding) is not in itself sufficient to induce voltage, nor is angular displacement with change in magnitude. The change of the quadrature-axis magnetizing flux linkage $\Delta\psi_m$ depends on the change of the quadrature-axis magnetizing current (Δi_{mQ}) and thus $\Delta\psi_m = L_{DQ}\,\Delta i_{mQ}$, where L_{DQ} is a cross-coupling inductance between the direct and quadrature axes which is zero under linear magnetic conditions.

It follows from Fig. 5.3 that if $\mu_m = 0$, there cannot be any cross-saturation coupling ($L_{DQ} = 0$) between the windings in space quadrature, since this corresponds to the case when only the direct-axis winding is excited and therefore there is no change in the resultant flux-density distribution. Furthermore, if $\mu_m = \pi/2$, this corresponds to the case when there is only quadrature-axis excitation, and therefore the resultant flux-density distribution is equal to the quadrature-axis flux-density distribution, and thus the cross-saturation coupling will be zero since there is no direct-axis excitation and the quadrature-axis flux density distribution does not change. Thus when $\mu_m = 0$ or $\mu_m = \pi/2$, $L_{DQ} = 0$, but when $\mu_m \neq 0$ it has to be proportional to $\cos\mu_m \cdot \sin\mu_m$. However, it has to be zero under linear magnetic conditions and has to be non-zero under saturated conditions. Thus L_{DQ} has to contain another term as well, and this is equal to $L_d - L_m$. In this term L_d is a dynamic inductance and it is equal to the slope of the magnetizing curve ($L_d = \mathrm{d}|\bar{\psi}_m|/\mathrm{d}|\bar{i}_m|$), where $|\bar{\psi}_m|$ and $|\bar{i}_m|$ are the moduli of the magnetizing flux-linkage and magnetizing-current space phasors respectively and L_m is the magnetizing inductance, $L_m = |\bar{\psi}_m|/|\bar{i}_m|$. Thus $L_{DQ} = (L_d - L_m)\sin\mu_m\cos\mu_m$ is obtained and a rigorous mathematical analysis yields the same expression for L_{DQ} [see the derivation of eqn (5.1-52)]. The change of the magnetizing flux linkage $\Delta\psi_m = L_{DQ}\,\Delta i_{mQ}$ yields the induced e.m.f. $L_{DQ}\,\mathrm{d}i_{mQ}/\mathrm{d}t$, which is the transformer voltage induced in the direct-axis winding as a result of cross-saturation. It follows that under saturated conditions, solely due to saturation, even sinusoidally distributed windings in space quadrature can become coupled. If the windings are not sinusoidally distributed, the cross-coupling described above also exists, but in addition, there are some other extra voltage terms which have to be considered.

The existence of the phenomenon of cross-saturation can also be proved by simple experiments, or by the application of numerical analysis techniques, e.g. the finite-element method. For this purpose it is very useful to employ the so-called method of frozen permeabilities. When this method is used, it can be shown that in a saturated induction machine, the obtained cross-saturation coupling inductance L_{DQ} agrees with the L_{DQ} obtained by the application of the simple expression $L_{DQ}=(L_d-L_m)\sin\mu_m\cos\mu_m$. Furthermore, it can also be shown that in a saturated salient-pole machine, in addition to the effects of the physically existing saliency, the effects of cross-saturation are also present. This will be discussed in Section 5.2.

The stator voltage equations, eqns (5.1-39) and (5.1-40) contain the quadrature-axis rotor current component (i_{ry}) and its derivative (di_{ry}/dt), but from eqn (5.1-43) i_{ry} can be substituted by the torque-producing stator current component (i_{sy}) and di_{ry}/dt can be expressed in terms of i_{sy}. The derivative di_{ry}/dt can be obtained directly by differentiation of eqn. (5.1-43). Under linear magnetic conditions this yields $-(L_m/L_r)\,di_{sy}/dt$, but under saturated conditions, owing to cross-saturation, it must also depend on the direct-axis magnetizing current (i_{mx}). The required derivative can be most simply obtained by utilizing eqn (2.1-151) and $L_r=L_{rl}+L_m$, where L_{rl} is the rotor leakage inductance, which is assumed to be constant, and thus the rotor flux-linkage space phasor in the rotor-flux oriented reference frame is obtained as

$$\bar{\psi}_{r\psi r}=|\bar{\psi}_r|=L_{rl}\bar{i}_{r\psi r}+\bar{\psi}_{m\psi r}, \tag{5.1-55}$$

where $\bar{i}_{r\psi r}$ has been defined in eqn (5.1-21) and $\bar{\psi}_{r\psi r}$ has been defined in eqn (5.1-24). The quadrature-axis component of eqn (5.1-55) yields

$$\psi_{ry}=0=L_{rl}i_{ry}+\psi_{my}, \tag{5.1-56}$$

where ψ_{my} has been defined by eqn (5.1-28). Eqn (5.1-56) together with eqn (5.1-35) can be used to obtain di_{ry}/dt in terms of i_{sx}, i_{sy} and i_{rx}. For this purpose eqn (5.1-56) is differentiated, and thus

$$0=L_{rl}\frac{di_{ry}}{dt}+\frac{d\psi_{my}}{dt}, \tag{5.1-57}$$

where $d\psi_{my}/dt$ has been given by eqn (5.1-35). Thus by substitution of eqn (5.1-35) into eqn (5.1-57),

$$di_{ry}/dt=-\left(L_{my}\frac{di_{sy}}{dt}+L_{xy}\frac{di_{sx}}{dt}+L_{xy}\frac{di_{rx}}{dt}\right)\Big/L_{ry}, \tag{5.1-58}$$

where $L_{ry}=L_{rl}+L_{my}$ is the self-inductance of the rotor along the quadrature axis of the rotor-flux-oriented reference frame and L_{my} and L_{xy} have been defined by eqns (5.1-36) and (5.1-38) respectively. It should be noted, that because of saturation, the self-inductance of the rotor winding along the direct axis (L_{rx}) differs from L_{ry}, and $L_{rx}=L_{rl}+L_{mx}$, where L_{mx} has been defined in eqn (5.1-37).

When eqns (5.1-58) and (5.1-43) are substituted into eqns (5.1-39) and (5.1-40) and eqn (5.1-41) is taken into account,

$$u_{sx}=R_si_{sx}+L_{sl}\frac{di_{sx}}{dt}+\left(L_{mx}-\frac{L_{xy}^2}{L_{ry}}\right)\left(\frac{di_{sx}}{dt}+\frac{di_{rx}}{dt}\right)+L_{xy}\left(1-\frac{L_{my}}{L_{ry}}\right)\frac{di_{sy}}{dt}-\omega_{mr}L_s'i_{sy} \tag{5.1-59}$$

and

$$u_{sy}=R_s i_{sy}+L_{sl}\frac{di_{sy}}{dt}+L_{my}\left(1-\frac{L_{my}}{L_{ry}}\right)\frac{di_{sy}}{dt}$$

$$+L_{xy}\left(1-\frac{L_{my}}{L_r}\right)\left(\frac{di_{sx}}{dt}+\frac{di_{rx}}{dt}\right)+\omega_{mr}[L_{sl}i_{sx}+L_m(i_{sx}+i_{rx})], \qquad (5.1\text{-}60)$$

where L'_s is the stator transient inductance, which in general varies with saturation, $L'_s=L_s-L_m^2/L_r$, and $L_r=L_{rl}+L_m$, $L_{ry}=L_{rl}+L_{my}$. Thus the stator voltage equations are coupled, and extra coupling (cross-saturation coupling) exists as a result of saturation. Under linear magnetic conditions ($L_{mx}=L_{my}=L_m$ and $L_{xy}=0$) the stator voltage equations are coupled owing to the existence of the rotational voltages $\omega_{mr}L'_s i_{sy}$ and $\omega_{mr}[L_{sl}i_{sx}+L_m(i_{sx}+i_{rx})]$.

The stator equations can be decoupled in the same way as that used in Section 4.1. Thus if an ideal drive is assumed, independent control of the stator currents i_{sx} and i_{sy} can be achieved by defining new voltages which directly control the stator current components. These new voltages can be defined as $\hat{u}_{sx}$ and $\hat{u}_{sy}$ and they are directly related to the stator currents i_{sx} and i_{sy} respectively,

$$\hat{u}_{sx}=\left[R_s+\left(L_{sx}-\frac{L_{xy}^2}{L_{ry}}\right)p\right]i_{sx}$$

$$=u_{sx}-\left(L_{mx}-\frac{L_{xy}^2}{L_{ry}}\right)pi_{rx}-L_{xy}\left(1-\frac{L_{my}}{L_{ry}}\right)pi_{sy}$$

$$+\omega_{mr}L'_s i_{sy} \qquad (5.1\text{-}61)$$

$$\hat{u}_{sy}=\left[R_s+\left(L_{sy}-\frac{L_{my}^2}{L_{ry}}\right)p\right]i_{sy}$$

$$=u_{sy}-L_{xy}\left(1-\frac{L_{my}}{L_{ry}}\right)pi_{sx}$$

$$-L_{xy}\left(1-\frac{L_{my}}{L_{ry}}\right)pi_{rx}-\omega_{mr}(L_s i_{sx}+L_m i_{rx}), \qquad (5.1\text{-}62)$$

where according to eqn (5.1-41) $L_{sx}=L_{sl}+L_{mx}$ and $p=d/dt$. Thus the required decoupling circuit will be relatively complicated.

Rotor voltage equations: As for the stator voltage equations derived in the previous section, the rotor voltage equatons can also be expressed in terms of L_m, which varies with the modulus of the magnetizing-current space phasor $|\bar{i}_m|$. For this purpose eqn (2.1-153) is used and it follows that in the rotor-flux-oriented reference frame ($\omega_g=\omega_{mr}$), by also considering eqn (5.1-21) and $L_r=L_{rl}+L_m$, where L_{rl} is the rotor leakage inductance which is assumed to be constant and L_m is the magnetizing inductance which varies due to saturation,

$$\bar{u}_{r\psi r}=0=R_r\bar{i}_{r\psi r}+L_{rl}\frac{d\bar{i}_{r\psi r}}{dt}+\frac{d\bar{\psi}_{m\psi r}}{dt}+j(\omega_{mr}-\omega_r)(L_{rl}\bar{i}_{r\psi r}+\bar{\psi}_{m\psi r}), \tag{5.1-63}$$

where $\bar{\psi}_{m\psi r}$ has been defined in eqn (5.1-24). Resolution of eqn (5.1-63) into its real- and imaginary-axis components gives the following voltage equations, if it is considered that in the rotor-flux-oriented reference frame the quadrature-axis rotor flux-linkage component is zero (and according to eqn (5.1-56) $\psi_{ry}=L_{rl}i_{ry}+\psi_{my}=0$),

$$0=R_ri_{rx}+L_{rl}\frac{di_{rx}}{dt}+\frac{d\psi_{mx}}{dt} \tag{5.1-64}$$

$$0=R_ri_{ry}+L_{rl}\frac{di_{ry}}{dt}+\frac{d\psi_{my}}{dt}+(\omega_{mr}-\omega_r)(L_{rl}i_{rx}+\psi_{mx}), \tag{5.1-65}$$

where $d\psi_{mx}/dt$ and $d\psi_{my}/dt$ are given in Eqns (5.1-29) and (5.1-30) respectively and ψ_{mx} and ψ_{my} have been defined in Eqns (5.1-27) and (5.1-28) respectively. In eqn (5.1-64) there is no term multiplied by $(\omega_{mr}-\omega_r)$, since the quadrature-axis rotor flux-linkage component is zero. In eqn (5.1-65) the derivative term di_{ry}/dt, which has been given by eqn (5.1-58) is present. Thus by substitution of eqns (5.1-29), (5.1-30), (5.1-27), (5.1-28) and (5.1-43) into eqns (5.1-64) and (5.1-65).

$$\begin{aligned}0=R_ri_{rx}+L_{rl}\frac{di_{rx}}{dt}&+\left(L_{mx}-\frac{L_{xy}^2}{L_{ry}}\right)\frac{di_{rx}}{dt}\\&+\left(L_{mx}-\frac{L_{xy}^2}{L_{ry}}\right)\frac{di_{sx}}{dt}+\left(L_{xy}-\frac{L_{xy}L_{my}}{L_{ry}}\right)\frac{di_{sy}}{dt}\end{aligned} \tag{5.1-66}$$

$$0=-R_r\frac{L_m}{L_r}i_{sy}+(\omega_{mr}-\omega_r)(L_ri_{rx}+L_mi_{sx}), \tag{5.1-67}$$

where $L_r=L_{rl}+L_m$, $L_{ry}=L_{rl}+L_{my}$. In eqn (5.1-67) $\psi_{rx}=(L_ri_{rx}+L_mi_{sx})$ and under linear magnetic conditions ($L_{mx}=L_{my}=L_m$, $L_{xy}=0$) eqns (5.1-66) and (5.1-67) yield eqns (4.1-25) and (4.1-26) respectively. This can be proved by substitution of $\psi_{rx}=|\bar{\psi}_r|=L_m|\bar{i}_{mr}|$ into eqn (5.1-67), where $|\bar{i}_{mr}|$ is the modulus of the rotor magnetizing-current space phasor, and by substitution of the expression for i_{rx} given by eqn (5.1-42) into eqn (5.1-66). Whilst the rotor voltage equations under linear magnetic conditions imply the decoupled control of the rotor flux and torque by the independent control of i_{sx} and i_{sy} respectively, it follows from eqns (5.1-66) and (5.1-67) that under saturated conditions this is not the case. It follows from eqn (5.1-66) that if i_{sx} is kept constant, as a result of the cross-saturation coupling a change in the torque-producing stator current (i_{sy}) will cause a change the direct-axis rotor current

i_{rx} and thus the rotor flux level (which depends on the rotor currents also) will change. It should be noted that eqn (5.1-66) is much more complicated than eqn (5.1-18). However, whilst eqn (5.1-18) has been derived on the basis that the magnetizing inductance is a function of the modulus of the rotor magnetizing-current space phasor, which has only a direct-axis component in the rotor-flux-oriented reference frame, eqn (5.1-66) has been obtained on the basis that the magnetizing inductance is a function of the modulus of the magnetizing-current space phasor, which does have direct-axis and quadrature-axis components in the same reference frame. Furthermore, it should also be noted that eqn (5.1-66) cannot be obtained from eqn (4.1-25) by simply replacing the constant magnetizing inductance by the variable magnetizing inductance which is a function of the modulus of the magnetizing-current space phasor.

For completeness, the rotor voltage equations are also given in the stationary reference frame. These can be most simply obtained by considering the real- and imaginary-axis components of eqn (5.1-63) and by utilizing that $\omega_{mr}=0$. Furthermore, in the stationary reference frame the rotor-current space phasor $\bar{i}_r'$ can be expressed in terms of its two-axis components as $\bar{i}_r'=i_{rd}+ji_{rq}$ and the two-axis components of the magnetizing flux-linkage space phasor are ψ_{mD} and ψ_{mQ}, which have been defined in eqns (5.1-46) and (5.1-47) respectively. Thus

$$u_{rd}=0=R_r i_{rd}+L_{rl}\frac{di_{rd}}{dt}+\frac{d\psi_{mD}}{dt}+\omega_r(L_r i_{rq}+L_m i_{sQ}) \tag{5.1-68}$$

$$u_{rq}=0=R_r i_{rq}+L_{sl}\frac{di_{rq}}{dt}+\frac{d\psi_{mQ}}{dt}-\omega_r(L_r i_{rd}+L_m i_{sQ}) \tag{5.1-69}$$

where $d\psi_{mD}/dt$ and $d\psi_{mQ}/dt$ are defined in eqns (5.1-48) and (5.1-49) respectively. Thus by the substitution of eqns (5.1-48) and (5.1-49) into eqns (5.1-68) and (5.1-69), the following rotor voltage equations are obtained for an induction machine when saturation of the main flux paths is present,

$$u_{rd}=0=R_r i_{rd}+L_{rd}\frac{di_{rd}}{dt}+L_{mD}\frac{di_{sD}}{dt}+L_{DQ}\left(\frac{di_{sQ}}{dt}+\frac{di_{rq}}{dt}\right)+\omega_r(L_r i_{rq}+L_m i_{sQ}) \tag{5.1-70}$$

$$u_{rq}=0=R_r i_{rq}+L_{rq}\frac{di_{rq}}{dt}+L_{mQ}\frac{di_{sQ}}{dt}+L_{DQ}\left(\frac{di_{sD}}{dt}+\frac{di_{rd}}{dt}\right)-\omega_r(L_r i_{rd}+L_m i_{sD}), \tag{5.1-71}$$

where $L_{rd}=L_{rl}+L_{mD}$ and $L_{rq}=L_{rl}+L_{mQ}$ are the self-inductances of the rotor in the direct and quadrature axes of the stationary reference frame and $L_r=L_{rl}+L_m$. As a result of saturation these are not equal, but under linear magnetic conditions $L_{rd}=L_{rq}=L_r$. In eqns (5.1-70) and (5.1-71) the effects of cross-saturation are also present and under saturated conditions in, say, the direct-axis rotor winding, a change in the quadrature-axis stator current will induce a transformer e.m.f. ($L_{DQ}\,di_{sQ}/dt$) in the direct-axis rotor winding.

If eqns (5.1-70) and (5.1-71) are combined with eqns (5.1-53) and (5.1-54), the stationary-axis model of a saturated smooth-air-gap machine is obtained as:

$$\begin{bmatrix} u_{sD} \\ u_{sQ} \\ u_{rd} \\ u_{rq} \end{bmatrix} = \begin{bmatrix} R_s+L_{sD}p & L_{DQ}p & L_{mD}p & L_{DQ}p \\ L_{DQ}p & R_s+L_{sQ}p & L_{DQ}p & L_{mQ}p \\ L_{mD}p & \omega_r L_m+L_{DQ}p & R_r+L_{rd}p & \omega_r L_r+L_{DQ}p \\ -\omega_r L_m+L_{DQ}p & L_{mQ}p & -\omega_r L_r+L_{DQ}p & R_r+L_{rq}p \end{bmatrix} \begin{bmatrix} i_{sD} \\ i_{sQ} \\ i_{rd} \\ i_{rq} \end{bmatrix}. \tag{5.1-72}$$

When eqn (5.1-72) is compared with eqn (2.1-123), which describes the linear model, it follows that as a result of saturation, all the inductance terms are modified and there will be 16 inductance terms in the impedance matrix of the saturated commutator model. Furthermore, in the saturated model the self-inductances of both the stator windings and the rotor windings are unequal, the magnetizing inductances along the two axes are unequal and all the windings in space quadrature are coupled owing to cross-saturation. The cross-saturation coupling terms are present in the transformer e.m.f.s. It is important to note that in eqn (5.1-72) only $\omega_r L_m i_{sQ}+\omega_r L_r i_{rq}$ and $-(\omega_r L_m i_{sD}+\omega_r L_r i_{rd})$ are rotational voltage components which correspond to $\omega_r\psi_{rd}$ and $-\omega_r\psi_{rq}$ respectively. Thus under saturated conditions there will be no cross-saturation or other extra terms present in the expression for the electromagnetic torque as compared with the expression valid under linear magnetic conditions.

5.1.2 Magnetizing-flux-oriented control

In this section the effects of main-flux saturation are discussed for the equations of the saturated single-cage induction machine which have to be used if magnetizing-flux-oriented control is employed. For this purpose, the stator and rotor voltage equations are formulated in the magnetizing-flux-oriented reference frame. It will be shown that relatively simple equations arise, since in this reference frame, because the magnetizing flux-linkage space phasor has only one component, the effects of cross-saturation will be absent.

Stator voltage equations

It follows from eqns (2.1-148) and (2.1-150) that in the magnetizing-flux-oriented reference frame, which rotates at the speed $\omega_m=d\mu_m/dt$, where μ_m is the angle of the magnetizing-current space phasor with respect to the real axis of the stationary reference frame (see Fig. 5.2 or Section 4.3.1), by considering that the self-inductance of the stator (L_s) is equal to the sum of the stator leakage inductance (L_{sl}) and the magnetizing inductance (L_m), which varies due to saturation, the space-phasor voltage equation of the stator can be put into the form:

$$\bar{u}_{sm} = R_s \bar{\imath}_{sm} + L_{sl}\frac{d\bar{\imath}_{sm}}{dt} + \frac{d(L_m|\bar{\imath}_{mm}|)}{dt} + j\omega_m(L_{sl}\bar{\imath}_{sm} + L_m|\bar{\imath}_{mm}|), \quad (5.1\text{-}73)$$

where L_{sl} is assumed to be constant and $|\bar{\imath}_{mm}|$ is the magnetizing-current space phasor in the rotor-flux-oriented reference frame,

$$|\bar{\imath}_{mm}| = |\bar{\imath}_m| = \bar{\imath}_{sm} + \bar{\imath}_{rm} = i_{sx} + ji_{sy} + (i_{rx} + ji_{ry}), \quad (5.1\text{-}74)$$

where $|\bar{\imath}_m|$ is the modulus of the magnetizing-current space phasor in the stationary reference frame. In eqn (5.1-73) in general L_m is a non-linear function of the modulus of the magnetizing-current space phasor and in eqn (5.1-74) $\bar{\imath}_{sm}$ and $\bar{\imath}_{rm}$ are the space phasors of the stator and rotor current respectively in the magnetizing-flux-oriented reference frame [they have been defined in eqns (4.3-3) and (4.3-4)]. $\bar{u}_{sm} = \bar{u}_s e^{-j\mu_m}$ is the space phasor of the stator voltages in the same reference frame, where $\bar{u}_s$ is the space phasor of the stator voltages in the stationary reference frame. Resolution of eqn (5.1-73) into its direct- and quadrature-axis components yields

$$u_{sx} = R_s i_{sx} + L_{sl}\frac{di_{sx}}{dt} + \frac{d(L_m|\bar{\imath}_m|)}{dt} - \omega_m L_{sl} i_{sy} \quad (5.1\text{-}75)$$

$$u_{sy} = R_s i_{sy} + L_{sl}\frac{di_{sy}}{dt} + \omega_m(L_{sl} i_{sx} + L_m|\bar{\imath}_m|). \quad (5.1\text{-}76)$$

Under linear magnetic conditions, the magnetizing inductance is constant and eqns (5.1-75) and (5.1-76) yield eqns (4.3-10) and (4.3-11) respectively. However, under saturated conditions when L_m is a function of $|\bar{\imath}_m|$, the derivative term, $d/dt(L_m|\bar{\imath}_m|) \neq L_m\, d|\bar{\imath}_m|/dt$, which is the magnetizing voltage and is equal to the rate of change of the magnetizing flux-linkage space phasor, can be expressed as

$$\frac{d(L_m|\bar{\imath}_m|)}{dt} = \frac{d|\bar{\psi}_m|}{dt} = \frac{d|\bar{\psi}_m|}{d|\bar{\imath}_m|}\frac{d|\bar{\imath}_m|}{dt} = L_d\frac{d|\bar{\imath}_m|}{dt}, \quad (5.1\text{-}77)$$

where L_d is a dynamic inductance (also used in the previous section), $L_d = d|\bar{\psi}_m|/d|\bar{\imath}_m|$, where $|\bar{\psi}_m|$ is the modulus of the magnetizing flux-linkage space phasor in the stationary reference frame (and is equal to $|\bar{\psi}_{mm}|$, which is the modulus of the magnetizing flux-linkage space phasor in the magnetizing-flux-oriented reference frame). Substitution of eqn (5.1-77) into eqn (5.1-75) and rearrangement of the resulting equation, together with the rearranged form of eqn (5.1-76), yields

$$T_{sl}\frac{di_{sx}}{dt}+i_{sx}=\frac{u_{sx}}{R_s}+\omega_m T_{sl}i_{sy}-\frac{L_d}{R_s}\frac{d|\bar{i}_m|}{dt} \tag{5.1-78}$$

$$T_{sl}\frac{di_{sy}}{dt}+i_{sy}=\frac{u_{sy}}{R_s}-\omega_m\left(T_{sl}i_{sx}+\frac{|\bar{i}_m|L_m}{R_s}\right), \tag{5.1-79}$$

where $T_{sl}=L_{sl}/R_s$ is the stator leakage time constant. Under linear magnetic conditions L_m=constant, $L_d=L_m$ and eqns (5.1-78) and (5.1-79) agree with eqns (4.3-9) and (4.3-10) respectively. However, it follows from eqns (5.1-78) and (5.1-79) that when the effects of main-flux saturation are also considered, the stator voltage equations are very similar to those obtained under linear magnetic conditions, but in the direct-axis voltage equation the magnetizing inductance is replaced by the dynamic inductance and of course in the quadrature-axis voltage equation instead of the constant magnetizing inductance the variable magnetizing inductance has to be used. The simple forms of the equations are due to the fact that the magnetizing-current space phasor has only one component in the magnetizing-flux-oriented reference frame and therefore there cannot be any terms due to cross-saturation. However, the stator equations are still coupled and the required decoupling circuits can be obtained in a way similar to that described in Section 4.3.2. When the machine is supplied by impressed stator currents, only the rotor voltage equations have to be considered, and these are described in the next section.

Rotor voltage equations

From eqn (2.1-153), which is the rotor-voltage space-phasor in the general reference frame, it follows that in the magnetizing-flux-oriented reference frame, which rotates at the speed ω_m, when the effects of main-flux saturation are considered with $L_r=L_{rl}+L_m$, where L_{rl} is the rotor leakage inductance and L_m is the magnetizing inductance which changes with saturation, $L_m=L_m(|\bar{i}_m|)$, the following rotor voltage equation for the induction machine is obtained:

$$0=R_r\bar{i}_{rm}+L_{rl}\frac{d\bar{i}_{rm}}{dt}+\frac{d(L_m|\bar{i}_m|)}{dt}+j(\omega_m-\omega_r)(L_{rl}\bar{i}_{rm}+L_m|\bar{i}_m|), \tag{5.1-80}$$

where $|\bar{i}_m|$ is defined in eqn (5.1-74). Owing to saturation of the main flux paths, in eqn (5.1-80) $d/dt(L_m|\bar{i}_m|)\neq L_m\, d|\bar{i}_m|/dt$ but is given by eqn (5.1-77). Thus by the substitution of eqn (5.1-77) into eqn (5.1-80)

$$0=R_r\bar{i}_{rm}+L_{rl}\frac{d\bar{i}_{rm}}{dt}+L_d\frac{d|\bar{i}_m|}{dt}+j(\omega_m-\omega_r)(L_{rl}\bar{i}_{rm}+L_m|\bar{i}_m|). \tag{5.1-81}$$

However, the space phasor of the rotor currents can be written in terms of the space phasors of the stator currents and the magnetizing current and it follows from eqn (5.1-74) that $\bar{i}_{rm}=|\bar{i}_m|-\bar{i}_{sm}$ and when this expression is substituted into eqn (5.1-81),

$$0=R_r(|\bar{i}_m|-\bar{i}_{sm})+(L_{rl}+L_d)\frac{d|\bar{i}_m|}{dt}-L_{rl}\frac{d\bar{i}_{sm}}{dt}+j(\omega_m-\omega_r)(L_r|\bar{i}_m|-L_{rl}\bar{i}_{sm}). \tag{5.1-82}$$

This equation is very similar to eqn (4.3-17), (which holds under linear magnetic conditions and it should be noted that $|\bar{i}_m|=|\bar{i}_{mm}|$), but whilst in eqn (4.3-17) the derivative of the modulus of the magnetizing current space phasor is multiplied by the rotor self-inductance ($L_r=L_{rl}+L_m$), in eqn (5.1-82) $d|\bar{i}_m|/dt$ is multiplied by the inductance ($L_{rl}+L_d$) and under saturated conditions this varies and is smaller than $L_{rl}+L_m$. Thus the resolution of eqn (5.1-82) into its direct- and quadrature-axis components in the magnetizing-flux-oriented reference frame gives

$$\left(|\bar{i}_m|+T\frac{d|\bar{i}_m|}{dt}\right)/T_{rl}=\frac{di_{sx}}{dt}+\frac{i_{sx}}{T_{rl}}-\omega_{sl}i_{sy} \tag{5.1-83}$$

$$\omega_{sl}\left(|\bar{i}_m|\frac{T_r}{T_{rl}}-i_{sx}\right)=\frac{di_{sy}}{dt}+\frac{i_{sy}}{T_{rl}}, \tag{5.1-84}$$

where the various rotor time constants are defined as $T=(L_{rl}+L_d)/R_r$, $T_r=(L_{rl}+L_m)/R_r$ and $T_{rl}=L_{rl}/R_r$ and ω_{sl} is the angular slip frequency, $\omega_{sl}=\omega_m-\omega_r$. Under linear magnetic conditions $L_d=L_m=$constant and eqns (5.1-83) and (5.1-84) yield eqns (4.3-18) and (4.3-19) respectively. It follows from eqns (5.1-83) and (5.1-84) that there is unwanted coupling between these equations and $|\bar{i}_m|$ is also a function of i_{sy} (it depends on all the currents of the machine). The required decoupling circuit can be obtained by utilizing a method which is similar to the one described in Section 4.3.3.

5.1.3 Stator-flux-oriented control

In this section the stator and rotor voltage equations of the saturated single-cage induction machine are formulated in the stator-flux-oriented reference frame, which rotates at the speed ω_{ms}, and the effects of main-flux saturation are considered.

Stator voltage equations

It follows from eqns (2.1-148) and (2.1-150) that, by considering $\omega_g=\omega_{ms}$,

$$\bar{u}_{s\psi s}=R_s\bar{i}_{s\psi s}+\frac{d\bar{\psi}_{s\psi s}}{dt}+j\omega_{ms}\bar{\psi}_{s\psi s}, \tag{5.1-85}$$

where $\bar{u}_{s\psi s}$ and $\bar{i}_{s\psi s}$ are the space phasors of the stator voltages and stator currents respectively and have been defined in eqns (4.1-180) and (4.1-175) respectively and

$$\bar{\psi}_{s\psi s} = |\bar{\psi}_s| = L_m|\bar{i}_{ms}| = \psi_{sx} + j\psi_{sy} \tag{5.1-86}$$

is the space phasor of the stator flux linkages in the stator-flux-oriented reference frame ($\psi_{sy}=0$). In eqn (5.1-86) $|\bar{i}_{ms}|$ is the modulus of the stator magnetizing-current space phasor and it is related to the space phasor of the stator and rotor currents in the stator-flux-oriented reference frame as given by eqn (4.2-9). Substitution of eqn (5.1-86) into eqn (5.1-85) yields

$$\bar{u}_{s\psi s} = R_s\bar{i}_{s\psi s} + \frac{\mathrm{d}(L_m|\bar{i}_{ms}|)}{\mathrm{d}t} + j\omega_{ms}L_m|\bar{i}_{ms}|. \tag{5.1-87}$$

Under linear magnetic conditions L_m = constant and eqn (5.1-87) agrees with eqn (4.2-10). However, under saturated conditions, when the magnetizing inductance is changing and is a non-linear function of $|\bar{i}_{ms}|$,

$$\frac{\mathrm{d}(L_m|\bar{i}_{ms}|)}{\mathrm{d}t} = \frac{\mathrm{d}|\bar{\psi}_s|}{\mathrm{d}|\bar{i}_{ms}|}\frac{\mathrm{d}|\bar{i}_{ms}|}{\mathrm{d}t} = L^{ds}\frac{\mathrm{d}|\bar{i}_{ms}|}{\mathrm{d}t}. \tag{5.1-88}$$

In eqn (5.1-88) L^{ds} is a dynamic inductance, $L^{ds} = \mathrm{d}|\bar{\psi}_s|/\mathrm{d}|\bar{i}_{ms}|$ and substitution of eqn (5.1-88) into eqn (5.1-87) yields

$$\bar{u}_{s\psi s} = R_s\bar{i}_{s\psi s} + L^{ds}\frac{\mathrm{d}|\bar{i}_{ms}|}{\mathrm{d}t} + j\omega_{ms}L_m|\bar{i}_{ms}|. \tag{5.1-89}$$

This equation is very similar to eqn (4.2-10), which holds under linear magnetic conditions, but instead of the constant L_m now L^{ds} is present in the stator voltage equation. Thus the resolution of eqn (5.1-89) into its direct- and quadrature-axis components yields

$$i_{sx} = \frac{u_{sx}}{R_s} - \frac{L^{ds}}{R_s}\frac{\mathrm{d}|\bar{i}_{ms}|}{\mathrm{d}t} \tag{5.1-90}$$

$$i_{sy} = \frac{u_{sy}}{R_s} - \frac{\omega_{ms}L_m|\bar{i}_{ms}|}{R_s}, \tag{5.1-91}$$

which are similar to eqns (4.2-11) and (4.2-12), which hold under linear magnetic conditions. Eqns (5.1-90) and (5.1-91) contain unwanted coupling terms and these can be removed by the application of the suitable decoupling circuit. It should be noted that cross-saturation coupling terms are not present in eqns (5.1-90) and (5.1-91) since in the stator-flux-oriented reference frame the stator magnetizing-current space phasor contains only a direct-axis component (the quadrature-axis component is zero) and the magnetizing inductance has been expressed in terms of the stator magnetizing-current space phasor. However, it is also possible to formulate the equations in the stator-flux-oriented reference frame, by expressing the magnetizing inductance in terms of the modulus of the magnetizing-current space phasor. In this case, under saturated conditions, the resulting equations will contain the cross-saturation coupling terms as well, similarly to the equations described in Section 5.1.1, since these will contain the rate of change of the magnetizing flux linkage space phasor expressed in the stator-flux-oriented reference frame ($\bar{\psi}_{\mathrm{m}\psi\mathrm{s}}$), and $\bar{\psi}_{\mathrm{m}\psi\mathrm{s}}$ is not coaxial with the stator-flux-oriented reference frame. Full details of this will not be given here, since the approach is very similar to that used in Section 5.1.1, but a cursory derivation of the equations will be given.

The magnetizing flux-linkage space phasor expressed in the stator-flux-oriented reference frame can be defined in a similar way to the flux-linkage space phasor given by eqn (5.1-24) and thus $\bar{\psi}_{\mathrm{m}\psi\mathrm{s}} = L_{\mathrm{m}}\bar{i}_{m\psi\mathrm{s}} = L_{\mathrm{m}}|\bar{i}_{\mathrm{m}}|e^{j(\mu_{\mathrm{m}}-\rho_{\mathrm{s}})} = \psi_{\mathrm{m}x} + j\psi_{\mathrm{m}y}$, where μ_{m} is the angle of the magnetizing flux-linkage space phasor with respect to the direct axis of the stationary reference frame and ρ_{s} is the angle of the stator magnetizing-current space phasor with respect to the real-axis of the stationary reference frame (see, for example, Fig. 2.17). It follows from the resolution of eqns (5.1-85) and (5.1-86) into their real- and imaginary-axis components that by utilizing $\psi_{\mathrm{s}y}=0$ and $\psi_{\mathrm{s}x}=L_{\mathrm{sl}}i_{\mathrm{s}x}+\psi_{\mathrm{m}x}$,

$$u_{\mathrm{s}x} = R_{\mathrm{s}}i_{\mathrm{s}x} + L_{\mathrm{sl}}\frac{\mathrm{d}i_{\mathrm{s}x}}{\mathrm{d}t} + \frac{\mathrm{d}\psi_{\mathrm{m}x}}{\mathrm{d}t}, \tag{5.1-92}$$

where as in eqn (5.1-34) the rate of change of the direct-axis component of the magnetizing flux-linkage space phasor in the stator-flux-oriented reference frame can be expressed as

$$\frac{\mathrm{d}\psi_{\mathrm{m}x}}{\mathrm{d}t} = L_{\mathrm{m}x}\frac{\mathrm{d}i_{\mathrm{m}x}}{\mathrm{d}t} + L_{xy}\frac{\mathrm{d}i_{\mathrm{m}y}}{\mathrm{d}t}, \tag{5.1-93}$$

where $i_{\mathrm{m}x}$ and $i_{\mathrm{m}y}$ are the magnetizing current components along the direct and quadrature axes of the stator-flux-oriented reference frame, $i_{\mathrm{m}x}=i_{\mathrm{s}x}+i_{\mathrm{r}x}$, $i_{\mathrm{m}y}=i_{\mathrm{s}y}+i_{\mathrm{r}y}$, $L_{\mathrm{m}x}$ is the magnetizing inductance in the direct-axis,

$$L_{\mathrm{m}x} = L_{\mathrm{d}}\cos^2(\mu_{\mathrm{m}}-\rho_{\mathrm{s}}) + L_{\mathrm{m}}\sin^2(\mu_{\mathrm{m}}-\rho_{\mathrm{s}}), \tag{5.1-94}$$

and L_{xy} is the cross-coupling inductance between the direct and quadrature axes of the stator-flux-oriented reference frame,

$$L_{xy} = \tfrac{1}{2}(L_{\mathrm{d}}-L_{\mathrm{m}})\sin[2(\mu_{\mathrm{m}}-\rho_{\mathrm{s}})]. \tag{5.1-95}$$

In eqns (5.1-94) and (5.1-95) L_{d} is a dynamic inductance, $L_{\mathrm{d}}=\mathrm{d}|\bar{\psi}_{\mathrm{m}}|/\mathrm{d}|\bar{i}_{\mathrm{m}}|$, where $|\bar{\psi}_{\mathrm{m}}|$ and $|\bar{i}_{\mathrm{m}}|$ are the moduli of the magnetizing flux-linkage and magnetizing-current space phasors respectively. Substitution of eqn (5.1-93) into eqn (5.1-92) gives the direct-axis voltage equation and it follows that under saturated conditions this contains cross-saturation coupling terms.

A similar derivation could be performed for the quadrature-axis stator voltage equation. However, it should be noted that both the direct- and quadrature-axis stator equations contain derivatives of the rotor currents (di_{rx}/dt, di_{ry}/dt) and, according to eqn (4.2-2), the rotor currents are related to the stator currents as

$$i_{rx}=|\bar{\imath}_{ms}|-i_{sx}\frac{L_s}{L_m} \tag{5.1-96}$$

$$i_{ry}=-i_{sy}\frac{L_s}{L_m}. \tag{5.1-97}$$

Rotor voltage equations

The rotor voltage equations of the saturated induction machine in the stator-flux-oriented reference frame can be obtained by considering the general form of the rotor voltage equation, eqn (2.1-186). These equations can be used for the various implementations of stator-flux-oriented control of the saturated induction machine with impressed stator currents. The same method is followed as in Section 4.2.3, but it is assumed that the magnetizing inductance varies with saturation. Thus by using the same method as for the derivation of eqn (4.2-15), the following rotor-voltage space-phasor equation is obtained in the stator-flux-oriented reference frame:

$$0=R_r[|\bar{\imath}_{ms}|-(L_s/L_m)\bar{\imath}_{s\psi s}]+\frac{d(L_r|\bar{\imath}_{ms}|)}{dt}-\frac{d[(L_s'L_r/L_m)\bar{\imath}_{s\psi s}]}{dt}$$
$$+j\omega_{sl}[L_r|\bar{\imath}_{ms}|-(L_s'L_r/L_m)\bar{\imath}_{s\psi s}], \tag{5.1-98}$$

where

$$\omega_{sl}=\omega_{ms}-\omega_r \tag{5.1-99}$$

is the angular slip frequency. In eqn (5.1-98) there are two derivative terms which contain the variable magnetizing inductance (which is a function of the modulus of the stator magnetizing-current space phasor, $|\bar{\imath}_{ms}|$) and these can be expanded by the application of the chain differentiation rule. Thus the first term can be expanded as

$$\frac{d(L_r|\bar{\imath}_{ms}|)}{dt}=L_{rl}\frac{d|\bar{\imath}_{ms}|}{dt}+\frac{d(L_m|\bar{\imath}_{ms}|)}{dt}=(L_{rl}+L^{ds})\frac{d|\bar{\imath}_{ms}|}{dt}, \tag{5.1-100}$$

where $L^{ds}=d|\bar{\psi}_s|/d|\bar{\imath}_{ms}|$ is a dynamic inductance, $|\bar{\psi}_s|$ and $|\bar{\imath}_{ms}|$ are the moduli of the stator flux-linkage space phasor and stator magnetizing-current space phasor respectively and under linear magnetic conditions $L^{ds}=L_m$. L_{rl} is the leakage inductance of the rotor. The second term of eqn (5.1-98) can be expanded as follows, if it is considered that the stator transient inductance can be expressed as $L_s'=L_s-L_m^2/L_r$:

$$\frac{d[(L_s'L_r/L_m)\bar{\imath}_{s\psi s}]}{dt}=\frac{L_s'L_r}{L_m}\frac{d\bar{\imath}_{s\psi s}}{dt}+\bar{\imath}_{s\psi s}\frac{d(L_sL_r/L_m-L_m)}{dt}. \tag{5.1-101}$$

The chain differentiation rule can be applied to the second term on the right-hand side of eqn (5.1-101), since it is a function of L_m which varies with saturation, and finally eqn (5.1-101) will take the following form,

$$\frac{\mathrm{d}[(L_s'L_r/L_m)\bar{\imath}_{s\psi s}]}{\mathrm{d}t}=\frac{L_s'L_r}{L_m}\frac{\mathrm{d}\bar{\imath}_{s\psi s}}{\mathrm{d}t}-\bar{\imath}_{s\psi s}\frac{L_{sl}L_{rl}}{L_m^2}\frac{\mathrm{d}L_m}{\mathrm{d}t}, \tag{5.1-102}$$

where L_{sl} is the leakage inductance of the stator. In eqn (5.1-102) the derivative $\mathrm{d}L_m/\mathrm{d}t$ is present; under linear magnetic conditions this is zero, but when the magnetizing inductance changes with saturation, then as in eqn (5.1-5)

$$\frac{\mathrm{d}L_m}{\mathrm{d}t}=\frac{L^{ds}-L_m}{|\bar{\imath}_{ms}|}\frac{\mathrm{d}|\bar{\imath}_{ms}|}{\mathrm{d}t}. \tag{5.1-103}$$

Thus by substitution of eqn (5.1-103) into eqn (5.1-102)

$$\frac{\mathrm{d}[(L_s'L_r/L_m)\bar{\imath}_{s\psi s}]}{\mathrm{d}t}=\frac{L_s'L_r}{L_m}\frac{\mathrm{d}\bar{\imath}_{s\psi s}}{\mathrm{d}t}-\frac{\bar{\imath}_{s\psi s}}{|\bar{\imath}_{ms}|}(L^{ds}-L_m)\frac{L_{sl}L_{rl}}{L_m^2}\frac{\mathrm{d}|\bar{\imath}_{ms}|}{\mathrm{d}t} \tag{5.1-104}$$

is obtained and substitution of eqns (5.1-104) and (5.1-100) into eqn (5.1-98) yields the final form of the space-phasor form of the rotor voltage equation:

$$0=R_r\left[|\bar{\imath}_{ms}|-\frac{L_s}{L_m}\bar{\imath}_{s\psi s}\right]+(L_{rl}+L^{ds})\frac{\mathrm{d}|\bar{\imath}_{ms}|}{\mathrm{d}t}-\frac{L_s'L_r}{L_m}\frac{\mathrm{d}\bar{\imath}_{s\psi s}}{\mathrm{d}t}$$

$$+\frac{\bar{\imath}_{s\psi s}}{|\bar{\imath}_{ms}|}(L^{ds}-L_m)\frac{L_{sl}L_{rl}}{L_m^2}\frac{\mathrm{d}|\bar{\imath}_{ms}|}{\mathrm{d}t}+j\omega_{sl}\left[L_r|\bar{\imath}_{ms}|-\frac{L_s'L_r}{L_m}\bar{\imath}_{s\psi s}\right] \tag{5.1-105}$$

and resolution into real- and imaginary-axis components yields

$$\frac{\mathrm{d}i_{sx}}{\mathrm{d}t}+\frac{i_{sx}}{T_r'}-\omega_{sl}i_{sy}=\left[\frac{L_m}{L_s'}A+\frac{i_{sx}}{|\bar{\imath}_{ms}|}B(L^{ds}-L_m)\right]\frac{\mathrm{d}|\bar{\imath}_{ms}|}{\mathrm{d}t}+\frac{L_m}{L_sT_r'}|\bar{\imath}_{ms}| \tag{5.1-106}$$

and

$$\omega_{sl}\left(\frac{L_m|\bar{\imath}_{ms}|}{L_s'}-i_{sx}\right)=i_{sy}\left[\frac{1}{T_r'}-\frac{(L^{ds}-L_m)B}{|\bar{\imath}_{ms}|}\right]\frac{\mathrm{d}|\bar{\imath}_{ms}|}{\mathrm{d}t}+\frac{\mathrm{d}i_{sy}}{\mathrm{d}t}, \tag{5.1-107}$$

where

$$A=\frac{L_{rl}+L^{ds}}{L_r}$$

$$B=\frac{L_{sl}L_{rl}}{L_mL_s'L_r} \tag{5.1-108}$$

and $L_s'=L_s-L_m^2/L_r$, $L_r=L_{rl}+L_m$ depend on the magnetizing inductance, which varies with saturation. Under linear magnetic conditions, eqns (5.1-106), (5.1-107) yield eqns (4.2-17) and (4.2-18) respectively, since $L^{ds}=L_m=$constant. For stator-flux- oriented control of the saturated induction machine with impressed stator currents eqns (5.1-106) and (5.1-107) have to be considered. However, there is unwanted coupling between the torque-producing stator current component (i_{sy}) and the stator magnetizing-current component (i_{sx}). This coupling can be removed by the application of a suitable decoupling circuit. This can be derived in a similar way to the decoupling circuit described in Section 4.2.3.

In the equations derived above the magnetizing inductance has been assumed to be a non-linear function of the modulus of the stator magnetizing-current space phasor and the rotor voltage equations have been obtained in the stator-flux-oriented reference frame. Since the stator magnetizing space phasor has only a direct-axis component in this reference frame, no cross-saturation terms appear in the equations. However, if the magnetizing inductance is expressed in terms of the modulus of the magnetizing-current space phasor and the rotor voltage equations are expressed in the stator-flux-oriented reference frame, under saturated conditions cross-saturation terms will be present in the equations, since the magnetizing-current space phasor is not coaxial with the direct axis of the reference frame and thus there is magnetization of both axes which will result in cross-saturation transformer e.m.f.s under saturated conditions. The derivation of the resulting rotor voltage equations is similar to the derivation of eqns (5.1-66) and (5.1-67) and thus it will not be discussed in detail. However, like eqn (5.1-63), the space-phasor voltage equation in the stator-flux-oriented reference frame can be expressed as

$$\bar{u}_{r\psi s}=0=R_r\bar{i}_{r\psi s}+L_{rl}\frac{d\bar{i}_{r\psi s}}{dt}+\frac{d\bar{\psi}_{m\psi s}}{dt}+j(\omega_{ms}-\omega_r)(L_{rl}\bar{i}_{r\psi s}+\bar{\psi}_{m\psi s}), \tag{5.1-109}$$

where the magnetizing flux-linkage space phasor in the stator-flux-oriented reference frame has been defined in the previous section as $\bar{\psi}_{m\psi s}=L_m\bar{i}_{m\psi s}=L_m|\bar{i}_m|e^{j(\mu_m-\rho_s)}=\psi_{mx}+j\psi_{my}$. Resolution of eqn (5.1-109) into its real- and imaginary-axis components gives the following equations, if it is considered that in the stator-flux-oriented reference frame the quadrature-axis stator flux-linkage component is zero ($\psi_{sy}=L_{sl}i_{sy}+\psi_{my}=0$):

$$0=R_ri_{rx}+L_{rl}\frac{di_{rx}}{dt}+\frac{d\psi_{mx}}{dt}-(\omega_{ms}-\omega_r)(L_{rl}i_{ry}+L_{sl}i_{sy}) \tag{5.1-110}$$

$$0=R_ri_{ry}+L_{rl}\frac{di_{ry}}{dt}+\frac{d\psi_{my}}{dt}+(\omega_{ms}-\omega_r)(L_{rl}i_{rx}+\psi_{mx}), \tag{5.1-111}$$

where, by considering eqn (5.1-93), which is repeated here for convenience,

$$\frac{d\psi_{mx}}{dt}=L_{mx}\frac{di_{mx}}{dt}+L_{xy}\frac{di_{my}}{dt}, \tag{5.1-112}$$

and as for eqn (5.1-35), the rate of change of the quadrature-axis magnetizing flux linkage component can be expressed as

$$\frac{d\psi_{my}}{dt} = L_{my}\frac{di_{my}}{dt} + L_{xy}\frac{di_{mx}}{dt}. \tag{5.1-113}$$

In these equations L_{mx} and L_{my} are the magnetizing inductances along the direct (x) and quadrature (y) axes respectively in the stator-flux-oriented reference frame. The inductance L_{mx} has been defined by eqn (5.1-94) and L_{my} can be defined in a similar way to eqn (5.1-37) and therefore

$$L_{my} = L_d \sin^2(\mu_m - \rho_s) + L_m \cos^2(\mu_m - \rho_s). \tag{5.1-114}$$

In eqns (5.1-112) and (5.1-113) L_{xy} is the cross-coupling inductance between the x and y axes, and it has been given in eqn (5.1-95). Substitution of eqns (5.1-112) and (5.1-113) into eqns (5.1-110) and (5.1-111) yields the rotor voltage equations in the stator-flux-oriented reference frame, and the effects of main-flux saturation are also incorporated in these. They will contain the cross-saturation terms as well, in agreement with the discussion presented above. The rotor voltage equations thus obtained can be used for the implementation of stator-flux-oriented control of an induction machine supplied by impressed stator currents, in a similar way to that described in Section 4.2.3. It follows from eqns (5.1-94), (5.1-114) and (5.1-95) that under linear magnetic conditions when $L_d = L_m$, $L_{mx} = L_{my} = L_m$ and $L_{xy} = 0$, and thus by also considering eqns (5.1-96) and (5.1-97) it can be shown that the rotor voltage equations yield eqns (4.2-17) and (4.2-18).

5.2 VECTOR CONTROL OF SALIENT-POLE MACHINES BY CONSIDERING THE EFFECTS OF MAIN-FLUX SATURATION

In this part of the Section the effects of main-flux saturation are incorporated into the five-winding model of the salient-pole synchronous machine discussed in Section 3.2. All the other assumptions are those used for the saturated smooth-air-gap machine. Only a brief discussion is given since the various forms of vector control technique can be developed in ways similar to those described in Sections 3.1. and 3.2.

5.2.1 Stator voltage equations

It follows from eqns (3.2-14) and (3.2-15) that in the reference frame fixed to the rotor the stator equations take the form,

$$u_{sd} = R_s i_{sd} + \frac{d\psi_{sd}}{dt} - \omega_r \psi_{sq} \tag{5.2-1}$$

$$u_{sq} = R_s i_{sq} + \frac{d\psi_{sq}}{dt} + \omega_r \psi_{sd}, \tag{5.2-2}$$

where the two-axis components of the stator flux linkages are

$$\psi_{sd} = L_{sl} i_{sd} + \psi_{md} \tag{5.2-3}$$

and

$$\psi_{sq} = L_{sl} i_{sq} + \psi_{mq}. \tag{5.2-4}$$

ψ_{md} and ψ_{mq} are the direct- and quadrature-axis components of the magnetizing flux linkages in the reference frame fixed to the rotor and from eqns (3.2-1) and (3.2-2) that they can be defined as

$$\psi_{md} = L_{md} i_{md} = L_{md}(i_{sd} + i_{r\alpha} + i_{rF}) \tag{5.2-5}$$

$$\psi_{mq} = L_{mq} i_{mq} = L_{mq}(i_{sq} + i_{r\beta}). \tag{5.2-6}$$

i_{md} and i_{mq} are the magnetizing current components along the real and imaginary axes of the reference frame fixed to the rotor and L_{md} and L_{mq} are the magnetizing inductances in the direct and quadrature axes respectively. As a result of physical saliency these are different even under linear conditions. However, under saturated conditions they vary with the currents. This will now be discussed.

In the salient-pole machine, because of the physical saliency, in general, the magnetizing flux-linkage space phasor is not coaxial with the magnetizing-current space phasor (the resultant flux wave is not collinear with the resultant m.m.f. wave). In special cases, when the m.m.f. lies completely in one axis only (the magnetizing current space phasor has only a direct-axis or a quadrature-axis component) then the magnetizing flux-linkage space phasor is coaxial with the magnetizing-current space phasor. Thus it is more problematic to characterize the saturation level in the saturated salient-pole machine than in the saturated smooth-air-gap machine, where the magnetizing flux-linkage space phasor is coaxial with the magnetizing-current phasor and where the amplitude of the magnetizing flux-linkage space phasor is a non-linear function of the magnetizing current, as discussed in Section 5.1. In the salient-pole machine, because of symmetry, the magnetizing flux linkage along the direct axis (ψ_{md}) must be zero if the direct-axis magnetizing current is zero; thus under saturated conditions $\psi_{md} = L_{md}(i_{md}, i_{mq}) i_{md}$, where L_{md} is a non-linear function of the magnetizing current components, and similarly $\psi_{mq} = L_{mq}(i_{md}, i_{mq}) i_{mq}$, where L_{mq} is another non-linear function of the magnetizing currents. However, it is assumed that the pole tips and/or the teeth are the most saturated parts of the magnetic circuit, and therefore, as for the smooth-air-gap machine, it is assumed that in each axis the saturation level is determined by the amplitude of the magnetizing-current space phasor $|\bar{i}_m| = (i_{md}^2 + i_{mq}^2)^{1/2}$ Thus ψ_{md}/i_{md} and ψ_{mq}/i_{mq} depend only on $|\bar{i}_m|$ and

$$\psi_{md} = L_{md}(|\bar{i}_m|) i_{md} \tag{5.2-7}$$

$$\psi_{mq} = L_{mq}(|\bar{i}_m|) i_{mq}. \tag{5.2-8}$$

The magnetizing inductances L_{md} and L_{mq} are chord-slope, static inductances and can be measured or computed by considering the direct- and quadrature-axis magnetizing curves respectively, with excitation in the corresponding axis only (when $i_{md}=|\bar{i}_m|$ and $i_{mq}=|\bar{i}_m|$ respectively). If the flux linkages are the state variables of the model, then it is only the chord-slope inductances which are present in the equations (together with other parameters), but if such a model is used, where the currents are the state variables, the 'tangent-slope' dynamic inductances in the two axes are also present. The direct-axis dynamic inductance can be defined as $L^d=d\psi_{md}/di_{md}$, but again there is only excitation along the direct axis and thus $|i_{md}|=|\bar{i}_m|$. Similarly, the quadrature-axis dynamic inductance can be defined as $L^q=d\psi_{mq}/di_{mq}$ and there is only excitation along the quadrature axis, thus $|i_{mq}|=|\bar{i}_m|$. The inductances L_{md}, L_{mq}, L_d and L_q will now be utilized in the expanded forms of the voltage equations.

Eqns (5.2-1)–(5.2-6) describe the stator voltage equations in the presence of saturation of the main flux paths and can be used for various implementations of vector control. It is also possible to expand these equations, and by the substitution of eqns (5.2-3) and (5.2-4) into eqns (5.2-1) and (5.2-2),

$$u_{sd}=R_s i_{sd}+L_{sl}\frac{di_{sd}}{dt}+\frac{d\psi_{md}}{dt}-\omega_r(L_{sl}i_{sq}+\psi_{mq}) \qquad (5.2\text{-}9)$$

$$u_{sq}=R_s i_{sq}+L_{sl}\frac{di_{sq}}{dt}+\frac{d\psi_{mq}}{dt}+\omega_r(L_{sl}i_{sd}+\psi_{md}). \qquad (5.2\text{-}10)$$

R_s and L_{sl} are the resistance and leakage inductance of a stator winding respectively and L_{sl} is assumed to be constant. Eqns (5.2-9) and (5.2-10) contain the magnetizing voltages along the direct and quadrature axes respectively and by the application of the chain differentiation rule, they can be expanded in a similar way to that shown in eqns (5.1-34) and (5.1-35), so that

$$d\psi_{md}/dt=\frac{d(L_{md}\,di_{md})}{dt}=L_{mD}\frac{di_{md}}{dt}+L_{dq}\frac{di_{mq}}{dt} \qquad (5.2\text{-}11)$$

$$d\psi_{mq}/dt=\frac{d(L_{md}\,di_{mq})}{dt}=L_{mQ}\frac{di_{mq}}{dt}+L_{qd}\frac{di_{md}}{dt}, \qquad (5.2\text{-}12)$$

$$L_{mD}=L^d\cos^2\mu'+L_{md}\sin^2\mu' \qquad (5.2\text{-}13)$$

$$L_{mQ}=L^q\sin^2\mu'+L_{mq}\cos^2\mu'. \qquad (5.2\text{-}14)$$

In eqns (5.2-13) and (5.2-14) the angle $\mu'=\mu_m-\theta_r$ is the angle of the space phasor of the magnetizing currents with respect to the direct axis of the reference frame fixed to the rotor, μ_m is the angle of the same space phasor with respect to the direct axis of the stationary reference frame and θ_r is the rotor angle. In eqns (5.2-11) and (5.2-12)

$$L_{dq} = (L^d - L_{md}) \sin \mu' \cos \mu' \tag{5.2-15}$$

$$L_{qd} = (L^q - L_{mq}) \sin \mu' \cos \mu' \tag{5.2-16}$$

and L_{dq} and L_{qd} are the cross-coupling (mutual) inductances between the d and q, and the q and d axes respectively. In the smooth-air-gap machine $L_{dq} = L_{qd}$ due to symmetry, but in the salient-pole machine this reciprocity only holds if certain criteria are satisfied. Namely, it can be shown, that as for the reciprocity conditions of two stationary, non-linear inductors (denoted by a, b respectively), where the reciprocity condition for the mutual inductances ($M_{ab} = M_{ba}$) holds only if the inductances are assumed to be lossless or if the losses are modclled separately, in the salient-pole machine the reciprocity condition $L_{dq} = L_{qd}$ is only satisfied if the inductances of the machine are considered to be lossless (or if the losses are modelled separately).

Substitution of eqns (5.2-11) and (5.2-12) into eqns (5.2-9) and (5.2-10) respectively yield the following stator voltage equations, if eqns (5.2-5) and (5.2-6) are also substituted into eqns (5.2-9) and (5.2-10),

$$u_{sd} = R_s i_{sd} + L_{sl} \frac{di_{sd}}{dt} + L_{mD} \frac{d(i_{sd} + i_{r\alpha} + i_{rF})}{dt} + L_{dq} \frac{d(i_{sq} + i_{r\beta})}{dt}$$
$$- \omega_r [L_{sl} i_{sq} + L_{mq}(i_{sq} + i_{r\beta})] \tag{5.2-17}$$

$$u_{sq} = R_s i_{sq} + L_{sl} \frac{di_{sq}}{dt} + L_{mQ} \frac{d(i_{sq} + i_{r\beta})}{dt} + L_{qd} \frac{d(i_{sd} + i_{r\alpha} + i_{rF})}{dt}$$
$$+ \omega_r [L_{sl} i_{sd} + L_{md}(i_{sd} + i_{r\alpha} + i_{rF})]. \tag{5.2-18}$$

It follows that in general, due to saturation, in the stator voltage equations four inductances, L_{md}, L_{mq}, L^d and L^q have to be considered together with the angle of the magnetizing current space phasor μ'. The equations also contain the effects of cross-saturation and thus there exists cross-saturation coupling between all the windings which are in space quadrature, resulting in cross-saturation transformer e.m.f.s. Under linear magnetic conditions $L_{mD} = L^d = L_{md}$ and $L_{mQ} = L^q = L_{mq}$, $L_{dq} = L_{qd} = 0$ and the well-known equations are obtained.

If $i_{rF} = 0$ and a smooth air-gap is considered, the static inductances are equal in the two axes and $L_{md} = L_{mq} = L_m$. Furthermore, the dynamic inductances are also equal and $L^d = L^q = L_d$. It follows from eqns (5.2-13), (5.2-14) and (5.2-15) that under these conditions, in the stationary reference frame the expressions for L_{mD}, L_{mQ} and L_{dq} agree with eqns (5.1-50), (5.1-51) and (5.1-52) respectively and eqns (5.2-17) and (5.2-18) yield the stator voltage equations given by eqns (5.1-53) and (5.1-54) respectively.

5.2.2 Rotor voltage equations

The rotor voltage equations of the saturated five-winding models of the salient-pole machine can be obtained in a similar way to the equations derived in the previous section. It follows from eqns (3.2-18) and (3.2-19) that the voltage equation for the field winding can be put into the following form in the reference frame fixed to the rotor:

$$u_{rF}=R_{rF}i_{rF}+L_{Fl}\frac{di_{rF}}{dt}+\frac{d\psi_{md}}{dt}, \qquad (5.2\text{-}19)$$

where R_{rF} is the resistance of the field winding and; L_{Fl} is the leakage inductance of the field winding. Similarly, it follows from eqns (3.2-20) and (3.2-21) that the voltage equations of the direct- and quadrature-axis damper windings are

$$0=R_{r\alpha}i_{r\alpha}+L_{r\alpha l}\frac{di_{r\alpha}}{dt}+\frac{d\psi_{md}}{dt} \qquad (5.2\text{-}20)$$

$$0=R_{r\beta}i_{r\beta}+L_{r\beta l}\frac{di_{r\beta}}{dt}+\frac{d\psi_{mq}}{dt}, \qquad (5.2\text{-}21)$$

where $R_{r\alpha}$ and $R_{r\beta}$ are the resistances of the damper windings along the direct and quadrature axes respectively and $i_{r\alpha}$ and $i_{r\beta}$ are the currents in the two damper windings. Eqns (5.2-19)–(5.2-21) contain the derivatives of the direct- and quadrature-axis magnetizing flux-linkage components, and they have been given by eqns (5.2-11) and (5.2-12) respectively. Thus when eqns (5.2-11) and (5.2-12) are substituted into eqns (5.2-19), (5.2-20) and (5.2-21), the following rotor voltage equations are obtained, if eqns (5.2-5) and (5.2-6) are also considered,

$$u_{rF}=R_{rF}i_{rF}+L_{Fl}\frac{di_{rF}}{dt}+L_{mD}\frac{d(i_{sd}+i_{r\alpha}+i_{rF})}{dt}+L_{dq}\frac{d(i_{sq}+i_{r\beta})}{dt} \qquad (5.2\text{-}22)$$

$$0=R_{r\alpha}i_{r\alpha}+L_{r\alpha l}\frac{di_{r\alpha}}{dt}+L_{mD}\frac{d(i_{sd}+i_{r\alpha}+i_{rF})}{dt}+L_{dq}\frac{d(i_{sq}+i_{r\beta})}{dt} \qquad (5.2\text{-}23)$$

$$0=R_{r\beta}i_{r\beta}+L_{r\beta l}\frac{di_{r\beta}}{dt}+L_{mQ}\frac{d(i_{sq}+i_{r\beta})}{dt}+L_{qd}\frac{d(i_{sd}+i_{r\alpha}+i_{rF})}{dt}. \qquad (5.2\text{-}24)$$

In these equations the inductances L_{mD}, L_{mQ} and L_{dq}, L_{qd} are present, and these have been defined in eqns (5.2-13)–(5.2-16) respectively. Because of saturation of the main flux paths, these equations contain the effects of cross-saturation as well. Under linear magnetic conditions, the well-known forms of these equations are obtained. In general, the cross-saturation coupling terms (L_{dq}, L_{qd}) only disappear under linear magnetic conditions or if the magnetizing-current space phasor lies along the direct axis or the quadrature axis (when $\mu_m = 0$ or $\mu_m = \pi/2$).

BIBLIOGRAPHY

Anvari, H. A. and Faucher, J. (1988). Simulation considering the cross-magnetization effect. In *Conference Proceedings of International Conference on Electrical Machines, Pisa*, pp. 519–24.

Anvari, H. A., Faucher, J., and Trennoy, B. (1986). On the cross-coupling saturation effect and its computation by the finite difference solution. In *Conference Proceedings of International Conference on Electrical Machines, München*, pp. 451–4.

Boldea, I. and Nasar, S. A. (1987). Unified treatment of core losses and saturation in the orthogonal-axis model of electrical machines. *IEE Proc.* **B 134**, 355–63.

Brown, J. E., Kovács, K. P., and Vas, P. (1982). A method of including the effects of main flux saturation in the generalized equations of a.c. machines. *IEEE Transactions on PAS*, **PAS-102**, 96–103.

Delfino, B., Denegri, G. B., Pinceti, P., and Schiappacasse, A. (1985). Dynamic analysis of turbogenerator performance based on saturated equivalent circuit models. In *Universities Power Engineering Conference, Huddersfield*, pp. 234–7.

Hallenius, K. E., Vas, P., and Brown, J. E. (1988). The analysis of a saturated self-excited asynchronous generator. *IEEE Power Engineering Society Winter Meeting, New York*, Paper 88 WM 017–6.

He, Y-K. and Lipo, T. A. (1985). Computer simulation of induction machine with spatially dependent saturation. *IEEE Transactions on Industry Applications*, **IA-21**, 226–34.

Jack, A. G., Vas, P., and Brown, J. E. (1980). A finite element study of the phenomenon of cross-magnetization. In *Conference Proceedings of International Conference on Electrical Machines, München*, pp. 462–5.

Jing-Qiu, Q. and Yun-Qiu, T. (1988). Cross-magnetization effect and d,q axis coupling reactances. In *Conference Proceedings of International Conference on Electrical Machines, Pisa*, pp. 109–12.

Khater, F. M. H., Lorenz, R. D., Novotny, D. W., and Tang, K. (1987). Selection of flux level in field-oriented induction machine controllers with consideration of magnetic saturation effects. *IEEE Transactions on Industry Applications*, **IA-23**, 276–82.

Macdonald, A. C. and Turner, P. J. (1983). Turbine generator reactances in the steady-state. IEE Colloquium on 'Finite element methods applied to electrical machines' IEE, London, March, also 18 Universities Power Engineering Conference, Surrey.

Melkebeek, J. A. (1983). Magnetizing field saturation and dynamic behaviour of induction machines, Part I: An improved calculation method for induction machine dynamics. IEE Proc. **B. 130**, 1–9.

Melkebeek, J. E., Vas, P., and Brown, J. E. (1985). Dynamic analysis of synchronous machines incorporating the effects of main flux saturation. *20 Universities Power Engineering Conference, Huddersfield*, pp. 238–41.

Melkebeek, J. A. and Willems, J. L. (1988). Reciprocity relations for the mutual inductances of orthogonal axis windings in saturated electrical machines. *IEEE IAS Annual Meeting*, pp. 104–10.

Minnich, S. H., Schulz, R. P., Baker, D. H., Sharma, D. K., Farmer, R. G., and Fish, J. H. (1987). Saturation function for synchronous generators from the finite elements. *IEEE Transactions on Energy Conversion*, **2**, 680–92.

Nehl, T. W., Fouad, F. A., and Demerdash, N. A. (1982). Determination of saturated values of rotating machinery incremental and apparent inductances by energy perturbation method. *IEEE Transactions on PAS*, **PAS-101**, 4441–52.

Robert, J. (1987). A simplified method for the study of saturation in a.c. machines. *International Association for Mathematics and Computers*. In *Simulation, Symposium on Modelling and Simulation of Electrical Machines, Converters and Power Systems, Quebec*, pp. 119–26.

Rácz, I. and Vas, P. (1985). Oscillatory behaviour of induction machines employed in railway traction. *2nd International Conference on Electrical Machines—Design and Applications, London*, pp. 168–75. IEE, London.

Ramshaw, R. S. and Xie, G. (1984). Nonlinear model of nonsalient synchronous machines. *IEEE Transactions on PAS*, **PAS-103**, 1809–15.

Richter, E. and Neumann, T. W. (1984). Saturation effects in salient-pole synchronous machines with permanent magnet excitation. In *Proceedings of International Conference on Electrical Machines, Lausanne*, pp. 603–6.

Shackshaft, G. (1979). Model of generator saturation for use in power system studies. *Proc. IEE*, **126**, 759–63.

Sneyers, B., Novotny, D. W., and Lipo, T. A. (1985). Field-weakening in buried permanent magnet a.c. drives. *IEEE Transactions on Industry Applications*, **IA-21**, 398–407.

Stiebler, M. H. and Ritter, C. (1985). Synchronous machine modelling including d/q flux saturation. *IEE Second International Conference on Electrical Machines and Drives, London*, pp. 108–12.

Vas, P. (1981). Generalized analysis of saturated a.c. machines. *Archiv für Elektrotechnik*, **63**, 57–62.

Vas, P. (1987). Various models for saturated electrical machines. *International Conference on Saturated Electrical Machines, Paper 1, Liege, 2 December 1987*.

Vas, P. and Hallenius, K. E. (1989). A three-phase and a quadrature-phase slip-ring model for saturated smooth-air-gap electrical machines. *Archiv für Elektrotechnik*, **72**, 59–68.

Vas, P. and Alaküla, M. (1989). Field-oriented control of saturated induction machines. *IEEE Power Engineering Society Summer Meeting, Long Beach, California*, Paper 89SM 738–6 EC.

Vas, P., Brown, J. E., and Hallenius, K. E. (1984). Cross-saturation effect in smooth-air-gap electrical machines. In *Proceedings of International Conference on Electrical Machines, Lausanne*, pp. 261–4.

Vas, P., Deleroi, W., and Brown, J. E. (1984). Transient analysis of smooth-air-gap machines incorporating the effects of main and leakage flux saturation. In *Proceedings of International Conference on Electrical Machines, Lausanne*, pp. 269–72.

Vas, P., Hallenius, K. E., and Brown, J. E. (1985). Transient analysis of saturated smooth-air-gap synchronous machines with and without damper windings. *11*

International Association for Mathematics and Computers in Simulation World Congress, Oslo, Vol. 3, pp. 289–92.

Vas, P., Hallenius, K. E., and Brown, J. E. (1985). Computer simulation of saturated cylindrical rotor synchronous machines. *11 International Association for Mathematics and Computers in Simulation World Congress, Oslo*, Vol. 3, pp. 303–6.

Vas, P., Hallenius, K. E., and Brown, J. E. (1986). Cross-saturation in smooth-air-gap electrical machines. *IEEE Transactions on Energy Conversion*, **EC-1**, 103–12.

6 Monitoring of the machine parameters and the rotor speed

For the various vector-controlled drives described in previous chapters it is necessary to know different parameters of the machine subjected to vector control. In this chapter the determination of those required parameters obtained prior to starting the machine and by utilizing on-line measurements for a single-cage induction machine will first be presented. Then an observer-based rotor estimator is described which utilizes the monitored rotor angle and the developed electromagnetic torque of the machine. Finally, load rejection tests will be described for the determination of the direct- and quadrature-axis synchronous reactances, transient reactances and subtransient reactances of a salient-pole synchronous machine. These tests make use of measured values of the terminal voltage, the field current and the load angle.

In many industrial applications, the machine and converter are sold by separate manufacturers and the parameters of the machine are not known. However, when the machine is subjected to vector control, as shown in previous chapters, certain parameters of the machine have to be known prior to starting the drive. Normally, this requires extensive testing of the machine and the tuning of the controllers, which can be an expensive and time-consuming process and requires specially trained staff. However, for the induction machine it is also possible to obtain the machine parameters and to tune the controllers by an automated process where the required electrical parameters (stator and rotor resistances, rotor time constant, transient stator inductance, etc.) are obtained from on-line measurements of voltages and currents when the machine is at standstill and where the inverter of the drive itself is utilized to generate the signals required for the parameter estimation. Some of these methods will now be discussed and it will also be shown how the inertia of the drive can be obtained by utilizing monitored values of the torque-producing stator current and rotor speed during a test run of the induction machine. A rotor speed observer will also be described which gives high precision and which can be used in drives containing induction or synchronous machines.

6.1 DETERMINATION OF THE ROTOR TIME CONSTANT OF AN INDUCTION MACHINE WITH IMPRESSED STATOR CURRENTS AND SUBJECTED TO INDIRECT ROTOR-FLUX-ORIENTED CONTROL

It follows from Section 4.1.1 that when indirect rotor-flux-oriented control of an induction machine is used, it is very important to have an accurate representation of the rotor time constant of the machine. This is a function of the rotor resistance and the rotor self-inductance. The former can increase by up to 100% of its nominal value for an increase of 170–180° in temperature. In systems which employ indirect rotor-flux-oriented control of an induction machine, where an encoder or a resolver is used, the position of the rotor flux-linkage space phasor with respect to the direct axis of the stationary reference frame (ρ_r, which is shown in Fig. 4.26), is obtained by the addition of the slip angle (θ_{sl}) and the monitored rotor angle (θ_r). The slip angle is obtained with a slip angle estimator which contains the effect of the open-circuit rotor time constant T_r [see eqn 4.1-26)]. When an incorrect value of the rotor time constant is used for the setting the gain of the slip controller, the performance of the machine will degrade under both steady-state and transient conditions. It is therefore important to utilize accurate values of the rotor time constant in the slip-angle estimator.

In Section 4.1.1 Model Reference Adaptive Control was used to obtain a circuit which continuously monitors the rotor resistance. In this section, an alternative technique is described for the determination of the rotor time constant.

6.1.1 Obtaining the rotor time constant prior to starting the machine

Various methods are discussed for obtaining the rotor time constant of the induction machine; these utilize measurements made prior to starting the machine. Traditionally, the rotor time constant can be obtained by performing the locked-rotor and no-load tests. However, this also requires mechanical blocking of the rotor and therefore it is more difficult to automatize these tests than tests which avoid mechanical blocking of the rotor. In the next section, such a rotor time-constant measurement system is described, which can be used for the automatic determination of the rotor time constant prior to starting the induction machine. The blocking of the rotor is avoided by applying single-phase excitation to the stator. However, the no-load test is still used.

Measurement which utilizes single-phase excitation of the stator

It is possible to obtain the rotor time constant from measurements made on the induction machine prior to start-up and by utilizing single-phase excitation of the stator windings, by measuring one of the line-to-line stator voltages of the machine. This voltage is then processed to yield the rotor time constant of the machine. The test utilizes the inverter which supplies the induction machine.

It is assumed that the induction machine is supplied by a current-controlled PWM inverter (described in Section 3.1.1) and at instant t_0, single-phase sinusoidal current $i_s(t)=I_{sref}\cos(\omega_1 t)$ is supplied to two of the stator phases of the induction machine with the third stator current controlled to zero. The time variation of the applied stator current of the induction machine is shown in Fig. 6.1.

As a result of the single-phase excitation, the rotor will remain at standstill and therefore it does not have to be locked during the tests. After switching on the a.c. current, there is a transient state followed by the steady state. In the steady state, at an appropriate instant t_1, where t_1 is much greater than the rotor time constant, the stator current is switched to a d.c. current, $i_s(t \geq t_1) = I_{sxref}$. This is also shown in Fig. 6.1. The instant t_1 is selected so that at this instant the instantaneous value of the stator current is equal to the reference value of the rotor flux-producing current component, which is a magnetizing current. The value of the magnetizing current can be obtained from a no-load test and it is dependent on the flux level in the machine.

Since the rotor speed is zero, the applied inverter angular frequency (ω_1) is equal to the angular slip frequency ($\omega_{sl} = \omega_1$). As a result of switching from a.c. current to d.c. current, a transient stator voltage will arise when the reference value of the angular slip frequency is not equal to the correct value. A physical analysis will be presented below. This transient voltage can be processed to yield the correct value of the rotor time constant. The inverter frequency is adjusted until this voltage transient is a minimum and when this is accomplished, from eqn (4.1-26), and $\omega_{sl} = \omega_1$, the rotor time constant can be obtained as

$$T_r = \frac{1}{\omega_1} \frac{I_{syref}}{I_{sxref}} = \frac{1}{\omega_1} \frac{(I_{sref}^2 - I_{sxref}^2)^{1/2}}{I_{sxref}}. \tag{6.1-1}$$

I_{syref} is the reference value of the torque-producing stator current, and the amplitude of the a.c. stator current can be expressed as $I_{sref} = (I_{sxref}^2 + I_{syref}^2)^{1/2}$.

It should be noted that, at standstill, the equivalent circuit shown in Fig. 4.10(c) is also valid under transient conditions, since for zero rotor speed

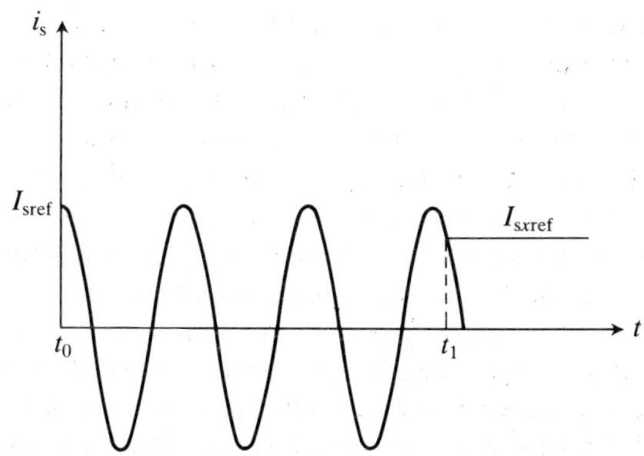

Fig. 6.1. Switching of the applied stator current.

all the speed voltages are zero and the induction machine can be described in a similar way to a transformer. This equivalent circuit can be used to give some physical considerations of the stator terminal voltages for two cases, when the reference value of the angular slip frequency is larger or smaller than the correct value of the angular slip frequency.

If the reference value of the angular slip frequency is larger than the correct value, both the actual magnetizing current and the rotor flux will be smaller than their reference values. Therefore the magnetizing current and the rotor flux are too small before the switching of the stator current. After the switching of the stator current to the d.c. value, the instantaneous value of the flux-producing stator current (i_{sx}) will be smaller than the instantaneous value of the stator current i_s. Since a current which flows across an inductor cannot be changed instantaneously, some trapped rotor current must flow across the rotor branch which contains the rotor resistance and its direction will be from point B to D in Fig. 4.10(c). The stator terminal voltage is the sum of three component voltages, the stator ohmic drop, the voltage across the transient inductance of the stator and the voltage across the magnetizing branch, which contains the inductance (L_m^2/L_r). The second component can be neglected, since it corresponds to a leakage drop, which is small compared with the other two components. It follows from Fig. 4.10(c) that the voltage across the magnetizing branch is determined by the current across the rotor resistance $R_r L_m^2/L_r^2$. Eventually, the trapped rotor current will flow across the magnetizing branch since it represents a short circuit at zero frequency, and in the steady-state, the rotor flux will be equal to the reference value. Initially, the stator terminal voltage will be higher than its steady-state value and this will decay.

If the reference value of the angular slip frequency is smaller than the correct value, then both the actual magnetizing current and the rotor flux are larger than their reference values before the switching of the stator current. After the switching of the stator current to the d.c. value, the instantaneous value of the rotor flux-producing stator current wil be higher than the instantaneous value of the stator current and thus the trapped rotor current will be reversed compared with the previous case and eventually this current will again decay to zero as the rotor flux decays to the reference value. Initially, the stator terminal voltage will be smaller than its steady-state value, since the transient voltage across the inductance L_m^2/L_r shown in Fig. 4.10(c) has a different polarity to the polarity of the ohmic stator drop.

It follows that the trapped rotor current will only be zero when the reference value of the angular slip frequency is equal to its correct value. Under this condition, both before and after the switching of the stator current, the current which flows across the magnetizing branch is equal to the reference value of the rotor flux-producing stator current. Thus no voltage will arise across the inductance L_m^2/L_r and the stator terminal voltage will be equal to the sum of the voltage across the stator resistance and the voltage across the stator transient inductance; the latter component can be neglected.

It follows from eqn (6.1-1) that for the determination of the rotor time constant it is necessary to utilize the two reference currents I_{sxref} and I_{syref}. However, to obtain high torque response, it is necessary to keep the flux-producing component constant. Thus the flux-producing current component must be held constant to yield the required flux in the machine when the reference value of the angular slip frequency is correct. It is preferable to perform the test at a constant value of I_{sxref}, which will produce the required flux in the machine. It follows from eqn (6.1-1) that the rotor time constant can be obtained with greater accuracy if the ratio I_{syref}/I_{sxref} is large. However, with a large ratio there will be large ohmic losses in the machine. If I_{syref} or I_{syref}/I_{sxref} are chosen to result in a small rotor flux error, then by calculating the rotor flux error which is due to the presence of the incorrect value of the rotor time constant it follows from the rotor voltage equations presented in Section 4.1.1 that an acceptable level of the rotor flux error can be obtained even if the current ratio (I_{sxref}/I_{syref}) is not too large.

To summarize the measurement procedure, it follows that by the application of a current controlled PWM inverter, a single-phase a.c. current is supplied to the stator windings and this is switched to an appropriate d.c. value in the steady state. This d.c. value is equal to the reference value of the rotor flux-producing component of the stator current. Its value is obtained from a no-load test and is dependent on the required flux level. One of the line-to-line voltages is monitored and the frequency of the inverter is varied until a minimum of this voltage is reached and, under this condition, eqn (6.1-1) is used to obtain the rotor time constant. It follows that when this technique is used together with the indirect rotor-flux-oriented control system described in Section 4.1.2, the only extra signal to be monitored is a line-to-line stator voltage. The experiment is performed with a rated value of the flux reference, but the actual flux is equal to its rated value only when the correct value of the angular slip-frequency reference is used.

In the case of an induction machine supplied by a current-controlled PWM inverter, the measured terminal voltage transients must be low-pass filtered. When the induction machine is supplied by a hysteresis current-controlled PWM inverter, the undesirable time-harmonic components will be spread over the entire frequency domain and this property complicates the design of the filter. However, when a current-controlled PWM with sine–triangle modulation is used, the harmonic frequencies are fixed at the harmonic frequencies of the triangular wave and thus it is easier to design the required filter than in the previous case. The filter can be a moving-average filter; such a filter can easily be implemented in a microprocessor or by utilizing a shift register and Sample-and-Hold (S/H) circuits.

If the input signal to the moving-average filter is $x(t)$, the output signal is

$$y(t) = \frac{1}{T} \int x(t)\,\mathrm{d}t, \tag{6.1-2}$$

where T is the sampling rate. Thus this filter yields the mean value of the input signal in the time interval $t-T, t$. The integral given by eqn (6.1-2) can be replaced by a summation and, if the input signal is sampled k times in the period T, then the output signal at instant t_i is

$$y(t_i) = \frac{1}{k} \sum x(t_j). \tag{6.1-3}$$

By utilizing eqn (6.1-3), Fig. 6.2 shows an implementation of the moving average filter. The input to the circuit shown in Fig. 6.2 is the signal $x(t)$, which is a line-to-line voltage in the experiment described above. The sample instants are determined by a rotating pulse in a Shift Register (SR). The sampling pulses are $s_1, s_2, s_3 \ldots s_k$. The frequency of the input pulse train of the shift register is $f = k/T$. Each SH circuit holds one sample during the period T. The k samples of the line-to-line voltage are added, and are then divided by k to yield the moving-average value.

The accuracy of the test described above can be checked by either comparing the obtained value of the rotor time constant with the value obtained with other tests, or, and this is simpler, by utilizing the measured drive to check the accuracy. For this purpose it is possible to use the relationship between the output torque and the reference torque in the steady-state. This will only be linear if the correct value of the rotor time constant is

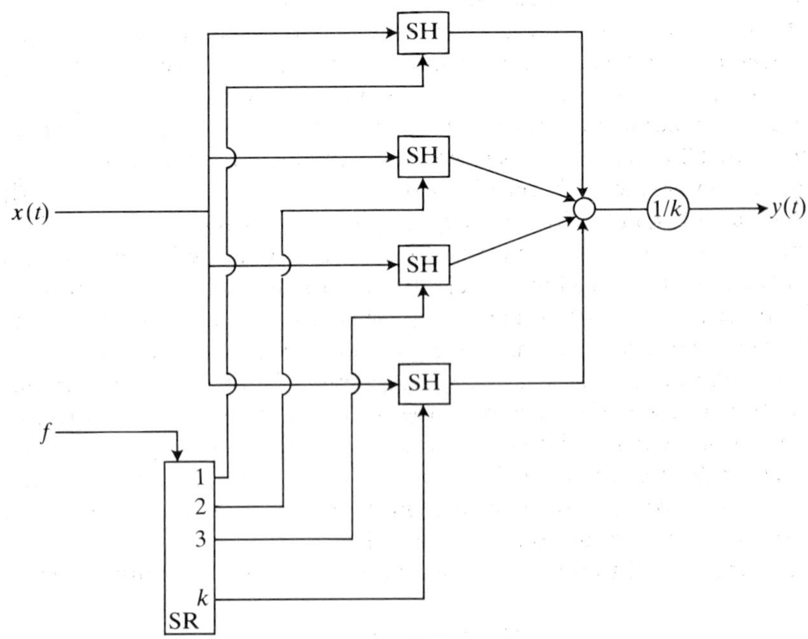

Fig. 6.2. An implementation of the moving average filter.

used in the calculation block of the angular rotor slip frequency. Since the rotor resistance can change significantly with temperature, during the operation of the drive, the initial value has to be corrected. For this purpose it is also possible to use a simplified thermal model of the machine.

6.2 DETERMINATION OF ELECTRICAL AND MECHANICAL PARAMETERS OF AN INDUCTION MACHINE WITH IMPRESSED STATOR VOLTAGES AND SUBJECTED TO ROTOR-FLUX-ORIENTED CONTROL

It is assumed that the induction machine is supplied by a voltage source inverter, as discussed in Section 4.1.1, and rotor-flux-oriented control is employed. It is assumed that all control tasks (inverter control, rotor-flux-oriented control) can be performed by a computer. The stator current in stator phase sA is measured and its instantaneous value is i_{sA}. The electrical parameters of the machine are obtained when the machine is at standstill by utilizing measured values of appropriately formed stator currents and measured terminal voltages. For this purpose the voltage equations of the machine expressed in the stationary reference frame are utilized, but with the two-axis components of the stator currents (i_{sD}, i_{sQ}) and the two axis rotor magnetizing-current components (i_{mrD}, i_{mrQ}) chosen as state-variables.

It follows from eqns (4.1-34) and (4.1-29) that the space-phasor form of the stator and rotor voltage equations in the stationary reference frame are as follows if the machine is at standstill ($\omega_r=0$) and if the effects of magnetic saturation are neglected:

$$\bar{u}_s = R_s\bar{i}_s + L_s'\frac{d\bar{i}_s}{dt} + (1-\sigma)L_s\frac{d\bar{i}_{mr}}{dt} \qquad (6.2\text{-}1)$$

$$\frac{d\bar{i}_{mr}}{dt} = \frac{\bar{i}_s - \bar{i}_{mr}}{T_r}. \qquad (6.2\text{-}2)$$

In eqn (6.2-1) the derivative of the rotor magnetizing-current space phasor is present. When its expression [eqn 6.2-2)] is substituted into eqn (6.2-1), the stator voltage equation can be expressed as

$$\bar{u}_s = (R_s + R_{rref})\bar{i}_s + L_s'\frac{d\bar{i}_s}{dt} - \frac{(1-\sigma)L_s\bar{i}_{mr}}{T_r}. \qquad (6.2\text{-}3)$$

By introducing the referred value of the rotor flux-linkage space phasor,

$$\bar{\psi}_{rref} = (1-\sigma)L_s\bar{i}_{mr} = \frac{L_m}{L_r}\bar{\psi}_r', \qquad (6.2\text{-}4)$$

where $\bar{\psi}_r'$ is the space phasor of the rotor flux linkages and L_m/L_r is a referring factor (whose role has been described in Section 4.1.1), eqns (6.2-3) and (6.2-2) can be put into the form,

$$\bar{u}_s = (R_s + R_{rref})\bar{i}_s + L'_s \frac{d\bar{i}_s}{dt} - \frac{\bar{\psi}_{rref}}{T_r} \tag{6.2-5}$$

$$\frac{d\bar{\psi}_{rref}}{dt} = R_{rref}\bar{i}_s - \frac{\bar{\psi}_{rref}}{T_r}, \tag{6.2-6}$$

where R_{rref} is the referred value of the rotor resistance,

$$R_{rref} = \left(\frac{L_m}{L_r}\right)^2 R_r \tag{6.2-7}$$

and has been shown in the equivalent circuit given in Figs 4.10(a) and 4.10(c).

Equations (6.2-5)–(6.2-7) are used to obtain several parameters (stator transient time constant, stator resistance, rotor resistance, rotor time constant, etc.) of the induction machine supplied by the voltage-source PWM inverter by employing tests performed at standstill.

6.2.1 Determination of the stator transient inductance

First the stator transient inductance (L'_s) is determined. For this purpose appropriate short voltage impulses are generated by the inverter which supplies the induction machine itself. It follows from eqns (6.2-5)–(6.2-7) that if the direct-axis stator current expressed in the stationary reference frame is zero, $i_{sD}=0$, and the direct-axis component of the referred value of the rotor flux-linkage space phasor expressed in the stationary reference frame is zero, $\psi_{rref\alpha}=0$, then the direct-axis component of the equations yields

$$L'_s = u_{sD}/(di_{sD}/dt), \tag{6.2-8}$$

where u_{sD} is the direct-axis component of the stator-voltage space phasor in the stationary reference frame. Equation (6.2-8) is used to obtain the stator transient inductance. For this purpose first two appropriate stator terminals (SR, TR) are supplied by the d.c. voltage U_d, using the inverter itself, where U_d is the d.c. link voltage. From the definition of the space phasor of the stator voltages, eqn (2.1-126), $u_{sD}=\frac{2}{3}U_d$ is obtained and this is the value of which has to be used in eqn (6.2-8). Fig. 6.3 shows the terminal voltage u_{SR} and the current response in phase sA.

The duration of the pulses in the terminal voltage is selected to be much shorter than the open-circuit rotor time constant (T_r), and it follows from the voltage equations that, with respect to $i_{sA}=i_{sD}$, the induction machine behaves like a first-order delay element and the initial rise of the stator current i_{sD} is determined by the stator transient inductance. In Fig. 6.3 at instant t_1 the

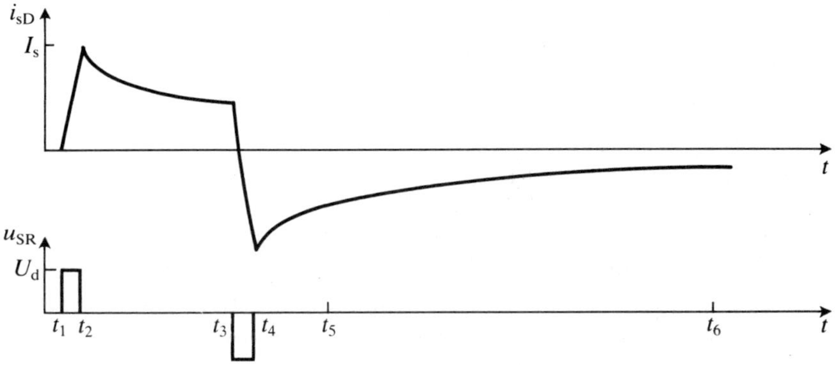

Fig. 6.3. Variation of u_{SR} and i_{sA}.

voltage U_d appears at the two stator terminals indicated above. Thus the stator current starts to rise and at instant t_2 it reaches its peak value (peak of rated stator current I_s). At this instant the stator terminals are short circuited with the aid of the inverter valves and thus the stator current starts to decrease. At instant t_3, where the phase current is approximately equal to half the peak value of the rated current, a negative value of the d.c. link voltage is applied and thus the stator current will further decrease and will become negative, as shown in Fig. 6.3. At instant t_4 the stator current reaches its negative peak value and at this instant the stator terminals are again short-circuited. Thus the stator current will start to rise.

In eqn (6.2-8) the derivative $\mathrm{d}i_{sD}/\mathrm{d}t$ is present. This can be computed by considering the monitored values of the stator current shown in Fig. 6.3 in either the interval t_1–t_2 or the interval t_3–t_4. However, since $t_4-t_3>t_2-t_1$ and $i_{sD}(t_3)-i_{sD}(t_4)<i_{sD}(t_2)-i_{sD}(t_1)$, more accurate results are obtained if the values of the current at the instants t_3 and t_4 are used for the calculation of the derivative of the stator current. Furthermore, the average value of the stator current in the second interval is closer to zero than in the first interval and this also leads to a more accurate value of L_s'. It follows that L_s' is computed by considering the following expression,

$$L_s'=\tfrac{2}{3}U_d\,\frac{t_4-t_3}{i_{sD}(t_3)-i_{sD}(t_4)}. \qquad (6.2\text{-}9)$$

Equation (6.2-9) has been derived by considering the voltage equations, eqns (4.1-33) and (4.1-29), which have been obtained by assuming ideal conditions, e.g. by neglecting the effects of magnetic saturation, eddy currents and current displacement. Thus when the stator transient inductance is determined by utilizing eqn (6.2-9), the value obtained will be smaller than the actual value. However, an improvement can be obtained, if instead of the measured value of the stator current at instant t_4, $i_{sD}(t_4)$, a computed value $i_{sDc}(t_4)$ is used, which is obtained by fitting an exponential current in the interval t_5–t_6 shown in Fig. 6.3 and by using this to compute the value of the stator current at instant t_4. This value, $i_{sDc}(t_4)$, is smaller than the measured value $i_{sD}(t_4)$, and thus $i_{sD}(t_3)-i_{sD}(t_4)>i_{sD}(t_3)-i_{sDc}(t_4)$. The stator transient inductance can be computed from

$$L_s' = \frac{2}{3} U_d \frac{t_4 - t_3}{i_{sD}(t_3) - i_{sDc}(t_4)}. \tag{6.2-10}$$

When the stator transient inductance is determined by using eqn (6.2-10) it will be larger than the value obtained by using eqn (6.2-9). It is possible to use a similar technique to obtain the variation of the other two stator currents and to determine the resulting transient stator inductance. The average value of the results will give a more accurate representation of the actual value of the stator transient inductance.

6.2.2 Determination of the stator resistance

The stator resistance, the referred value of the rotor resistance and the rotor time constant can be obtained by impressing different values of direct current on the machine. This could be performed by a current controller and a PWM modulator. The current controller can be implemented by software. Figure 6.4 shows the employed stator current and also the output of the current controller, (u_c), the average value of which is proportional to the terminal voltage.

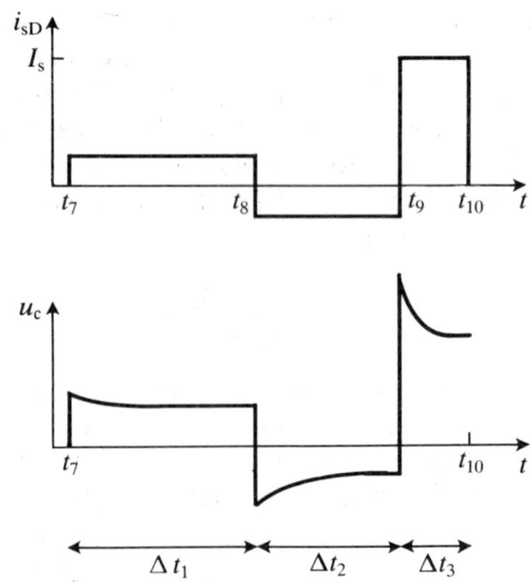

Fig. 6.4. The impressed stator current and the output voltage of the current controller.

It should be noted that the interval t_1–t_2 in Fig. 6.3 is in the millisecond region, and the interval t_7–t_{10} shown in Fig. 6.4 can be a continuation of the interval t_1–t_6 shown in Fig. 6.3. However, interval t_7–t_{10} is much longer than t_1–t_6, and can be a few seconds long. At instant t_7 a d.c. current is applied to the stator winding sA, which is approximately one third of the peak value of the rated current (I_s). This value is chosen so that it should be less than the magnetizing current of the machine to ensure that the effects of magnetic saturation do not affect the measured parameters. Since the value of the magnetizing current is unknown prior to the measurements, this current can be reduced to a lower level if it is established that saturated conditions are obtained when the d.c. current is equal to approximately one third of the peak value of the rated current. As a result of the applied d.c. current, a voltage (u_c) will appear on the output of the current controller and will start to decrease until it reaches the steady state, as shown in Fig. 6.4. In the steady state, at instant t_8, a negative value of the previously impressed d.c. current is applied and thus the voltage on the output of the current controller will be negative and will start to increase until the steady state is reached. Thus at instant t_9 the applied d.c. current is equal to the peak of the rated current and a large voltage will arise on the output of the current controller; this will start to decrease and at instant t_{10} the steady state is reached.

It follows from eqns (6.2-5)–(6.2-7) that, in the steady state, the direct-axis stator voltage in the stationary reference frame is

$$u_{sD}=R_s i_{sD}. \qquad (6.2\text{-}11)$$

As mentioned above, the average value of the terminal voltage is proportional to the voltage on the output of the current controller and can be determined if the d.c. voltage is known. However, the voltage drops due to the semiconductor valves cause a small deviation. To obtain a correct value of the stator resistance, the following expression can be used:

$$R_s=\frac{u_{sD}(\Delta t_3)-u_{sD}(\Delta t_1)}{i_{sD}(\Delta t_3)-i_{sD}(\Delta t_1)}, \qquad (6.2\text{-}12)$$

where $u_{sD}(\Delta t_3)$ and $u_{sD}(\Delta t_1)$ are the average values of the terminal voltage in intervals (t_9–t_{10}) and (t_7–t_8) respectively, and $i_{sD}(\Delta t_3)$, $i_{sD}(\Delta t_1)$ are the direct-axis stator current components in the same two intervals.

The stator transient inductance obtained in the previous section, together with the stator resistance, can be used for the estimation of the stator transient time constant, $T_s'=L_s'/R_s$.

6.2.3 Determination of the rotor time constant and the referred value of the rotor resistance

It follows from eqns (6.2-5)–(6.2-7) that in interval t_8–t_9 of Fig. 6.4, the direct-axis component of the referred rotor flux $\psi_{rref\alpha}$ (and of course the unreferred rotor flux as well) changes exponentially from $\psi_{rref\alpha}(\Delta t_1)=$

$T_r R_{rref} i_{sD}(\Delta t_1)$ to $\psi_{rref\alpha}(\Delta t_2)=T_r R_{rref} i_{sD}(\Delta t_2)$, and the time constant of the exponential change is equal to the rotor time constant. Since, according to eqn (6.2-5), the stator voltage also contains the change of the rotor flux, the rotor time constant can be obtained from the output voltage of the controller in the interval Δt_2. In practice, when the controller is realized digitally, the output voltage of the controller is not smooth because of the sampling and thus an exponential function is fitted to it in the interval Δt_2 and the rotor time constant is obtained from this fitted curve. It should be noted that the third interval Δt_3 shown in Fig. 6.4 cannot be used for the estimation of the rotor time constant, since in this region the stator current is high (it is equal to the peak rated value) and thus the rotor flux very quickly reaches its saturated value and this would result in too small a value of the rotor time constant.

From eqns (6.2-5)–(6.2-7), in the third interval shown in Fig. 6.4, the initial value of the terminal voltage can be expressed as

$$u_{sD0}(\Delta t_3)=R_s i_{sD}(\Delta t_3)+R_{rref}[i_{sD}(\Delta t_3)-i_{sD}(\Delta t_2)]. \qquad (6.2\text{-}13)$$

Thus it follows that the referred value of the rotor resistance can be obtained as

$$R_{rref}=\frac{u_{sD0}(\Delta t_3)-R_s i_{sD}(\Delta t_3)}{i_{sD}(\Delta t_3)-i_{sD}(\Delta t_2)}. \qquad (6.2\text{-}14)$$

Equation (6.2-14) can be used to obtain R_{rref}, and the stator resistance can be obtained from eqn (6.2-12). Similarly to above, instead of the terminal voltage the output of the controller is used, and thus $u_{c0}(\Delta t_3)$ is used instead of $u_{sD0}(\Delta t_3)$, where $u_{c0}(\Delta t_3)$ is the value of the output voltage of the controller at the beginning of the third interval shown in Fig. 6.4. Furthermore, to avoid the effects of current displacement, a similar technique can be used to that presented above in connection with the determination of the stator transient inductance (where computed values are used instead of measured ones and these are obtained from fitted curves).

6.2.4 Determination of the stator self-inductance and the resultant leakage factor

Since T_r and R_{rref} are known and according to eqn (6.2-7) the referred rotor resistance can be expressed as $R_{rref}=(L_m/L_r)^2 R_r$, it follows that $R_{rref}=L_m^2/(L_r T_r)$ and thus

$$L_m^2/L_r=(1-\sigma)L_s=T_r R_{rref}, \qquad (6.2\text{-}15)$$

where σ is the resultant leakage factor, $\sigma=1-L_m^2/(L_s L_r)$, and $\sigma L_s=L_s'$ is the stator transient inductance. It should be noted that the factor L_m^2/L_r is also present in the expression for the electromagnetic torque, eqn (2.1-197).

Since L_s' can be determined from eqn (6.2-10), it follows from eqn (6.2-15) that the stator self-inductance L_s can be estimated by using

$$L_s = L'_s + T_r R_{rref} \tag{6.2-16}$$

and, since L_s and L'_s are now known quantities, the leakage factor can be determined by considering

$$\sigma = \frac{L'_s}{L_s}. \tag{6.2-17}$$

It should be noted that, in practice, some of the machine parameters can change during the operation of the drive. For example, the rotor resistance can change considerably with temperature. Thus during the operation of the drive some of the machine parameters derived on the basis of the method described above must be corrected. A background computer program can take care of this updating. When the drive is closed down, the required parameters can be saved in RAM for reuse.

6.2.5 Determination of the inertia of the drive

It is also possible to obtain the inertia of the drive by using on-line measurements, but standstill tests cannot be used. The inertia can also be computed from suitable measurements of the currents and speed of the induction machine during a test run.

For the purpose of obtaining the expression for the inertia as a function of machine currents and rotor speed, the equation of motion is considered. For simplicity a two-pole induction machine is considered. It follows from eqn (2.1-109) that if $D=0$, the equation of motion can be expressed as

$$t_e - t_l = J\frac{d\omega_r}{dt}, \tag{6.2-18}$$

where t_e is the developed electromagnetic torque of the machine, t_l is the load torque, J is the inertia and ω_r is the rotor speed. Since it is assumed that rotor-flux-oriented control is employed, it is useful to express the electromagnetic torque in terms of the torque-producing stator current (i_{sy}) and the modulus of the rotor magnetizing current space phasor $|\bar{i}_{mr}|$ and, by considering eqn (2.1-197), it follows that

$$t_e = \frac{3}{2}\frac{L_m^2}{L_r}|\bar{i}_{mr}|i_{sy}. \tag{6.2-19}$$

It is possible to determine the inertia by considering the response of the drive to changes in the reference speed and to obtain an expression for the inertia which does not contain the load torque. This allows the technique to be used for a large range of drive applications. If the machine runs at a given rotor speed $\omega_r(\Delta t_1)$, approximately 30 per cent of the base speed, during interval Δt_1

of the test run, and then the speed reference is changed (interval Δt_2) and in interval Δt_3 the rotor speed is $\omega_r(\Delta t_3)$, and, say, the speed can rise to approximately 60 per cent of the base speed in this region, then it follows from eqns (6.2-18)–(6.2-19) that if the mean values of the load torque in intervals Δt_1 and Δt_3 are the same, $t_{lm}=t_{lm}(\Delta t_1)=t_{lm}(\Delta t_2)$, then

$$t_{em}(\Delta t_1)-t_{lm}=J\Delta\omega_r(\Delta t_1)T \tag{6.2-20}$$

and

$$t_{em}(\Delta t_3)-t_{lm}=J\Delta\omega_r(\Delta t_3)T, \tag{6.2-21}$$

where $t_{em}(\Delta t_1)$ and $t_{em}(\Delta t_3)$ are the mean values of the electromagnetic torque in intervals Δt_1 and Δt_3 respectively and $\Delta\omega_r(\Delta t_1)$ and $\Delta\omega_r(\Delta t_3)$ represent the changes in rotor speed in the intervals Δt_1 and Δt_3 respectively. Furthermore, $T=\Delta t_1=\Delta t_2$ is the duration of the intervals where the measurement of the speed is performed, together with the measurement of the torque-producing stator current in the same two intervals, as will be discussed below. It follows from eqn (6.2-19) that for a fixed value of the rotor flux, the electromagnetic torque is the product of a constant $c=\frac{3}{2}(L_m^2/L_r)|\bar{i}_{mr}|$ and the quadrature-axis stator current, which is expressed in the rotor-flux-oriented reference frame but is accessible in the drive system. In the two intervals Δt_1 and Δt_3 the mean values of the electromagnetic torque can thus be expressed as $ci_{sym}(\Delta t_1)$ and $ci_{sym}(\Delta t_3)$ respectively, where $i_{sym}(\Delta t_1)$ and $i_{sym}(\Delta t_3)$ are the mean values of the torque-producing stator current component in the same two intervals. Thus

$$ci_{sym}(\Delta t_1)-t_{lm}=J\Delta\omega_r(\Delta t_1)T \tag{6.2-22}$$

and

$$ci_{sym}(\Delta t_3)-t_{lm}=J\Delta\omega_r(\Delta t_3)T \tag{6.2-23}$$

are obtained and, when eqn (6.2-23) is subtracted from eqn (6.2-22), finally the following expression is obtained for the inertia of the drive,

$$J=\frac{cT[i_{sym}(\Delta t_3)-i_{sym}(\Delta t_1)]}{\Delta\omega_r(\Delta t_3)-\Delta\omega_r(\Delta t_1)}. \tag{6.2-24}$$

It should be noted that it has been assumed that the mean values of the load torque in the intervals Δt_3 and Δt_1 are equal, and thus eqn (6.2-24) gives an accurate value of the inertia only if this condition is fulfilled. In practice, the load torque can be constant and this condition is automatically satisfied or it can change with the speed or time. However, even in these two cases, the error in the determination of the inertia will be small if T is chosen to be small.

It should be noted that if prior to starting the drive the parameters of the speed controller are not known, initially, these parameters can be set to small values which yield safe operation and during the interval $\Delta t_1+\Delta t_2+\Delta t_3$ these parameters can be held at their initial values. In the next interval the speed-controller parameters can be adjusted by using the known value of the inertia.

During the commissioning of the drive the output of the speed controller is monitored in order to identify closed-loop oscillations. In the case of oscillations the parameters of the speed controller are reduced.

It also follows from eqn (6.2-24) that when $i_{sym}(\Delta t_3) - i_{sym}(\Delta t_1)$ is small, i.e. when the acceleration torque is small, an incorrect value of the inertia will be obtained. This can be the case if the change of the reference speed is too small, or if the torque limit is very near to the load torque or if the speed ramp is too slow. If torsionally flexible coupling exists between the machine and the load, then eqn (6.2-24) cannot be used.

6.3 HIGH-RESOLUTION ROTOR SPEED ESTIMATOR

In Section 4.1.1 a rotor speed estimator was described for use in the rotor-flux-oriented control scheme of a voltage-source inverter-fed induction machine. The rotor speed estimator was obtained by using the technique of Model Reference Adaptive Control and the inputs of the rotor speed estimator circuit was the two-axis components of the stator voltage references expressed in the rotor-flux-oriented reference frame, together with the actual values of the rotor flux- and torque-producing stator current components and the speed of the rotor flux-linkage space phasor.

However, it is possible to obtain the rotor speed by using a rotor speed transducer, which can be expensive, or a rotor position transducer (incremental and absolute optical encoders, electromagnetic resolvers, synchros, inductosyns, etc.). For synchronous machines, absolute encoders must be used. The application of electromagnetic devices (resolvers, synchros, etc.) for position sensing increases the inertia. They can also be susceptible to electromagnetic interference. The resolver is a more rugged device than the encoder. When conventional optical encoders are used, the optical emitters, detectors and the signal processing cannot operate in high-temperature environments. It is also possible to obtain the rotor angle by utilizing a reflective/non-reflective code band installed on the rotor and illuminated by optical fibres. The reflected light can be transmitted by return optical fibre lines which are connected to photodetectors. Such a system is immune to electromagnetic interference, radio-frequency interference, etc.

When a rotor position transducer is used for the estimation of the rotor speed, generally some type of recursive algorithm is used which also utilizes the previous value of the rotor position angle. Such algorithms can be implemented either by hardware (analog or digital) or by software. It is a drawback of the application of such techniques that they consider the reference torque as if it was an unknown disturbance torque and thus they will generally lag the actual rotor speed during acceleration or deceleration. Other problems can arise due to modulation errors caused by asynchronous sampling (e.g. if the sampling period of the rotor speed loop is not synchronous with the carrier frequency and a resolver is used), whereby low-frequency beat patterns can be produced even if the rotor speed is actually constant. Then, to offset the modulation errors, actual position errors will be produced.

6.3.1 Estimation of the rotor speed by utilizing the monitored rotor position

When only the rotor angle (θ_r) is monitored, the simplest algorithm for the estimated rotor speed is as follows,

$$\hat{\omega}_r = \frac{\theta_r(k) - \theta_r(k-1)}{T_s}, \tag{6.3-1}$$

where $\theta_r(k)$ and $\theta_r(k-1)$ are the kth and $(k-1)$th sampled values respectively of the monitored value of the rotor angle and T_s is the sampling time interval for the speed loop. Thus the average value of the rotor speed is obtained in the last sample interval. However, when the estimation of the rotor speed is based on eqn (6.3-1), if there is a quick change in the reference value of the torque and thus the machine accelerates, incorrect results can be obtained. Furthermore, the rotor speed resolution is limited by the resolution of the position transducer and also by the sampling time. An improvement in the rotor speed resolution can be obtained by the application of a moving-average filter, which has been discussed in Fig. 6.2.

6.3.2 Estimation of the rotor speed by utilizing an optimal rotor speed observer

It is possible to obtain the rotor speed by using a rotor speed observer. If an open-loop observer is used, this will only yield a correct value of the rotor speed if the model of the observer and the parameters of this model are identical to those of the actual system. However, the parameter sensitivity of the open-loop observer can be reduced by using a closed-loop observer. Such an observer is now presented for the estimation of the rotor speed and this uses as its inputs the electromagnetic torque, t_e (which is present in all the vector-controlled schemes described earlier) and the monitored rotor angle ω_r. An implementation of the continuous-domain rotor speed observer can be based on the following two equations, which contain a corrective signal obtained from the position error and utilize the equation of motion, eqn (2.1-109) and the fact that the rotor speed is equal to the first time derivative of the rotor angle:

$$t_e + k_1(\theta_r - \hat{\theta}_r) + k_2 \int (\theta_r - \hat{\theta}_r)\,dt = J\frac{d\hat{\omega}_r}{dt} + D\hat{\omega}_r \tag{6.3-2}$$

$$\hat{\omega}_r + k_3(\theta_r - \hat{\theta}_r) = \frac{d\hat{\theta}_r}{dt}. \tag{6.3-3}$$

θ_r is the quantized rotor angle, $\hat{\theta}_r$ is the estimated value of the rotor angle, and thus $\Delta\theta_r = (\theta_r - \hat{\theta}_r)$ is the rotor position error. k_1, k_2, and k_3 are suitably chosen scalar observer gain parameters which are used to force the estimated rotor angle to converge to its monitored value, J is the inertia, D is the damping factor, and $\hat{\omega}_r$ is the estimated value of the rotor speed. It follows that when the estimated value of the rotor angle is equal to its actual value, $\Delta\theta_r = 0$ and the well-known equations are obtained. An implementation of eqns (6.3-2) and (6.3-3) is shown in Fig. 6.5 where $1/p$ represents an integrator.

The rotor speed observer shown in Fig. 6.5 can be implemented by analog circuits. However, it is more convenient to have a software implementation. Figure 6.6 shows the corresponding implementation of the digital rotor speed observer, which can, for example, be implemented by using a signal processor.

In Fig. 6.6 T_s is the sampling time and z is a complex variable related to the Laplace operator p by $z = e^{pT_s}$. It should be noted that the role of the

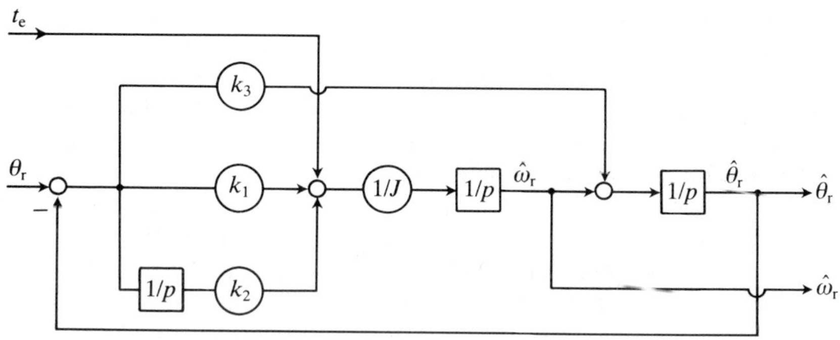

Fig. 6.5. An implementation of the rotor speed observer, if D=0.

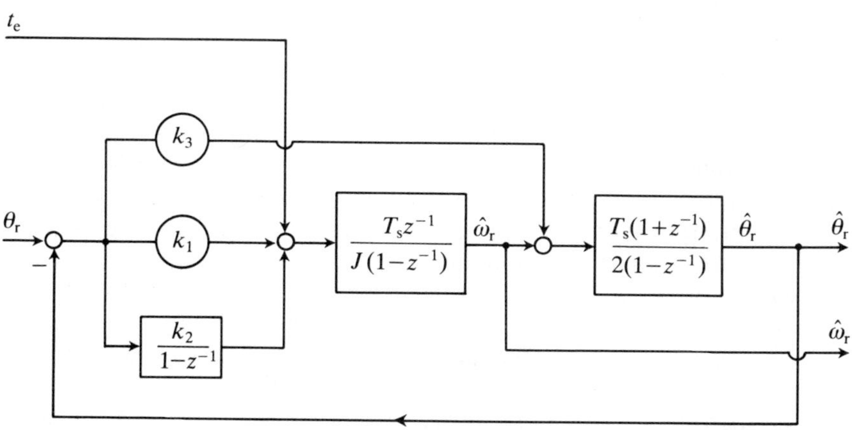

Fig. 6.6. An implementation of the digital rotor speed observer.

z-transform in a digital system is similar to the role of the Laplace transform in a continuous-domain system. The gains should be selected to yield good dynamic performance without any overshoot. When compared with a classical d.c. tachometer, improved dynamic performance is obtained when such a rotor speed observer is used.

6.4 DETERMINATION OF SYNCHRONOUS MACHINE PARAMETERS BY LOAD-REJECTION TESTS

There are many techniques which can be used for the identification of the parameters of salient-pole synchronous machines. However, it is very convenient to perform tests which utilize measured voltages and currents of the machine, together with the measured load angle, and which involve simple experiments and simple estimation processes. For the derivation of the direct- and quadrature-axis reactance parameters (synchronous reactances, transient reactances, subtransient reactances) it is very convenient to perform load-rejection tests. These can be the so-called direct-axis and quadrature-axis load-rejection tests or the so-called arbitrary load-rejection test. Compared with the short-circuit and slip tests, which are inconvenient and where there is a risk of damaging the machine under test, these methods can be implemented more easily. Furthermore, the determination of the parameters by the load-rejection test compares well with frequency-response testing. The latter is ideal for laboratory-type machines, but is more complicated and more expensive than load-rejection tests, and with frequency-response testing it is more difficult to obtain the parameters under saturated conditions.

The salient-pole machine can be described by the model given in Section 3.2. It is now assumed that saturation can be present along the direct-axis. In the case of the direct-axis load-rejection test, a synchronous generator is initially assumed to be connected to the supply, to have zero active power, but to draw or supply reactive power. Thus there is only direct-axis flux. The machine is assumed to be underexcited; so saturation does not play any role. The machine is suddenly disconnected from the supply and it is assumed that at this instant the armature current is i_{sd0}. The decaying terminal voltage and the field current are then monitored and the direct-axis synchronous reactance (X_{d}), together with the direct-axis transient and subtransient reactances (X'_{d}, X''_{d}) and the direct-axis no-load transient and subtransient time constants (T'_{d0}, T''_{d0}), are estimated. It should be noted that the subtransient reactances and time constants are only given here for generality, since they are not required in the vector-control schemes described for the synchronous machine. Figure 6.7 shows the decaying terminal voltage after the load rejection.

It follows from the equations of the machine that X_{d}, X'_{d}, and X''_{d} can be determined from Fig. 6.7 as

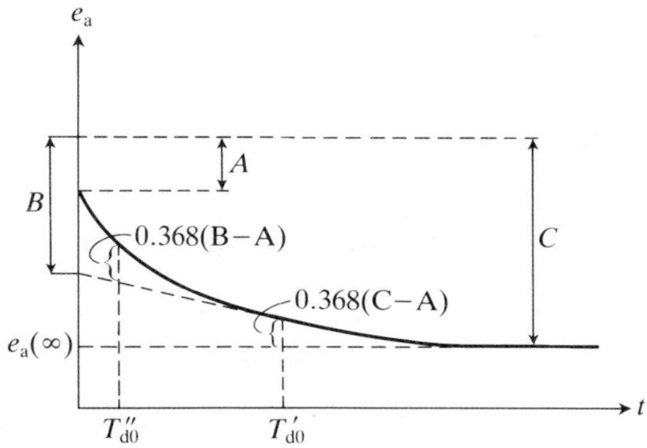

Fig. 6.7. The decaying terminal voltage after the load rejection.

$$X_d = \frac{C}{i_{sd0}} \tag{6.4-1}$$

$$X'_d = \frac{B}{i_{sd0}} \tag{6.4-2}$$

$$X''_d = \frac{A}{i_{sd0}}, \tag{6.4-3}$$

where A, B, and C are also shown in Fig. 6.7. The no-load direct-axis transient and subtransient time constants (T'_{d0} and T''_{d0}) can also be obtained from Fig. 6.7. These two time constants could also be derived if the field current was measured. To obtain parameters under saturated conditions, the same tests have to be performed as before, but with overexcited initial conditions.

For the determination of the quadrature-axis parameters, the quadrature-axis load-rejection test is performed. For this purpose it would be desirable if the armature current had only a quadrature-axis component (i_{sq0}). Thus the sudden switch-off is performed when the armature current has only a quadrature-axis component (i_{sq0}) and the reactive power is adjusted so that the absolute value of the phase angle between the armature current and the terminal voltage should be equal to the load angle. Figure 6.8 shows the variation of the terminal voltage after the load rejection.

From the equations of the machine, the following expressions for X_q and X''_q can be obtained by using the measured values of A, B, and C shown in Fig. 6.8:

$$X_q = \frac{(A^2 - C^2)^{1/2}}{i_{sq0}} \tag{6.4-4}$$

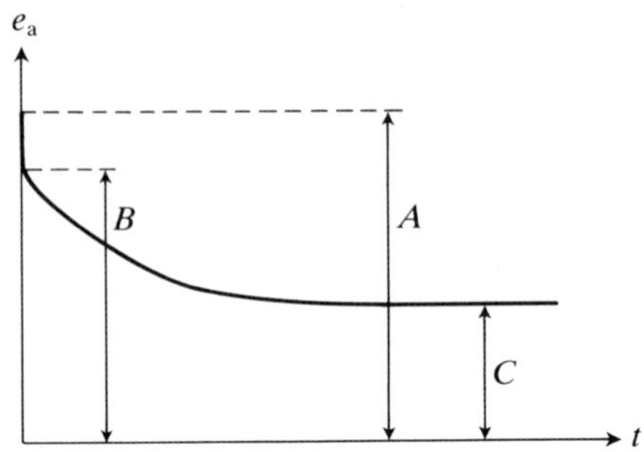

Fig. 6.8. Variation of the terminal voltage after the load rejection.

$$X_q'' = \frac{[(A^2 - C^2)^{1/2} - (B^2 - C^2)^{1/2}]}{i_{sq0}}. \tag{6.4-5}$$

Since it is difficult to adjust the load to the ideal quadrature-axis load-rejection condition, it is preferable to use the so-called arbitrary-axis load-rejection method. In this case the machine is suddenly switched off while the armature current has both (direct- and quadrature-axis) components. The variation of the load angle (δ) is also monitored, and the terminal voltage and armature current are separated into direct- and quadrature-axis components ($e_a \sin\delta$, $e_a \cos\delta$, i_{sd0}, i_{sq0}). Figure 6.9 shows the variation obtained in the voltage $e_a \sin\delta$, where a logarithmic scale has been used for the voltage.

From the machine equations it can be shown that the quadrature-axis synchronous reactance can be obtained as

$$X_q = \frac{e_a \sin\delta}{i_{sq0}} \tag{6.4-6}$$

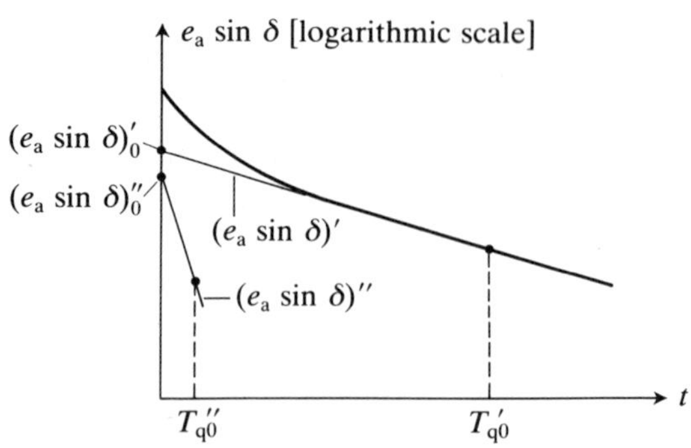

Fig. 6.9. Variation of $e_a \sin\delta$.

and the quadrature-axis transient and subtransient reactances can be obtained as

$$X'_q = X_q - \frac{(e_a \sin \delta)'_0}{i_{sq0}} \tag{6.4-7}$$

$$X''_q = X'_q - \frac{(e_a \sin \delta)''_0}{i_{sq0}} \tag{6.4-8}$$

In eqns (6.4-7) and (6.4-8) $(e_a \sin \delta)'_0$ and $(e_a \sin \delta)''_0$ are present; these are also shown in Fig. 6.9 and correspond to the initial values of the direct-axis component of the transient and subtransient terminal voltages respectively. In Fig. 6.9 the quadrature-axis no-load transient and subtransient time constants are also shown (T'_{q0}, T''_{q0}).

BIBLIOGRAPHY

Canay, I. M. (1983). Determination of model parameters of synchronous machines. *IEE Proc.* **B 130,** 86–93.

Harms, K. and Leonhard, W. (1987). Parameter adaptive control of induction motor based on steady-state model. *2nd European Conference on Power Electronics and Applications, Grenoble*, pp. 2.57–2.62.

Hindmarsh, J. (1977) *Electrical machines and their applications*. Pergamon Press, Oxford.

Loeser, F. and Sattler, P. K. (1985). Identification and compensation of the rotor temperature of a.c. drives by an observer. *IEEE Transactions on Industry Applications* **IA-21**, 1387–93.

Lorenz, R. D. and Patten, K. V. (1988). High resolution velocity estimation for all digital, a.c. servo-drives. *IEEE IAS Annual Meeting, Pittsburgh*, pp. 363–8.

Matsuo, T. and Lipo, T. A. (1985). A rotor parameter identification scheme for vector-controlled induction motor drives. *IEEE Transactions on Industry Applications* **IA-21**, 624–32.

Mello, F. P. de, and Ribeiro, J. R. (1977). Derivation of synchronous machine parameters from tests. *IEEE Transactions on PAS*, PAS-96, 1211–18.

Namba, M., Hosoda, J., Doi, S., and Udo, M. (1981). Development for measurement of operating parameters of synchronous generator and control systems. *IEEE Transactions on PAS* **PAS-100**, 618–28.

Say, M. G. (1976). *Alternating current machines*. Pitman, London.

Schierling, H. (1988). Fast and reliable commissioning of a.c. variable speed drives by self-commissioning. *IEEE IAS Annual Meeting, Pittsburgh*, pp. 489–92.

Wang, C., Novotny, D. W., and Lipo, T. A. (1986). An automated rotor time constant measurement system for indirect field oriented drives. *IEEE IAS Annual Meeting, Denver*, pp. 140–6.

Index